Advances in Design and Disaster Mitigation of Engineering Structures

Advances in Design and Disaster Mitigation of Engineering Structures

Editors

Liqiang Jiang
Jihong Ye
Wei Guo

MDPI

Basel • Beijing • Wuhan • Barcelona • Belgrade • Novi Sad • Cluj • Manchester

Editors

Liqiang Jiang
Civil Engineering
Central South University
Changsha
China

Jihong Ye
Civil Engineering
China University of Mining and Technology
Xuzhou
China

Wei Guo
Civil Engineering
Central South University
Changsha
China

Editorial Office
MDPI
St. Alban-Anlage 66
4052 Basel, Switzerland

This is a reprint of articles from the Special Issue published online in the open access journal *Buildings* (ISSN 2075-5309) (available at: https://www2.mdpi.com/journal/buildings/special_issues/Disaster_Structures).

For citation purposes, cite each article independently as indicated on the article page online and as indicated below:

Lastname, A.A.; Lastname, B.B. Article Title. *Journal Name* **Year**, *Volume Number*, Page Range.

ISBN 978-3-7258-0244-9 (Hbk)
ISBN 978-3-7258-0243-2 (PDF)
doi.org/10.3390/books978-3-7258-0243-2

Contents

buildings

MDPI

Article

Fatigue Damage Mitigation for Welded Beam-to-Column Connections in Steel High-Rise Buildings Using Passive Vibration Control

Zhao Fang [1], Jianshao Zhang [2,3], Fan Yang [3] and Aiqun Li [3,4,*]

1 School of Architecture Engineering, Nanjing Institute of Technology, Nanjing 211167, China
2 Guang Dong Architectural Design & Research Institute Co., Ltd., Guangzhou 510010, China
3 School of Civil Engineering, Southeast University, Nanjing 210096, China
4 Beijing Advanced Innovation Center for Future Urban Design, Beijing University of Civil Engineering and Architecture, Beijing 100044, China
* Correspondence: aiqunli@seu.edu.cn

Abstract: To investigate the use of vibration control systems in fatigue damage mitigation for welded beam-to-column connections in steel high-rise buildings, cases of both a single connection under constant amplitude cyclic loading and multi-connections in a high-rise building under the stochastic wind, with and without the fluid viscous damper (VFD) and the tuned mass damper (TMD), are discussed respectively. The finite element analysis and the fatigue assessment are conducted so that the mitigation effect, the effect of technical parameters and the conditions of both high-cycle fatigue and low-cycle fatigue are all discussed. The results show that the VFD and the TMD systems are both effective in the mitigation of local fatigue damage along with the structural displacement for both cases. The VFD generally has a better mitigation effect than the TMD and it starts to take effect instantly with the external loading, but it causes a phase difference in structural responses, while the situation of the TMD is quite the opposite. The displacement and the local stress show similar and synchronous mitigation trends so that the damping systems can be designed based on either of them. The VFD should be designed with a smaller damping exponent and a larger damping coefficient in a braced installation form, while the TMD can be designed using the optimal parameters. The optimized VFD layout plan is that VFDs are placed between the two connections with large relative displacement and relative velocity on higher floors and these two connections with VFDs should be near to the targeted connection. The negative fatigue damage mitigation mainly stems from insufficient lateral support force so that the direct installation of VFDs may result in a negative fatigue damage mitigation effect in the connections with limited lateral support.

Keywords: fatigue damage mitigation; fluid viscous damper; tuned mass damper; welded beam-to-column connection; steel high-rise building

Citation: Fang, Z.; Zhang, J.; Yang, F.; Li, A. Fatigue Damage Mitigation for Welded Beam-to-Column Connections in Steel High-Rise Buildings Using Passive Vibration Control. *Buildings* **2022**, *12*, 1898. https://doi.org/10.3390/buildings12111898

Academic Editors: Liqiang Jiang, Jihong Ye and Wei Guo

Received: 7 September 2022
Accepted: 3 November 2022
Published: 5 November 2022

1. Introduction

Welded beam-to-column connections in steel high-rise buildings are susceptible to fatigue damage under cyclic loading, in the form of high-cycle fatigue damage under wind and low-cycle fatigue damage under strong earthquakes. The fatigue damage may even lead to the collapse of the whole building [1]. Under such circumstances, it is indispensable to think of ways to decrease fatigue damage under wind and earthquakes during the design of welded connections. Traditionally, engineers tend to do it by increasing the fatigue damage resistance, but the disadvantages are obvious, in that the increase in the resistance may lead to high costs and a reduction in the structural ductility. Thus, the use of passive vibration control becomes another alternative.

The passive vibration control technique is mainly classified into three types: vibration isolation, energy dissipation and vibration absorbers. Vibration isolation is typically placed at the foundation of a structure and partially absorbs external input energy before this energy can be transmitted to the structure; therefore, it is limited to the case of a structure under earthquakes and not much applicable to a structure under wind. Energy dissipation is applicable to a structure under both wind and earthquakes. The basic function of energy dissipation devices is that they can consume a portion of the input energy and thereby reduce the structural vibration when they are incorporated into a structure [2]. They include a wide range of damping devices and the most common ones are metallic dampers, viscoelastic dampers and viscous fluid dampers (VFDs), etc. The metallic dampers consume energy based on plastic yielding of metal materials so that they are more applicable to strong external excitations such as earthquakes and typhoons, which inevitably make the metal yield. Viscoelastic dampers consume energy based on the shear deformation of viscoelastic solids; they provide both additional damping and additional stiffness to the structure and thereby change the structural stiffness. It indicates that they are more applicable to structures requiring additional stiffness to resist lateral deformation. Compared with the two above-mentioned energy dissipation devices, VFDs have the advantage of being applicable to both strong external excitations such as earthquakes/typhoons and weak ones such as moderate wind. Moreover, the use of VFDs does not provide additional stiffness to the structure, thereby avoiding the unexpected change of the structural stiffness and natural vibration frequency, which is important in structural designs. The most two common types of vibration absorbers are the tuned mass damper (TMD) and the tuned liquid damper (TLD). The former absorbs energy based on the inertial force of mass and the latter absorbs energy based on the dynamic pressure of liquid. Comparatively speaking, TMDs are much more widely used in engineering structures than TLDs, since the former is a well-developed technique that is easier to design and construct. From the above descriptions, the use of VFDs as typical energy dissipation devices and TMDs as typical vibration absorbers to mitigate fatigue damage of welded beam-to-column connections is discussed in this paper.

There are two main fatigue analysis methods from the perspective of analysis procedures, which are the frequency domain analysis and the time domain analysis. The frequency domain analysis based on mode superposition considers the random nature of external loading and proves to be a reliable option with low computational expense and a simple computing model. For example, Janbazi and Tabeshpour [3] discussed the application of several configurations of viscous dampers on jackets and its effect on damage and fatigue lifetime by introducing a spectral method for assessment and rehabilitation of the jacket platform structure; Allen et al. [4] presented a computationally efficient frequency domain model for of a floating wind turbine with multiple hull-based TMDs; Repetto and Solari [5] analyzed the fatigue collapse of two slender structures due to wind-induced vibrations using frequency domain analysis. However, two distinguished shortcomings must be pointed out for this method. The first one is that the frequency domain analysis based on mode superposition does not suit nonlinear analysis. It is revealed in Section 2 that the damping force output by a VFD is nonlinearly related to the speed of the piston movement and, moreover, plastic strain is found near the welding location in the case of low-cycle fatigue damage, which means nonlinear analysis is indispensable. The second shortcoming is that the frequency domain analysis is not as applicable as the time domain analysis to the description of the local stress near the welding location, due to the complex geometry properties and the resulting serious stress concentration and nonlinear stress gradient. Based on the above two shortcomings, frequency domain analysis is obviously not the optimum option. On the contrary, the time domain analysis has the advantages of being suitable to nonlinear cases and being capable of describing the local stress near the welding location more precisely, while it has the disadvantages of high computational expense and a complex computing model, since the tedious time-history analysis and complex finite element models are almost inevitable. Therefore, it is thought that frequency domain analysis provides elegant solutions but is difficult to use in engineering applications, while time domain approaches allow classical fatigue analyses but require considerable computational or experimental efforts [6]. From the above discussion, the time domain analysis proves to be a better option for the fatigue assessment of welded beam-to-column connections.

Based on the review of the related literature, some new real-time output-based solutions such as the real-time eigen perturbation strategies in structural health monitoring must be mentioned. These solutions can illustrate damage detection in an online mode with considerably less computational expense [7,8]. However, according to the latest literature, these solutions are mainly limited to damage detection including crack detection when dealing with fatigue damage problems [9] and have never been applied to the fatigue life assessment of engineering structures. Although they may have a promising future, the usability and availability of them in fatigue life assessment is still unclear and they are obviously not the main topic of this paper. Thus, the real-time output-based solutions are not discussed in the following paragraphs.

Vibration control is traditionally used in vibration response mitigation such as displacement and acceleration of buildings under wind and earthquakes. However, researchers have started to realize that passive vibration control can also be used in fatigue damage mitigation within the last two decades. Palmeri and Ricciardelli [10] estimated the fatigue life of structural components of high-rise buildings provided with viscoelastic dampers and found that viscoelastic dampers are effective in mitigating both the buffeting response and fatigue damage in medium-rise and high-rise buildings. Golafshani and Gholizad [11] evaluated the efficiency of an optimally designed tuned mass damper for fatigue damage mitigation in real steel jacket platforms and found that such dampers showed a great performance in this application. Ambrosio et al. [12] proposed an active control design to minimize fatigue damage on the structure from a theoretical, numerical and experimental point of view, which showed a significant improvement. Ripamonti et al. [13] proposed an adaptive vibration controller to increase the fatigue life of a smart structure made of composite material and actuated with piezoelectric patches. Andersson et al. [14] studied the use of passive and adaptive damping systems to mitigate vibrations in a railway bridge during resonance and to increase its fatigue service life.

However, there are still limitations in the abovementioned literature. Firstly, the above literature discussed the fatigue damage mitigation from a more global view of the whole structures, ranging from high-rise buildings and jacket platforms to railway bridges, but none of them discussed the fatigue damage mitigation of the more complex welded connections in structures, which are completely different from common components and whole structures. The welding part of a beam-to-column connection consists of three welded joints (the fillet welded joint connecting the beam web and the column flange, the butt welded joint connecting the top beam flange and the column flange and the butt welded joint connecting the bottom beam flange and the column flange) and two welding holes between them, as shown in Figure 1. The complex geometry properties lead to a serious stress concentration effect and nonlinear stress gradient near the local welding location. Consequently, a time domain fatigue analysis based on the computation of the local stress using a refined computational model is indispensable for a beam-to-column connection. On the contrary, the fatigue analysis of a common structure does not need to consider such local stress effects and the frequency domain fatigue analysis is more commonly used instead. Secondly, each of the above literature only focused on a single kind of vibration control system and lacks systematical comparison between common kinds of vibration control systems and, moreover, detailed discussions of the effect of technical parameters on the mitigation effect are still limited. Finally, when the external cyclic loading is large enough to result in plastic strain and low-cycle fatigue damage near the welding location, the elastoplastic analysis is required. However, the fatigue analysis of a global structure barely considers the effect of plastic strain, since in most cases, plastic strain is only found near the welding location and does not affect the behavior of the global structure. Therefore, the above literature mainly involves the use of vibration control in high-cycle fatigue damage mitigation, while low-cycle fatigue damage mitigation is barely discussed.

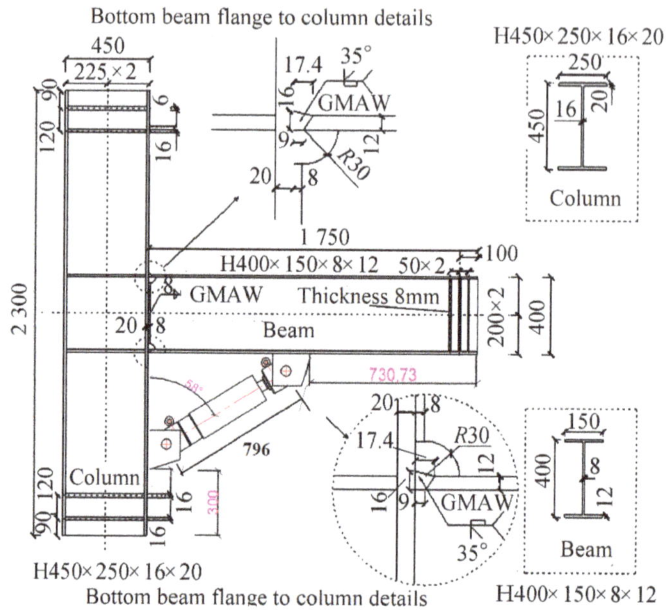

Figure 1. Welded beam-to-column connection.

The novelty of this research is to solve the above three current limitations. The fatigue damage mitigation of the welded beam-to-column connections in steel high-rise buildings is discussed and systematical comparisons of the two mitigation systems, including VFDs and TMDs, are made, with the effect of technical parameters on the mitigation effect being elaborated. Some original and unreported mitigation phenomena of the two vibration control systems are revealed and the use of them in low-cycle fatigue damage is also discussed for the first time. To the authors' knowledge, it is the first original research to discuss fatigue damage mitigation in welded beam-to-column connections.

In this paper, numerical study is conducted to research the use of two passive vibration control systems, VFDs and TMDs, in fatigue damage mitigation for welded beam-to-column connections in steel high-rise buildings. The fatigue damage mitigation is studied based on finite element analysis for the case of a single connection under constant amplitude cyclic loading in Section 2, which consists of three parts: the use of VFDs (Section 2.1), the use of TMDs (Section 2.2) and the condition of low-cycle fatigue damage (Section 2.3). In addition to the above fatigue assessment and the comparison of mitigation effects, a parametric study is also conducted to study the technical parameters of the two damping systems on the mitigation effect within Sections 2.1 and 2.2. The fatigue damage mitigation using the two damping systems for multi-connections in a steel high-rise building under the stochastic wind is introduced in Section 3, including the fatigue assessment, the comparison of mitigation effects and technical parametric study. In Section 4, discussion of important and unaddressed issues, including the optimization of damper positions and the negative mitigation effect, is made before the conclusions in Section 5 are drawn.

2. Fatigue Damage Mitigation for a Single Connection

2.1. Mitigation Using the VFD System

A VFD is a passive velocity-dependent damping device which works based on the damping force given by the interaction of the viscous medium (silicone oil) and the damper components. It is considered to only provide additional damping to the structure without changing its rigidity or natural vibration period dramatically. Generally speaking, the damping force output by a VFD is described in Equation (1):

$$F_D = CV^{\alpha} \tag{1}$$

where F_D is the damping force; C is the damping coefficient; α is the damping exponent; V is the speed of the piston movement.

2.1.1. Formulation of the Model

A welded beam-to-column connection made of high-strength low-alloy structural steel GB Q345B, which is approximate to S355 in BS code, is discussed. The material properties of the steel are that the elastic modulus is 206 GPa, the Poisson's ratio is 0.3, the density is 7850 kg/m^3, the yield strength is 345 MPa and the tensile strength is 620 MPa. The detailed dimensions of the connection and weld details are all shown in Figure 1. The welding is conducted by the gas metal arc welding (GMAW) process in the form of the full penetration groove weld. Meanwhile, the VFD is connected to the connection as shown in Figure 1.

The three-dimensional finite element model of the connection without the VFD is established using an eight-node 3D solid structural element SOLID185 in the commercial software ANSYS, as shown in Figure 2a. The mesh is more refined near the weld detail with a mesh size of about 5 mm, while it becomes more sparse at locations far away from the weld detail at a mesh size of 13 mm, as shown in Figure 2b. Meanwhile, a model of the connection with a VFD is also established, as shown in Figure 2c. The VFD is modeled by COMBIN37, a kind of unidirectional spring damper element, and it connects the bottom of the beam flange with the lateral surface of the column web. The damping coefficient is C = 750 kN·s/m and the damping index is α = 0.25. To consider the role of the potential gusset plate and to avoid the extreme local stress concentration brought by the two ends of the VFD, a rigid region is generated by degree-of-freedom constraint equations between relate nodes at the two locations that the VFD connects. The degree of freedom of nodes at the top and the bottom of the column are all constrained. The constant amplitude cyclic loading as a sinusoidal wave with a force amplitude of ΔF = 60 kN, a stress ratio of $R = -1$ and a period of T = 24 s is exerted on the far end of the beam, as shown in Figure 2a. The bilinear kinematic hardening rule using von Mises plasticity (BKIN) is assumed for the steel.

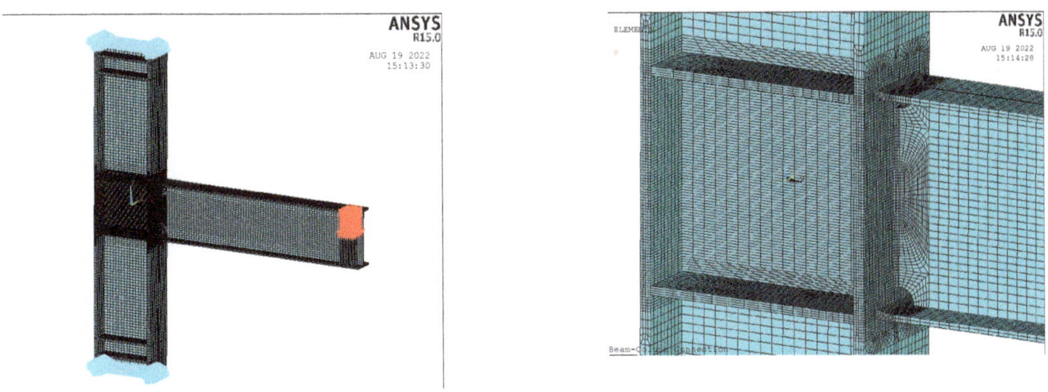

(a) (b)

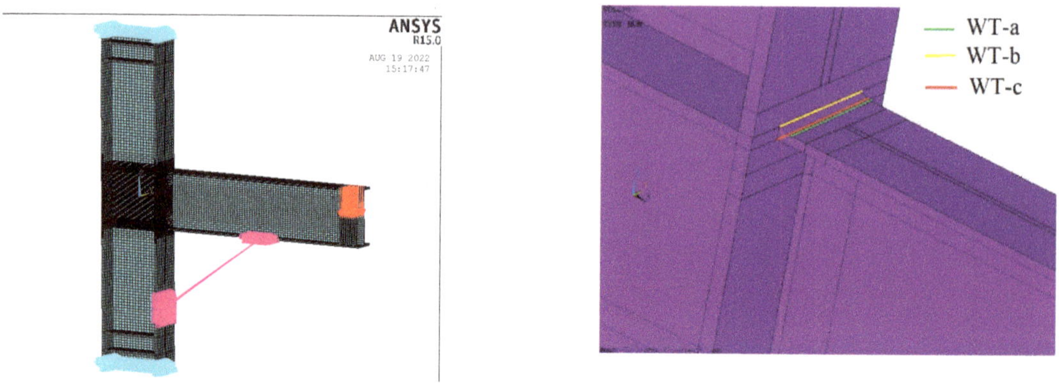

(c) (d)

Figure 2. Finite element model of the connection. (**a**) Connection without the VFD. (**b**) Local view of the weld detail. (**c**) Connection with the VFD. (**d**) Definition of weld toe lines.

2.1.2. Results and Analysis

The fatigue performance of this connection is strongly related to the stress amplitude at the weld detail. Thus, three weld toe lines of the weld seam connecting the top beam flange and the column are defined as shown in Figure 2d, where WT-a is the weld toe on the top beam flange, WT-b is the top weld toe on the column web and WT-c is the bottom weld toe on the column web. The maximum local principal stress $\sigma_{L\text{-max}}$ along the three weld toe lines in the connection with and without the VFD during the loading is, respectively, illustrated in Figure 3a,b. From the figures, it is found that the three weld toe lines show different stress distribution characteristics. The stress approximately presents an inverted V-shaped distribution along WT-a and the stress concentration is more serious at the center of the line, while it presents a W-shaped distribution along WT-b and WT-c, and the stress concentration is more serious at both the two sides and the center of the line. Moreover, it is found from the comparison of the figures that the maximum local principal stress along the weld toe decreases dramatically with the installation of the VFD for all the three weld toe lines, at a percentage of about 70%. It indicates that the VFD plays a great role in the mitigation of local stress near the weld detail.

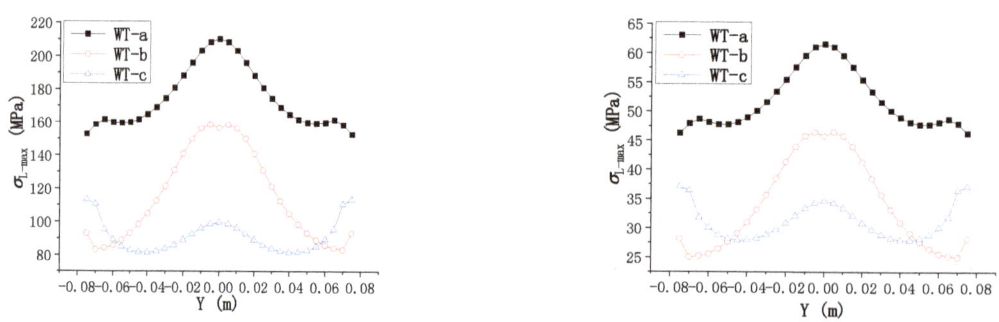

(a) (b)

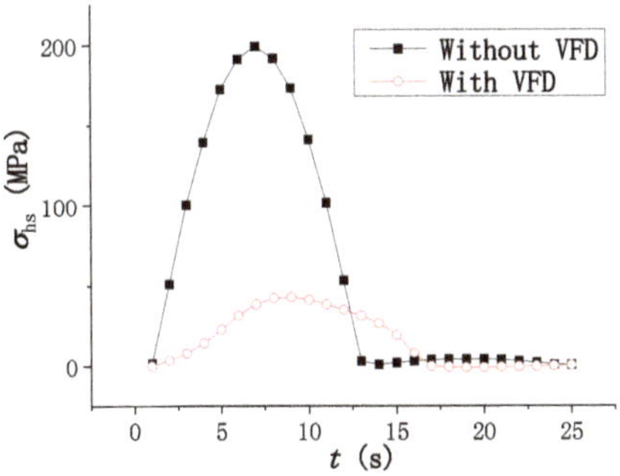

(c)

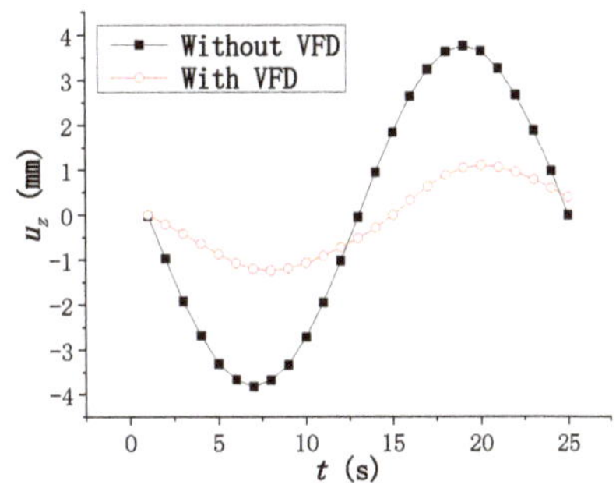

(d)

Figure 3. Stress and displacement of the connection with or without VFD. (**a**) Maximum local principal stress along the weld toe in the connection without the VFD. (**b**) Maximum local principal stress along the weld toe in the connection with the VFD. (**c**) Hotspot stress comparison. (**d**) Vertical displacement comparison.

The fatigue life of a component consists of fatigue crack initiation life and fatigue crack growth life. The fatigue crack initiation life is the life from the start of the cyclic loading to the appearance of visible cracks with a length of 1 mm [15], while the fatigue crack growth life starts from the appearance of these visible cracks to final ruptures of the component. Since the fatigue crack growth life is relatively short and it tends to be dangerous if final rupture is regarded as the failure criterion; the fatigue life discussed in this paper is the fatigue crack initiation life according to the suggestion made by the IIW recommendation (IIW2259-15) [16]. It is worth noting that instances of a sudden damage in a connection or a structure can lead to an abrupt change in the mass or stiffness. However, as soon as small visible cracks are found in welded beam-to-column connections, the fatigue analysis stops and there is no need to discuss the following events such as the abrupt change in the mass or stiffness, since visible cracks with a length of 1 mm in beam-to-column connections lead to little change in the structural mass or stiffness.

So far, the authors have conducted a series of research in fatigue assessment of welded beam-to-column connections using different fatigue assessment approaches, including the nominal stress approach, the hotspot stress approach, the equivalent structural stress approach, the notch stress approach and the theory of critical distances, etc. [17,18]. The detailed information and advantages/disadvantages of all these approaches can be found in the authors' previous publication [19]. Thus, it is necessary to decide proper fatigue assessment approaches for the connections in this paper. The finite element model of the single connection in this section is mainly established using SOLID elements and thus the nominal stress is not a good option, since the abstraction of the nominal stress from the analysis results is not as easy as that in a model established using BEAM elements. The notch stress approach and the theory of critical distances are two approaches based on local stress, so that they both require very refined mesh sizes in the finite element model and thus take a tedious amount of computing time and expends tremendous computing resources. Since there are numerous computing conditions involved and the computation of models with VFDs even requires time-consuming nonlinear analysis, these two approaches are also not the best options. The two approaches left are the equivalent structural stress approach and the hotspot stress approach. Although the former has the advantage of being insensitive to mesh size and requires only one S-N curve, it is more effective in finite element models using SHELL elements and the abstraction of the membrane stress and the bending stress require an integral calculation in models using SOLID elements, while the calculation using the latter model is more efficient, especially in the computation of numerous computing conditions. Thus, the hotspot stress approach is selected as the proper approach for the single connection in this section. On the contrary, all the approaches except the nominal stress approach are not very applicable to the multi-connections of a structure in Section 3, since these approaches all require models using SOLID or SHELL elements. It is true that the use of a multi-scale finite element model may be the solution to this problem. However, fatigue analysis needs to be conducted within numerous connections in Section 3 and it is nearly impossible to establish a multi-scale finite element model for so many connections. Therefore, the fatigue assessment for the multi-connections in a high-rise building is conducted using the nominal stress approach. It is worth noting that the above discussion is mainly for the connections under high-cycle fatigue damage. Plastic strain is found near the welding location under low-cycle fatigue damage; therefore, the local stress–strain approach is the best option.

The hotspot stress represents the structural stress at the hotspot location, i.e., the weld toe location, which includes all stress-raising effects of a structural detail. It is determined on the surface at the hotspot (weld toe) of the component to be assessed using two reference points, the location 0.4t and 1.0t away from the weld toe, to extrapolation to the weld toe location [16]. The hotspot stress σ_{hs} is calculated as Equation (2):

$$\sigma_{\text{hs}} = 1.67\sigma_{0.4t} - 0.67\sigma_{1.0t} \quad (2)$$

where $\sigma_{0.4t}$ and $\sigma_{1.0t}$ are the stresses of the location 0.4t and 1.0t away from the weld toe and t is the plate thickness. The stress results in these two locations can be easily abstracted from the finite element analysis results. After the hotspot stress range $\Delta\sigma_{\text{hs}}$ under the cyclic loading is obtained, the fatigue life is calculated using the S-N curve, as shown in Equation (3):

$$(\Delta\sigma_{\text{hs}})^{m_0} N = C_0 \quad (3)$$

where m_0 and C_0 are parameters of the S-N curve, dependent on the type of the welded connection. According to IIW recommendations (IIW2259-15) [16], FAT90 is the most suitable S-N curve for this connection, where $C_0 = 4.046 \times 10^{15}$ and $m_0 = 5$ when $\Delta\sigma_{\text{hs}} \leq 52.7$ MPa (knee point of the S-N curve), while $C_0 = 1.458 \times 10^{12}$ and $m_0 = 3$ when $\Delta\sigma_{\text{hs}} > 52.7$ MPa. From the brief introduction above to the hotspot stress approach, it is clear that this approach uses the stress obtained by extrapolation from the two reference points to the weld toe as the key stress index (hotspot stress) and thus the fatigue life is calculated from this key stress index based on an S-N curve specified by IIW recommendations.

The hotspot stress σ_{hs} near WT-a, with more serious stress concentration than the other two weld toe lines, within one cycle of loading is shown in Figure 3c. Meanwhile, the far-end vertical displacement of the beam u_z is illustrated in Figure 3d. It is found from the figures that the hotspot stress and the vertical displacement are both considerably decreased by about 70% during the cycle of loading and the two indexes show similar and synchronous mitigation trends that show a better vertical displacement mitigation effect along with a better stress mitigation effect. Moreover, it is found that the two indexes decrease dramatically from the beginning of the cyclic loading, which means that the VFD takes effect instantly with the external loading. Meanwhile, it is worth noting that the hotspot stress and the vertical displacement results of the connection with the VFD show a phase difference of about $\pi/12$ lagging behind those of the connection without the VFD. It indicates that the use of the VFD changes the phase property of the structural responses (output sinusoidal wave).

The fatigue life is then calculated based on the hotspot stress range by Equation (3) and a fatigue life ratio k is defined as shown in Equation (4):

$$k = N_e/N_0 = T_e/T_0 \quad (4)$$

where N_0 and N_e are the fatigue life of the connection without and with the passive control system, and T_0 and T_e are their counterparts expressed in the unit of a year, which is used in Section 3. It is obvious that a larger k value indicates a better fatigue damage mitigation effect and fatigue life prolonged effect. The result is listed in the first row of Table 1. It is found that a k value of 15 is obtained so that the fatigue life is prolonged considerably with the installation of the VFD.

2.1.3. Parametric Study

To better understand the effect of the two technical parameters, the damping coefficient C and the damping exponent α on the mitigation effect of both the hotspot stress and vertical displacement, simultaneously, the double Y-axis plot is established as shown in Figure 4a,b. The Y-axis on the left side indicates the hotspot stress range $\Delta\sigma_{hs}$ and that on the right side shows the absolute value of the far-end vertical displacement of the beam, $|u_z|$. The variation trend of both the two indexes with the damping coefficient and the damping exponent is plotted. It is found from Figure 4a,b that both the hotspot stress range and the far-end displacement decrease with the increase in the damping coefficient and they increase with the increase in the damping exponent. The two indexes show similar and synchronous mitigation trends for all the parameter conditions, which coincides with the phenomenon found in Figure 3c,d. It indicates that the two optimal technical parameters which contribute to the maximum structural vibration response mitigation effect also lead to the maximum fatigue damage mitigation effect. This conclusion is rather constructive to the design of passive control systems in high-rise buildings, since the optimal technical parameters for the damping systems simultaneously lead to the most effective mitigation of both the stress (fatigue damage) and the vibration responses (displacement) in a structure.

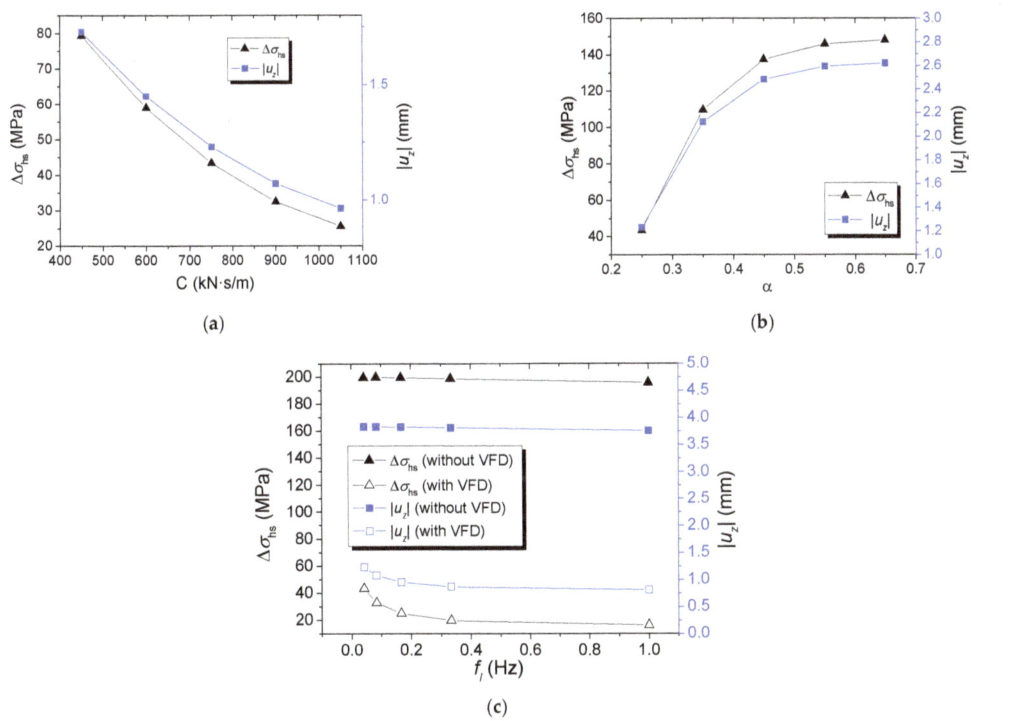

Figure 4. Effect of parameters on hotspot stress range results of the connection with and without the VFD. (**a**) The effect of the damping coefficient. (**b**) The effect of the damping exponent. (**c**) The effect of the loading frequency.

Table 1. Results of the connection with and without the VFD.

Condition No.	Parameters				Without the VFD			With the VFD			
	T (s)	f_l (Hz)	α	C (kN·s/m)	$\Delta\sigma_{hs}$ (MPa)	$\lvert u_z \rvert$ (mm)	N_0	$\Delta\sigma_{hs}$ (MPa)	$\lvert u_z \rvert$ (mm)	N_e	k
1	24	0.042	0.25	450				79.34	1.73	2.920×10^6	15
2	24	0.042	0.25	600				58.94	1.45	7.122×10^6	37
3	24	0.042	0.25	750				43.37	1.23	2.638×10^7	138
4	24	0.042	0.25	900				32.55	1.07	1.108×10^8	580
5	24	0.042	0.25	1050	196.86	3.83	1.911×10^5	25.56	0.96	3.705×10^8	1939
6	24	0.042	0.35	750				109.58	2.12	1.108×10^6	6
7	24	0.042	0.45	750				137.46	2.48	5.613×10^5	3
8	24	0.042	0.55	750				145.96	2.60	4.689×10^5	2
9	24	0.042	0.65	750				148.03	2.62	4.495×10^5	2
10	12	0.083	0.25	750	199.59	−3.828	1.834×10^5	43.37	1.230	2.638×10^7	144
11	6	0.167	0.25	750	199.72	−3.831	1.830×10^5	32.91	1.075	1.047×10^8	572
12	3	0.333	0.25	750	199.48	−3.826	1.837×10^5	24.94	0.953	4.193×10^8	2283
13	1	1.000	0.25	750	198.80	−3.813	1.856×10^5	19.73	0.868	1.352×10^9	7286

It is well-known to all that the effect of a VFD is strongly related to the relative moving velocity between its two ends and thus the mitigation effect with the loading frequency f_l is also discussed. The similar double *Y*-axis plot is illustrated in Figure 4c. It is found that the hotspot stress range and the far-end displacement of the connection without the VFD is not obviously affected by the loading frequency, while those of the connection with the VFD both decrease with the increase in the loading frequency nonlinearly. It can be explained by the fact that a larger loading frequency leads to larger relative velocity between the two ends of the VFD and finally gives rise to a better mitigation effect due to its velocity dependence.

The fatigue life and its prolonged effect for all the parameter conditions are listed in Table 1, where Conditions No. 1–5 show the effect of the damping coefficient, Conditions No. 6–9 give the effect of the damping exponent and Conditions No. 10–13 indicate the effect of the loading frequency. It is found that the k value varies dramatically in different conditions, which can even reach hundreds or thousands. It reveals that the use of a VFD is extremely effective in fatigue damage mitigation and it is more effective in the connection with a large damping coefficient and a small damping exponent under cyclic loading with a high frequency.

2.2. Mitigation Using the TMD System

The TMD system consists of a main structure and a substructure attached to it. The substructure contains masses, springs and dampers, so that it has the properties of mass, stiffness and damping. A typical SODF (Single Degree of Freedom) structure with a TMD is shown in Figure 5a. The natural frequency of the substructure is adjusted to be close to that of the main structure by changing its mass and stiffness. When the main structure vibrates under the external load, the substructure outputs an inertial force opposite to the vibration direction of the main structure and thus the kinetic energy of the main structure is dissipated through the damping of the substructure to obtain a mitigation effect.

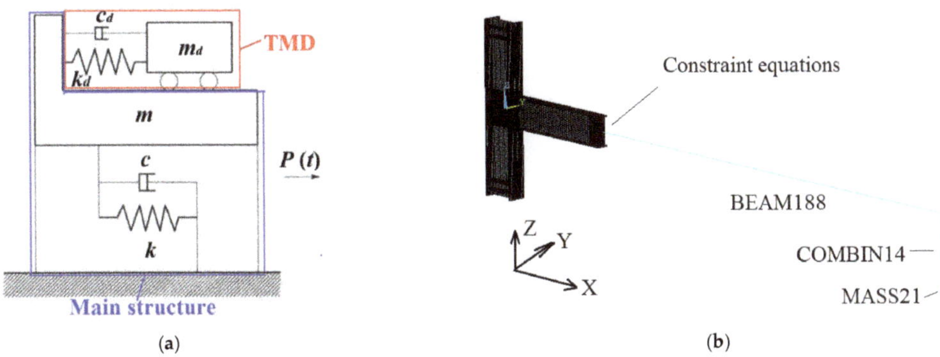

Figure 5. Mechanical model and finite element model of the structure or connection with a TMD. (**a**) Mechanical model of a SODF structure with a TMD. (**b**) Finite element model of the connection with a TMD.

2.2.1. Formulation of the Model

The finite element model of the connection without the TMD is defined as Model A and it keeps the same as that without the VFD in Section 2.1. A model analysis is conducted first to show the vibration modes of the connection and the results of the first three primary vibration modes are given in Table 2. It is found from Table 2 that the frequency of the third vibration mode is 104.79 Hz and this mode, as the first order of translation in the *Z*-axis, is the same vibration pattern as that under the cyclic loading. It is obvious that this vibration mode is too rigid for mitigation using the TMD, since it has been pointed out that a TMD works with better effect in flexible structures than in rigid structures [2,20].

Thus, the frequency of this connection is adjusted to make it more flexible by increasing the beam length by a factor of four so that the beam length is 1.65 × 4 = 6.6 m, while other dimensions and boundary conditions remain unchanged. Due to limited computing resources, a multi-scale finite element model defined as Model B is established as shown in Figure 5b. The lengthened part of the beam is simulated by BEAM188, a kind of 3D two-node beam element, and it is connected to the original beam using degree-of-freedom constraint equations. The mass part of the TMD simulated using MASS21, a kind of structural mass element with six degrees of freedom, is connected through COMBIN14, a kind of spring damper element, to the bottom of the beam far-end location where the largest vertical displacement of the third vibration mode is reached. The first three vibration modes are calculated and listed in Table 2, which are the first order of translation in the Y-axis, the first order of rotation around the X-axis and the first order of translation in the Z-axis. Thus, the cyclic loading is a sinusoidal wave with a force amplitude of ΔF = 6 kN, a stress ratio of $R = -1$ and a frequency of 6.31 Hz exerted on the far end of the beam in Model B, while the similar cyclic loading with a force amplitude of ΔF = 60 kN, a stress ratio of $R = -1$ and a frequency of 104.79 Hz is exerted on Model A, so that the frequency of the cyclic loading keeps almost equal to that of the third vibration mode of both Model A and Model B to make the TMD work with great efficiency.

Table 2. Frequency of models and parameters of the TMD.

Model	Modal Frequency (Hz)			Parameters of the TMD							
	1	2	3	m_d (kg)	m (kg)	μ	r_{fopt}	ξ_{opt}	f_d (Hz)	C_{opt} (N·s/m)	k_{opt} (N/m)
A	22.13	41.39	104.79	9.34	467.06	0.02	0.98	0.086	106.89	1075.87	4.21×10^6
B	1.30	4.67	6.31	15.99	799.47	0.02	0.98	0.086	6.44	110.96	2.62×10^4
C	0.396	0.417	0.468	2.305×10^7	1.152×10^6	0.05	0.952	0.134	0.416	8.040×10^5	7.854×10^6
D	0.390	0.427	0.452	-	-	-	-	-	-	-	-

Den Hartog [21] suggested the optimal technical parameters of a TMD such as the optimal frequency ratio r_{fopt}, the optimal damping ratio ξ_{opt}, the optimal spring stiffness k_{opt} and optimal damping coefficient C_{dopt}. They are calculated using Equations (5)–(8) [20,22]:

$$r_{fopt} = 1/(1+\mu) \tag{5}$$

$$\xi_{opt} = \sqrt{\frac{3\mu}{8(1+\mu)}} \tag{6}$$

$$k_{opt} = r_{fopt}^2 \omega^2 m_d \tag{7}$$

$$C_{dopt} = 2\xi_{opt} r_{fopt} \omega m_d \tag{8}$$

where the frequency ratio r_f is the ratio of the frequency of the TMD f_d to that of the structure f, as $r_f = f_d/f$, the mass ratio μ is the ratio of the mass of the TMD m_d to that of the structure m, as $\mu = m_d/m$, the damping ratio of TMD ξ is expressed as $\xi = c_d/(2m_d\omega_d)$ and c_d is the damping coefficient of the TMD and ω_d is the circular frequency of the TMD. According to Equations (5)–(8), the optimal parameters for the TMD in both models are calculated and listed in Table 2.

The time-history analysis is conducted, respectively, for Model A and Model B, and the hotspot stress of WT-a and the vertical displacement at the far end of the beam are illustrated in Figure 6. It is found that the TMD designed with the optimal parameters suggested by Den Hartog has little effect on the hotspot stress and vertical displacement results in Model A, while it decreases both the hotspot stress and displacement results distinctly in Model B, at a percentage rate of 10–15%. This phenomenon proves that the TMD works with greater efficiency in flexible structures than in rigid structures and it is the reason why Model A needs to be adjusted to Model B. It is also found that the hotspot stress and the far-end vertical displacement show similar and synchronous mitigation trends that show a better vertical displacement mitigation effect along with a better stress mitigation effect, which is the same as the phenomenon found in the connection with the VFD. Moreover, the equipment of the TMD causes little phase difference from the original phase in the two indexes, as shown in Figure 6c,d, and the TMD starts to take effect from the second cycle and reaches the greatest efficiency from the third cycle. This phenomenon is completely different from the equipment of the VFD, which causes phase difference from the original phase in the two indexes, and it starts to work instantly from the first cycle. It may stem from the fact that the mitigation of the TMD takes effect due to the inertial force generated by its vibration, and the vibration is relatively weak in the first cycle and becomes stronger and reaches a stable vibration stage at the second or third cycle when the TMD takes effect.

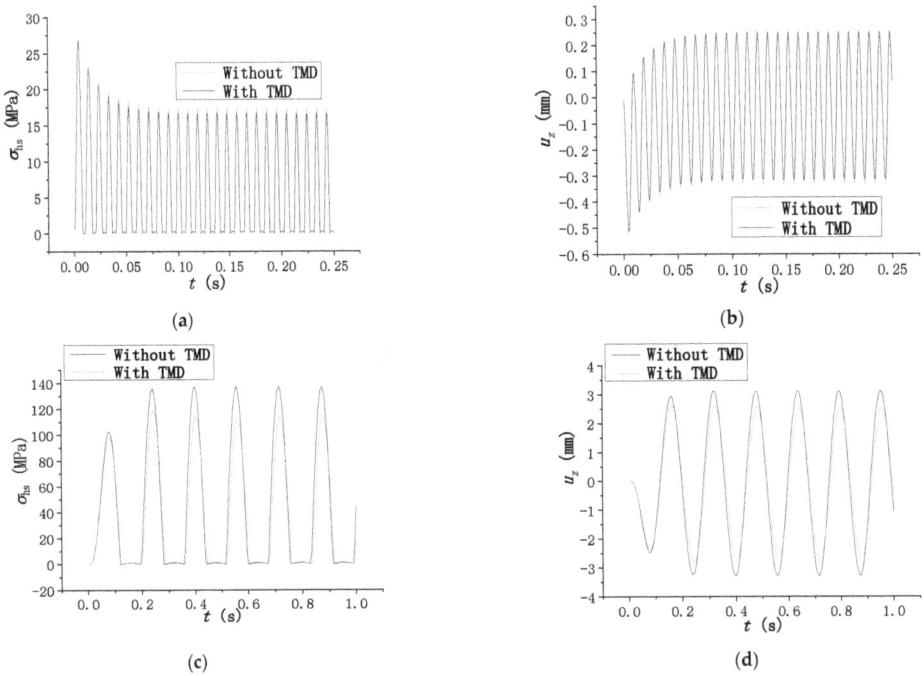

Figure 6. Time-history results of Model A and Model B. (**a**) Hotspot stress time-history of Model A. (**b**) Far-end vertical displacement of Model A. (**c**) Hotspot stress time-history of Model B. (**d**) Far-end vertical displacement of Model B.

2.2.2. Parametric Study

The effect of technical parameters such as the frequency ratio, the damping ratio and the mass ratio on the mitigation effect are all considered and the detailed results are illustrated in the double *Y*-axis plot in Figure 7 and listed in Table 3, where Conditions No. 1–7 show the effect of the frequency ratio, Conditions No. 8–12 indicate the effect of the damping ratio and Conditions No. 13–16 present the effect of the mass ratio.

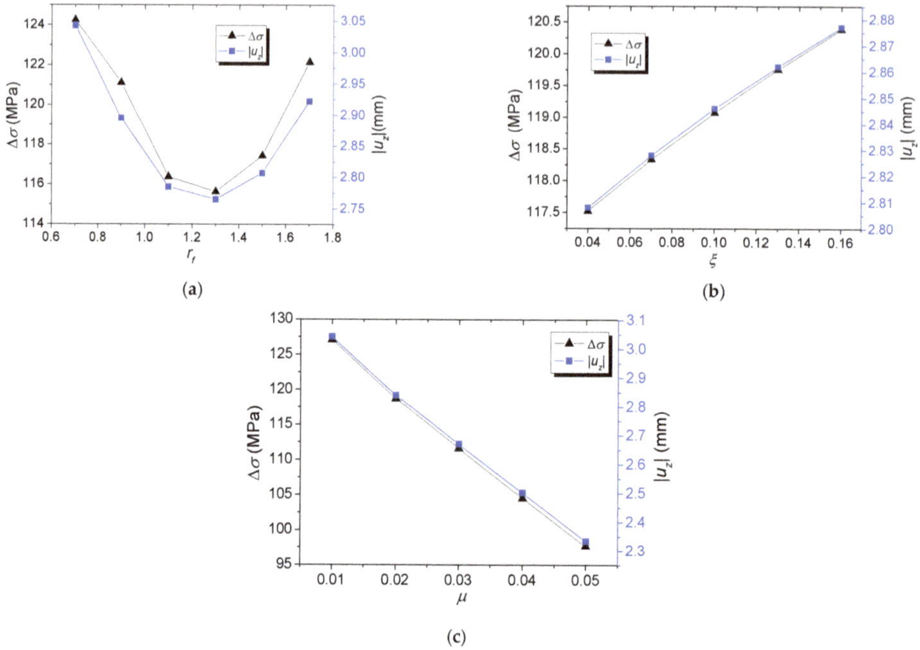

Figure 7. Effect of parameters on hotspot stress range results of the connection with and without the TMD. (**a**) Effect of the frequency ratio. (**b**) Effect of the damping ratio. (**c**) The effect of the mass ratio.

It is found that the mitigation effect decreases with the increase in frequency ratio until it reaches a minimum value; it starts to increase after that, when the damping ratio keeps constant at 0.086. It indicates that there is indeed an optimal frequency ratio for the TMD; therefore, the design of the TMD should be based on it. It is worth noting that the optimal frequency ratio for this TMD is about 1.3, which is only the optimal value for this specific loading frequency (6.31 Hz). It is completely different from the optimal frequency ratio of 0.98 proposed by Den Hartog, which is the optimal frequency ratio considering series of loading frequencies.

The hotspot stress increases almost linearly with the increases in the damping ratio when the frequency ratio keeps constant at 0.98, which indicates that the damping ratio leads to the decrease in the mitigation effect under such a frequency ratio. Meanwhile, the hotspot stress decreases almost linearly with the increase in the mass ratio when the damping ratio and the frequency ratio both keep constant, indicating that the increase in the TMD mass generally contributes to a better mitigation effect. However, it is impossible to increase it infinitely, since the overweight TMD increases the bending moment and shear force within the beam and causes potential failure danger. Thus, a mass ratio of 0.02–0.05 is often adopted by engineering designs.

Based on the hotspot stress results, the fatigue life and its ratio are calculated for all the parameter conditions of the TMD, as shown in Table 3. It is found that the fatigue life ratio generally ranges from 1 to 3. Although the connection with the TMD has a longer beam length and more flexible stiffness than that with the VFD and they are not the same connection; the fatigue life ratios of the connection with the VFD are still considerably larger than those with the TMD, based on the comparison of Tables 2 and 3. It indicates that the use of a VFD in a single welded beam-to-column connection generally has a better fatigue damage mitigation effect than the use of a TMD, and it has better usability and takes effect instantly, for the VFD is suitable for connections with any stiffness, while the TMD is suitable to more flexible connections with a relatively low natural frequency.

Table 3. Results of the connection with and without the TMD.

Condition No.	Parameters			Without the TMD			With the TMD			
	μ	r_f	ξ	$\Delta\sigma_{hs}$ (MPa)	$\|u_z\|$ (mm)	N_0	$\Delta\sigma_{hs}$ (MPa)	$\|u_z\|$ (mm)	N_e	k
1	0.02	0.98	0.086	137.326	3.28	5.630×10^5	118.729	2.837	8.711×10^5	1.5
2	0.02	0.70	0.086				124.254	3.040	7.600×10^5	1.3
3	0.02	0.90	0.086				121.105	2.900	8.209×10^5	1.5
4	0.02	1.10	0.086				116.362	2.790	9.254×10^5	1.6
5	0.02	1.30	0.086				115.626	2.770	9.432×10^5	1.7
6	0.02	1.50	0.086				117.425	2.810	9.005×10^5	1.6
7	0.02	1.70	0.086				122.144	2.920	8.001×10^5	1.4
8	0.02	0.98	0.040				117.521	2.808	8.983×10^5	1.6
9	0.02	0.98	0.070				118.331	2.828	8.800×10^5	1.6
10	0.02	0.98	0.100				119.074	2.846	8.636×10^5	1.5
11	0.02	0.98	0.130				119.757	2.862	8.489×10^5	1.5
12	0.02	0.98	0.160				120.387	2.877	8.356×10^5	1.5
13	0.01	0.98	0.086				127.076	3.040	7.105×10^5	1.3
14	0.03	0.98	0.086				111.636	2.669	1.048×10^6	1.9
15	0.04	0.98	0.086				104.534	2.501	1.276×10^6	2.3
16	0.05	0.98	0.086				97.699	2.334	1.563×10^6	2.8

2.3. *Low-Cycle Fatigue*

The above research mainly discusses the condition of high-cycle fatigue damage. However, welded beam-to-column connections may suffer damage under potential earthquakes, which can be classified as low-cycle fatigue damage. Since it was demonstrated that a VFD is more effective in fatigue damage mitigation than a TMD in the previous two sections, the use of the VFD is taken as an example to discuss its effectiveness in low-cycle fatigue damage mitigation.

The elastoplastic time-history analysis is conducted. Since the connection may suffer plastic deformation under low-cycle fatigue damage, the cyclic displacement loading with a displacement amplitude of $\Delta d = 6$ mm and a displacement ratio of $R = -1$ is exerted on the far end of the beam instead of the cyclic force loading, and the wave type and period remain the same. The local plastic strain near the weld detail of the connection with and without the VFD is shown as a contour map in Figure 8. It is found that the plastic strain near the weld detail decreases or even disappears with the equipment of the VFD.

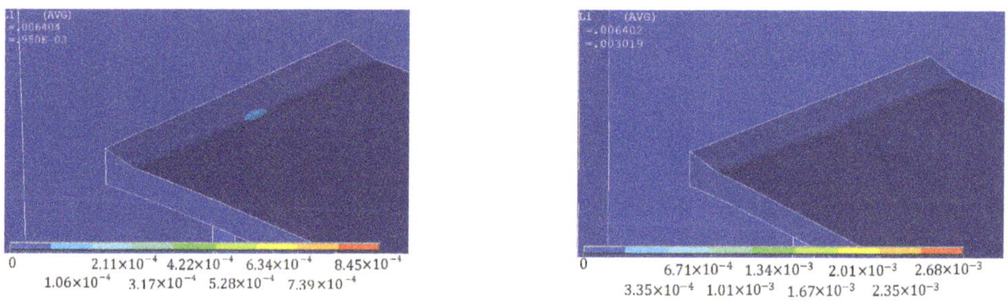

(**a**) (**b**)

Figure 8. Contour map of the local plastic strain. (**a**) Connection without the VFD. (**b**) Connection with the VFD.

The fatigue assessment of a welded connection under low-cycle fatigue needs to be conducted using the local stress–strain approach, instead of the hotspot stress approach, due to the existence of plastic strain. This approach is proposed by Socie [23], based on the assumption that the fatigue life of a welded specimen is the same as that of a smooth specimen if the stress–strain time history of the two specimens is the same. The fatigue life is calculated using the Coffin–Manson Equation, as shown in Equation (9):

$$2N = \left[\frac{\frac{\sigma'_f}{E}(2N)^{b-c} + \varepsilon'_f}{\frac{\Delta\varepsilon}{2}}\right]^{-\left(\frac{1}{c}\right)} \tag{9}$$

where σ'_f, ε'_f, b and c are four material-dependent parameters and they are 949 MPa, 0.159, −0.103 and −0.467 for the steel GB Q345B; $\Delta\varepsilon/2$ is the local strain amplitude. The results of the connections with and without the VFD using the damping coefficient C and the damping exponent α are listed in Table 4.

It is found from Table 4 that the fatigue damage mitigation of the connection with the VFD under the low-cycle fatigue condition is not as effective as that under the high-cycle fatigue condition, where the fatigue life ratio k is mainly about 2, while that in the high-cycle fatigue can reach even hundreds and thousands. The main reason may be that the increase in the stress slows down and reaches a stable stage in the stress–strain curve of the material due to the yield of the steel and the plastic strain, so that the local stress and strain in the connection with the VFD does not seem to vary a lot from that in the connection without the VFD. This may lead to the decreased fatigue life prolonging effect. Meanwhile, it is also found that the fatigue damage mitigation effect in the low-cycle fatigue is not greatly affected by the damping coefficient or the damping exponent, indicating that it is not sensitive to the change of technical parameters. This phenomenon is completely different from that in the high-cycle fatigue condition.

Table 4. Results of the connection with and without the VFD under low-cycle fatigue.

Condition No.	Parameters		Without the VFD			With the VFD			
	α	C (kN·s/m)	$\Delta\varepsilon$ (1×10^{-3})	u_z (mm)	N_0	$\Delta\varepsilon$ (1×10^{-3})	u_z (mm)	N_e	k
1	0.25	450	2.034	−6.35	58,256	1.840	−6.32	92,129	1.6
2	0.25	600				1.833	−6.32	93,801	1.6
3	0.25	750				1.828	−6.32	95,020	1.6
4	0.25	900				1.824	−6.32	96,011	1.6
5	0.25	1050				1.819	−6.32	97,269	1.7
6	0.35	750				1.844	−6.32	91,192	1.6
7	0.45	750				1.852	−6.32	89,355	1.5
8	0.55	750				1.854	−6.32	88,904	1.5
9	0.65	750				1.855	−6.32	88,679	1.5

3. Fatigue Damage Mitigation for Multi Connections in the Structure

3.1. Finite Element Model

A high-rise building located in a coastal city susceptible to severe wind disasters in China is taken as an example of a steel braced-frame structure. The building has an ellipse plane shape with a long axis of 59.4 m and a short axis of 22.02 m, as shown in Figure 9a. It has 27 floors above ground, with a building height of 96 m. The hot-rolled H-sectional beams and box-sectional columns made of structural steel GB Q345B are connected by welded connections, where all the welding is conducted in the form of the full penetration groove weld. The sections of the columns on the first floor are Z1(600 × 600 × 45 × 45), Z2(600 × 850 × 50 × 50), Z3(600 × 700 × 40 × 40), Z4(600 × 600 × 50 × 50), Z6(600 × 600 × 53 × 53) and Z9(700 × 700 × 65 × 65), as shown in Figure 9a (Unit: mm). The beams on the first floor are mainly with a section of H500 × 200 × 10 × 16 and H588 × 300 × 12 × 20 (Unit: mm). The sectional areas of beams and columns gradually decrease on higher floors. The braces are mainly of a section of H428 × 407 × 17.5 × 35 (Unit: mm). The floor plate, with a thickness of 120 mm, is made of the constructional concrete GB C30 with an elastic modulus of 30 GPa and a Poisson's ratio of 0.2. The material property of GB Q345B is referred to in Section 2.1.1. The finite element model is established in the ANSYS software, as shown in Figure 9b. The beams, columns and braces are all simulated using BEAM188, a type of 3D two-node beam element and floors are simulated by SHELL63, a type of four-node shell element, at a mesh size of 2 m. To better validate this finite element model, the model of the same structure is established in another finite element analysis software, ETABS, as shown in Figure 9c. The natural vibration frequency results of the first three vibration modes are listed in Table 2 as those of Model D and they are compared with those of the model established in ANSYS as Model C in the same table. The discrepancy of the frequency results of the first three vibration modes between the two models is 1.15%, 2.39% and 3.34%, respectively, and they are all within a discrepancy limit of 5%. It demonstrates that the established finite element model in ANSYS is valid in the establishment of the model.

Stochastic fluctuating wind time series with a duration of 50 s and a time interval of 0.1 s to calculate wind-induced fatigue are generated based on the Davenport power spectrum using the harmonic superposition method [24], as shown in Equations (10) and (11):

$$\frac{nS_v(n)}{\overline{v}_{10}^2} = \frac{4k_r x^2}{n(1+x^2)^{3/4}} \tag{10}$$

$$x = 1200\frac{n}{\overline{v}_{10}} \tag{11}$$

where $\overline{v}_{10}$ is the reference wind speed taken as 21.43 m/s, calculated from the reference wind pressure in this area suggested by Chinese code for structural design (GB50009-2019), with a reference height of 10 m, a time interval of 10 min and a return period of 50 years; n is the frequency of the fluctuating wind; k_r is the terrain roughness factor of this area, which is 0.22; $S_v(n)$ is the power spectrum of velocity fluctuation. The simulated wind speed time series are converted into the frequency domain and they are compared with the Davenport spectrum. The typical fluctuating wind speed time series at the height of the first floor is shown in Figure 9d and it was found that the spectrum of the simulated wind time series agrees with the Davenport spectrum well in most frequency bands, as illustrated in Figure 9e.

Wind loads are only applied in the Y-axis direction of the structure. The total wind speed time series at the location of the j-th connection on the i-th floor, as $V_{ji}(z,t)$, which consist of the mean wind speed $\overline{v}_{ji}(z)$ and fluctuating wind speed $v_{ji}(z,t)$, are converted into the wind pressure time series $W_{ji}(z,t)$ according to Bernoulli's theorem, where z is the height of the i-th floor and t is the time. Subsequently, the wind pressure time series are converted into the concentrated wind force time series $P_{ji}(t)$ by multiplying the loading area A_{ji} of each beam-to-column connection on the windward surface of the building. The definition of the loading area can be found in an earlier publication by the authors [17]. The above conversion of the wind speed time series to the concentrated wind force time series is summarized as Equation (12):

$$\begin{aligned} P_{ji}(t) = A_{ji}\mu_s W_{ji}(z,t) = \tfrac{1}{2}A_{ji}\mu_s\rho V_{ji}{}^2(z,t) = \tfrac{1}{2}A_{ji}\mu_s\rho\left[\overline{v}_{ji}(z) + v_{ji}(z,t)\right]^2 \\ \approx \tfrac{1}{2}A_{ji}\mu_s\rho\overline{v}_{ji}{}^2(z) + A_{ji}\mu_s\rho\overline{v}_{ji}(z)v_{ji}(z,t) \end{aligned} \tag{12}$$

where ρ is the air density; μ_s is the shape coefficient of the structure, suggested by Chinese code (GB50009-2019) as 1.1. Only the buffeting response is considered, since this building does not have a very slender structure.

3.2. Fatigue Analysis

The time-history analysis is conducted and the beam-to-column connections on the 11th floor with the maximum interlayer displacement ratio are the connections for the fatigue analysis. The calculation process and results of the interlayer displacement ratio of each floor are not detailed due to limited pages. Since the columns and beams are all simulated using the beam element, the fatigue assessment is conducted using the nominal stress approach. Nominal stress σ_{nom} is the sectional stress calculated based on the internal forces on the component section away from the welded joint, where the local stress-raising effects do not exist, as shown in Equation (13):

$$\sigma_{\text{nom}} = \frac{N_c}{A} + \frac{M_X}{W_X} + \frac{M_Z}{W_Z} \tag{13}$$

where N_c is the axial force, A is the sectional area of the beam in the Y-axis direction, M_X and M_Z are the moment around X-axis and Z-axis, and W_X and W_Z are the section modulus around X-axis and-Z axis. The directions of the axes are shown in Figure 9a.

The nominal stress spectrums (cumulative frequency diagrams), including the nominal stress ranges $\Delta\sigma$ and corresponding cycle numbers n, are obtained from the nominal stress time series by the rainflow-counting method, as shown in Figure 10. The effective nominal stress range $\Delta\sigma_e$ is calculated by Equation (14):

$$\Delta\sigma_e = \left[\frac{1}{D_N}\cdot\frac{\sum_{i=1}^{k} n_i(\Delta\sigma_{\text{nom}-i})^{m_0}}{\sum_{i=1}^{k} n_i}\right]^{1/m_0} \tag{14}$$

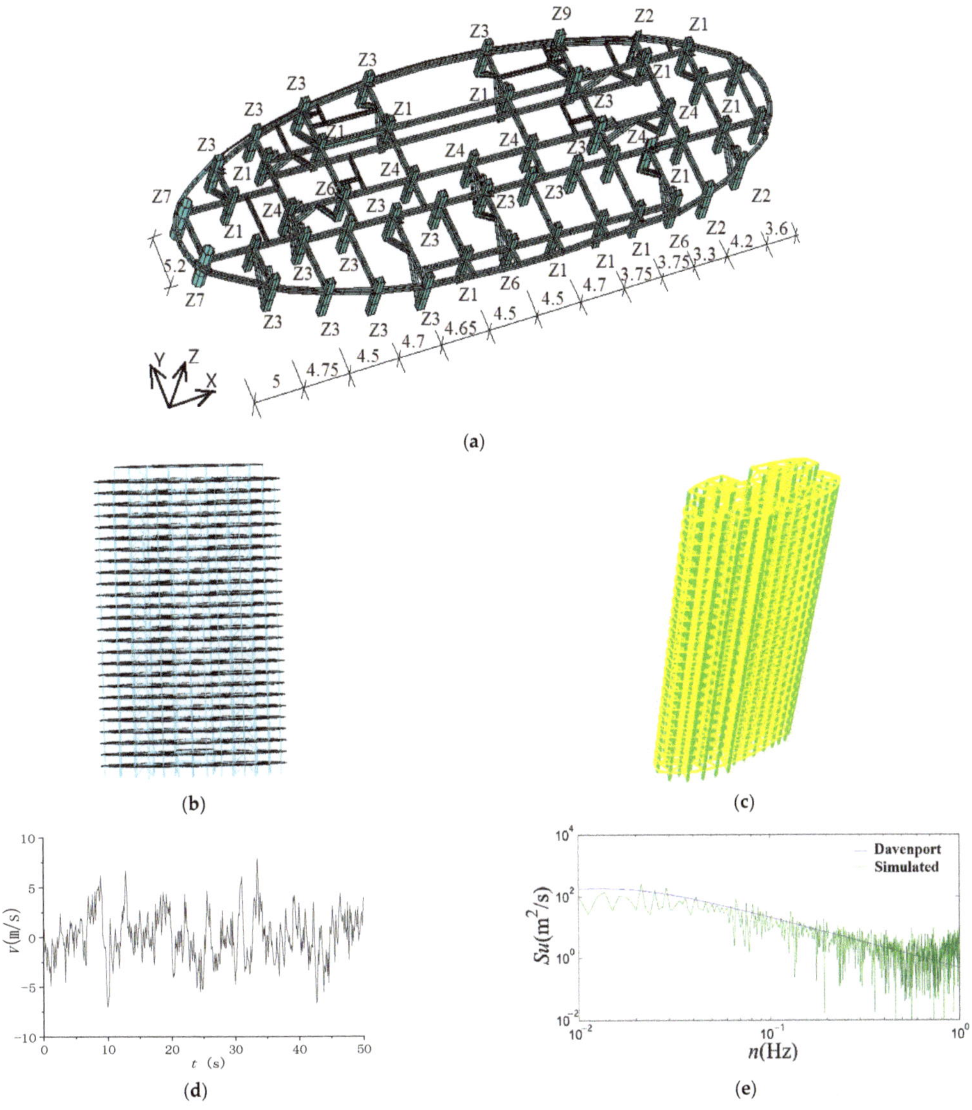

Figure 9. Summary of the building and the model. (**a**) Floor layout. (**b**) Finite element model in ANSYS. (**c**) Finite element model in ETABS. (**d**) Fluctuating wind speed time series. (**e**) Power spectrum comparison.

where $\Delta\sigma_{\mathrm{nom}-i}$ is the i-th nominal stress range causing fatigue damage in the cumulative frequency diagram; n_i is the cycle number under stress range $\Delta\sigma_{\mathrm{nom}-i}$; D_N is the specified Miner sum as 1. Based on the S-N curve of FAT36 suggested by IIW recommendation (IIW2259-15) for the nominal stress approach used in this connection, the fatigue damage within 50 s as D_t is calculated by Equation (15):

$$D_t = \frac{\Delta\sigma_e{}^{m_0} \sum_{i=1}^{k} n_i}{C_0} \tag{15}$$

where $C_0 = 4.143 \times 10^{13}$ and $m_0 = 5$ when $\Delta\sigma_{nom} \leq 21.1$ MPa (knee point of the S-N curve), while $C_0 = 9.331 \times 10^{10}$ and $m_0 = 3$ when $\Delta\sigma_{nom} > 21.1$ MPa [16]. The fatigue life T_0 (Unit: year) is calculated as shown in Equation (16):

$$T_0 = \frac{t_{50}}{D_t t_y} \tag{16}$$

where $t_{50} = 50$ s and t_y is the total time in one year, as 3.1536×10^7 s.

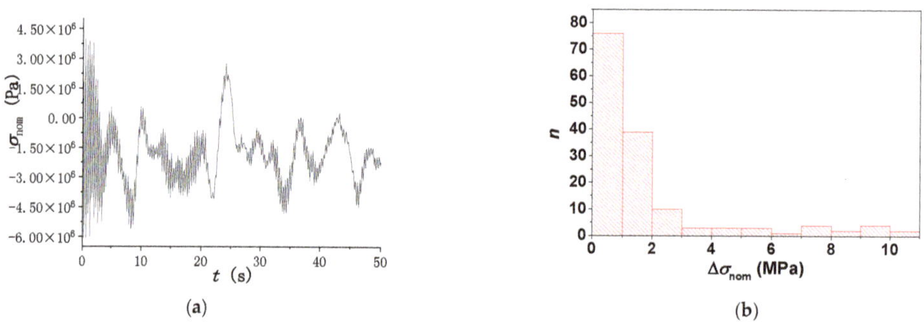

Figure 10. Nominal stress information of Connection 5561. (**a**) Time-history of the nominal stress. (**b**) Nominal stress range spectrum.

The results of eight typical connections, as shown in Figure 9a, are listed in Table 5. It is found from the table that Connection 5725 is the critical connection with the greatest fatigue damage, and its fatigue life is 93 years, which tends to be dangerous during its service life of 100 years. Thus, fatigue damage mitigation using the VFD and the TMD is necessary and is discussed, respectively.

3.3. Structure with VFD Systems

According to the results of the time-history analysis in the previous section, the VFDs are installed at specific locations, as shown in Figure 9a, in two ways, haunched and braced, as shown in Figure 11, from the 1st floor to the 20th floor. The damping coefficient of the VFD is C = 500 kN·s/m and the damping exponent is α = 0.25.

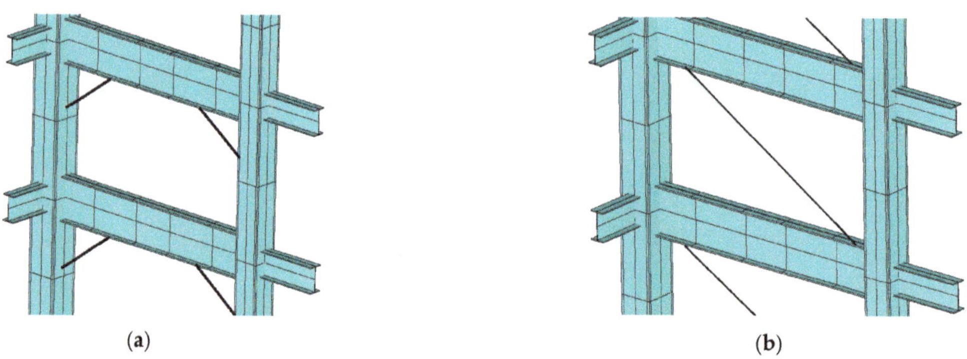

Figure 11. Installation of the VFD. (**a**) Haunched. (**b**) Braced.

Table 5. Fatigue analysis results.

No.	$\sum_{i=1}^{k} n_i$	Connection without Damping Systems		Connection with VFD						Connection with TMD		
				Haunched			Braced					
		D_t	T_0 (Year)	D_t	$T_{e\text{-HVFD}}$ (Year)	k	D_t	$T_{e\text{-BVFD}}$ (Year)	k	D_t	$T_{e\text{-TMD}}$ (Year)	k
5561	138	1.08×10^{-8}	146	7.73×10^{-10}	2052	14.1	5.80×10^{-10}	2735	18.7	2.07×10^{-9}	765	5.2
5585	137	4.85×10^{-9}	326	3.75×10^{-10}	4226	13.0	3.13×10^{-10}	5062	15.5	1.06×10^{-9}	1489	4.6
5626	137	3.59×10^{-9}	441	3.04×10^{-9}	522	1.2	9.79×10^{-9}	162	0.4	1.09×10^{-9}	1459	3.3
5788	138	1.39×10^{-8}	113	2.05×10^{-9}	773	6.8	1.27×10^{-9}	1249	11.1	6.98×10^{-9}	227	2.0
5622	137	5.93×10^{-9}	267	5.27×10^{-9}	301	1.1	1.17×10^{-8}	135	0.5	1.78×10^{-9}	893	3.3
5659	138	1.55×10^{-8}	102	1.49×10^{-9}	1065	10.4	1.01×10^{-9}	1574	15.4	4.86×10^{-9}	326	3.2
5725	138	1.70×10^{-8}	93	2.33×10^{-9}	680	7.3	1.51×10^{-9}	1047	11.3	6.89×10^{-9}	230	2.5
5811	139	2.52×10^{-10}	6285	6.98×10^{-11}	22,718	3.6	3.02×10^{-11}	52,434	8.3	1.26×10^{-10}	12,535	2.0

The simulation of the VFDs is kept the same as that in Section 2.1. The time-history analysis and fatigue analysis are conducted and the results are shown in Figure 12a and Table 5, where $T_{e\text{-HVFD}}$ and $T_{e\text{-BVFD}}$ are the fatigue life of the connections in the structure with haunched VFDs and braced VFDs. Figure 12a indicates that the nominal stress is reduced dramatically with the equipment of either haunched VFDs or bracing VFDs. However, the reduction effect is better in the structure with braced VFD systems, with less cost than that with haunched VFD systems. It is found from Table 5 that the fatigue life is prolonged for most typical connections except two connections, 5626 and 5622. This may be explained by the situation that these two connections are just the position where the VFD is installed, so that the damping force output by the VFD leads to the increase in the local stress within these two connections during the energy dissipation and thus decrease the fatigue life. This phenomenon is completely different from that of a single beam-to-column connection in Section 2, whose local stress is also reduced, even when the VFD is installed within this connection itself.

The two typical connections of 5561 and 5621 with the braced VFD systems are selected to demonstrate the effect of different damping coefficients (500 kN·s/m, 600 kN·s/m, 700 kN·s/m, 800 kN·s/m and 900 kN·s/m) and damping exponents (0.25, 0.35, 0.45, 0.55 and 0.7) on fatigue damage mitigation. The results of a total of 25 parameter conditions are shown in Figure 12b,c, where a fatigue life ratio k is defined as Equation (4).

It is found from Figure 12b that the fatigue life is significantly prolonged for the connection without the direct installation of the VFD, such as Connection 5561, and the prolonged effect tends to decrease with the increase in the damping exponent. This stems from the fact that the velocity response of the structure is relatively small under wind and the VFD designed with a low damping exponent is more sensitive to the small velocity response and leads to a larger damping force output. However, with the increase in the damping exponent, the influence of the damping coefficient on the mitigation effect becomes less. For example, the increase in the damping coefficient can significantly improve the mitigation effect when the damping exponent is 0.25, while it has little effect on the mitigation effect when the damping exponent reaches 0.7. It indicates that a small damping exponent is enough to reach a satisfying mitigation effect for the structure under wind.

Meanwhile, the mitigation effect is relatively bad or even becomes negative due to the increase in the local stress in the connection with the direct installation of the VFD, such as Connection 5626. Under such circumstances, the damping exponent should be designed to be greater than 0.55 to obtain a positive mitigation effect, according to Figure 12c. However, the VFDs are directly installed in two locations, as shown in Figure 9a, involving only four connections, so that the parameter design of the other connections can still refer to that of Connection 5561.

Moreover, the VFD systems can also contribute well to the mitigation of structural responses at the same time, as shown in Figure 12d. It is found that the structural top displacement response decreases dramatically and instantly after the equipment of the VFD systems, which is due to the phenomenon found in Section 2 that the stress and the displacement results show similar and synchronous mitigation trends.

3.4. Structure with TMD Systems

The modal analysis of this structure, defined as Model C, is conducted before the design of the TMD systems and the results of the first three vibration modes are listed in Table 2, which are the first order of translation in the Y axis, the first order of translation in the X axis and the first order of rotation around the Z axis. The maximum displacement for the first vibration mode mainly excited by the wind loading is reached at the top of the structure, so that this location is optimal for the installation of the TMD. Therefore, TMD is installed at the top of the structure, as shown in Figure 13a, simulated using MASS21, and it is connected to the structure through COMBIN14.

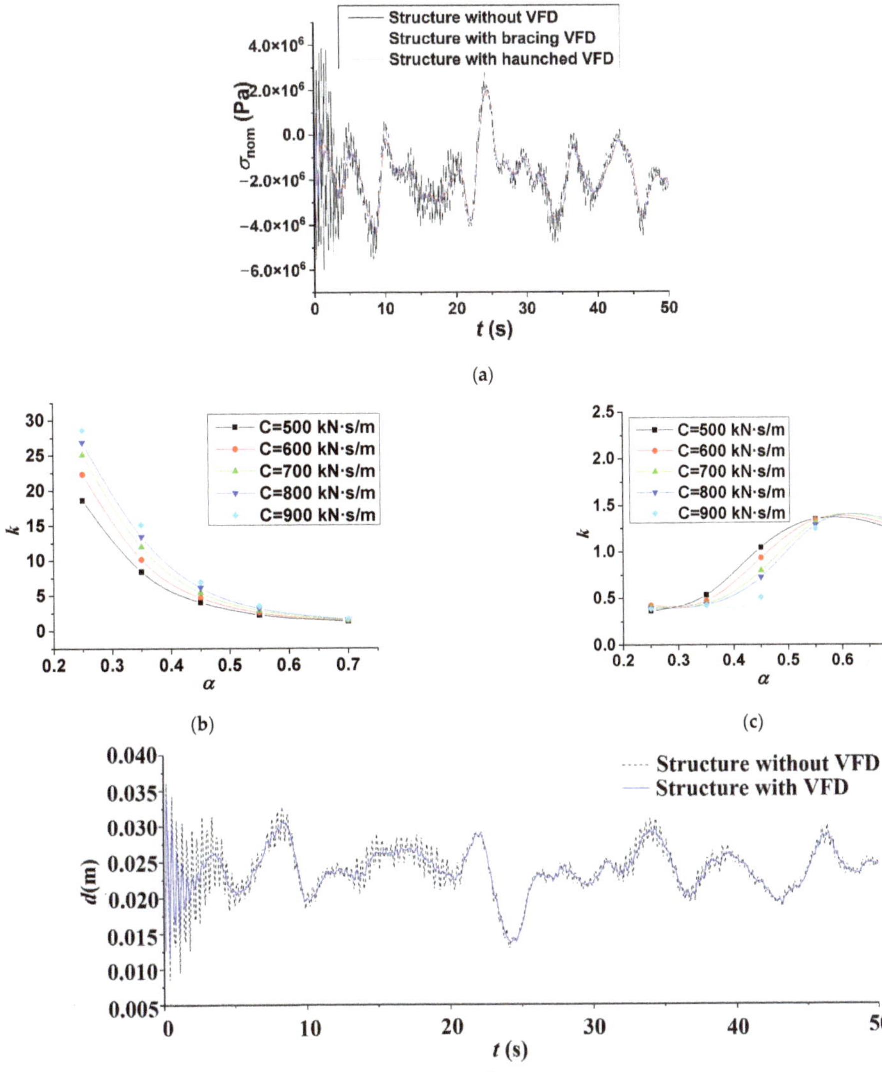

Figure 12. Fatigue mitigation effect of connections in structures with VFD systems. (**a**) Nominal stress time-history. (**b**) Parametric study of Connection 5561. (**c**) Parametric study of Connection 5626. (**d**) Time-history of structural top displacement response.

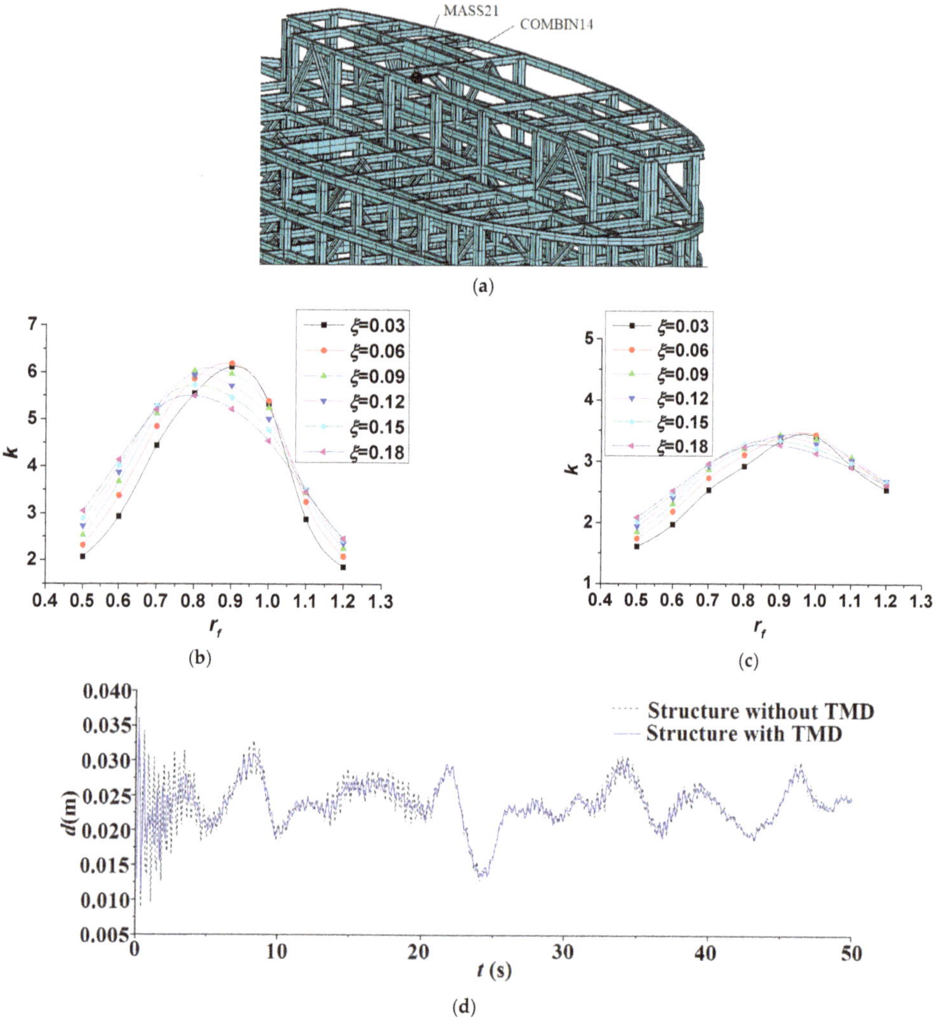

Figure 13. Fatigue mitigation effect of connections in structures with the TMD. (**a**) Installation of the TMD. (**b**) Parametric study of Connection 5561. (**c**) Parametric study of Connection 5626. (**d**) Time-history of structural top displacement response.

The mass ratio is set to be $\mu = 0.05$ and the optimal parameters for the TMD under wind loading suggested by Den Hartog are calculated and listed in Table 2. After the time-history analysis and the fatigue analysis are both conducted, the fatigue life and ratio results are listed in Table 5. It is found from Table 5 that the fatigue damage mitigation using the TMD with generally smaller k values is not as effective as that using the VFDs on the same connection in the same structure, which agrees with the phenomenon in Section 2. Thus, it comes to the conclusion that the VFD is generally more effective in fatigue damage mitigation than the TMD for welded beam-to-column connections in steel high-rise buildings. Regarding the economic effect, the VFD is more economical and easier to install for the mitigation of a small amount of connections, while the TMD is more economical and easier to install for the mitigation of all the connections in a high-rise building.

To better demonstrate the effect of technical parameters on the mitigation results, the influence of different frequency ratios (r_f = 0.5, 0.6, 0.7, 0.8, 0.9, 1.0, 1.1, 1.2) and damping ratios (ξ = 0.03, 0.06, 0.09, 0.12, 0.15, 0.18) on the fatigue performance of two typical connections, Connection 5561 and 5626, is studied. The variation trend of the k value with both the two parameters is shown in Figure 13b,c.

It is found from Figure 13b,c that the frequency ratio has a greater impact on the fatigue life results, compared with the damping ratio. With the increase in the frequency ratio at the early stage, the fatigue life ratio tends to increase and it reaches the maximum value when the frequency ratio is generally between 0.9–1.0, which indicates the optimal mitigation effect and it agrees with the optimal frequency ratio of 0.952 calculated based on the theory proposed by Den Hartog. After that, the fatigue life ratio shows a decreasing trend when the frequency ratio continues to increase.

The effect of the damping ratio is much more complex. The fatigue life ratio increases with the damping ratio when the frequency ratio is less than about 0.8. However, it starts to decrease with the damping ratio when the frequency ratio is between 0.8 and 1.05. When the frequency ratio is finally greater than 1.05, the fatigue life ratio starts to increase again or keeps almost constant with the increase in the damping ratio. It also explains the reason why the damping ratio of the TMD only leads to the decrease in the mitigation effect under a frequency ratio of 0.98 for the single connection in Section 2.

Moreover, TMD systems also contribute well to the mitigation of structural displacement responses at the same time, as shown in Figure 13d. It is found from the figure that, after the equipment of the TMD system, the fluctuation of the structural top displacement response d decreases dramatically. In addition to that, based on the comparison of Figures 12d and 13d, it is obvious that the structural top displacement response starts to decrease from the first fluctuating peak in the structure with the VFD, while it starts to decrease from the second and the third fluctuating peak, which agrees with the phenomenon found for the single connection in Section 2 that the VFD starts to take effect almost instantly, while the TMD starts to take effect after one or two cycles.

4. Discussion

4.1. Optimization of Damper Positions

In the previous sections, the fatigue damage mitigation using VFD systems or TMD systems is discussed. It is obvious that the layout positions of the dampers definitely affect the mitigation effect. Since the layout of numerous VFDs is much more complex than a single TMD, the VFD systems are taken as an example and the optimization of their layout positions is discussed in this section.

In theory, the most economical and efficient layout plan is to place VFDs in a single optimum position for each floor and thus the single position layout is discussed. The VFD is placed in ten typical positions (C1–C10) from the 1st floor to the 20th floor, as shown in Figure 14a. The fatigue life and the ratio k results for some typical beam-to-column connections on the 11th floor are shown in Table 6. For a better comparison, the k value results for these connections are illustrated in Figure 14b, where the maximum relative displacement and maximum relative velocity between the two positions which connect the two ends of the potential dampers in the model without VFDs from C1 to C10 are also listed in the legend part of Figure 14b.

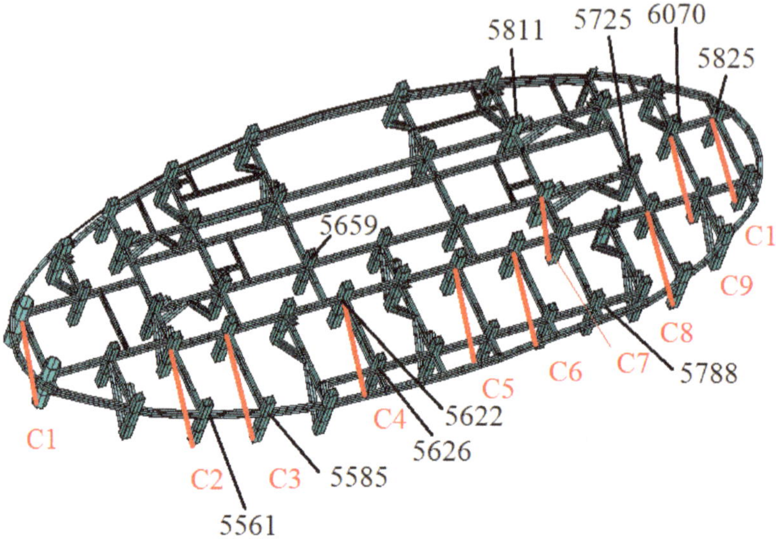
5811
5725
6070
5825
5659
C10
C9
C8
C7
5788
C6
C5
5622
C4
5626
C1
C2
C3
5585
5561

(a)

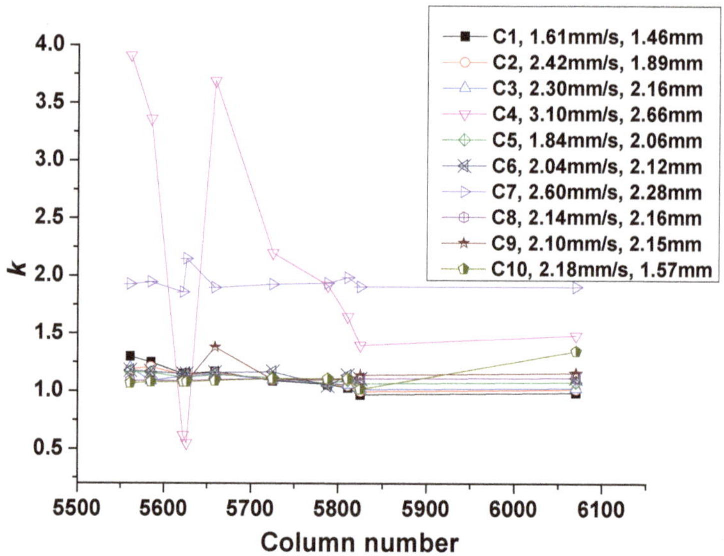
C1, 1.61mm/s, 1.46mm
C2, 2.42mm/s, 1.89mm
C3, 2.30mm/s, 2.16mm
C4, 3.10mm/s, 2.66mm
C5, 1.84mm/s, 2.06mm
C6, 2.04mm/s, 2.12mm
C7, 2.60mm/s, 2.28mm
C8, 2.14mm/s, 2.16mm
C9, 2.10mm/s, 2.15mm
C10, 2.18mm/s, 1.57mm
k
4.0
3.5
3.0
2.5
2.0
1.5
1.0
0.5
5500
5600
5700
5800
5900
6000
6100
Column number

(b)

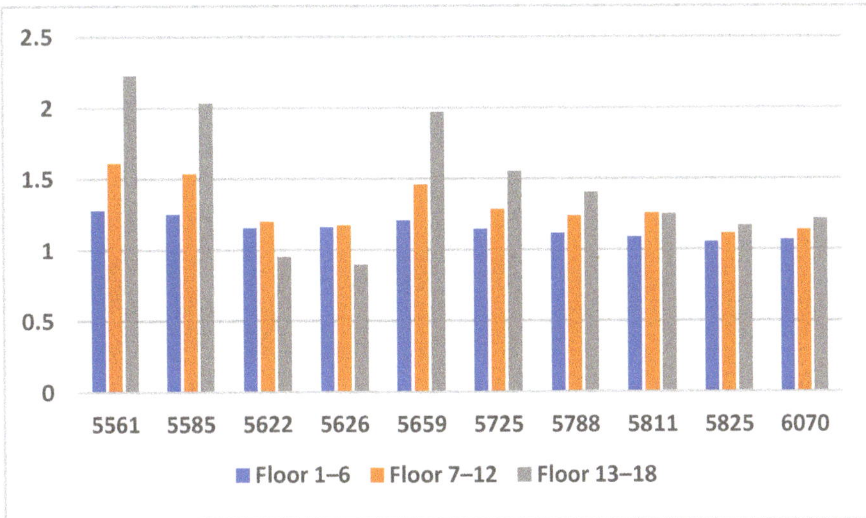

(c)

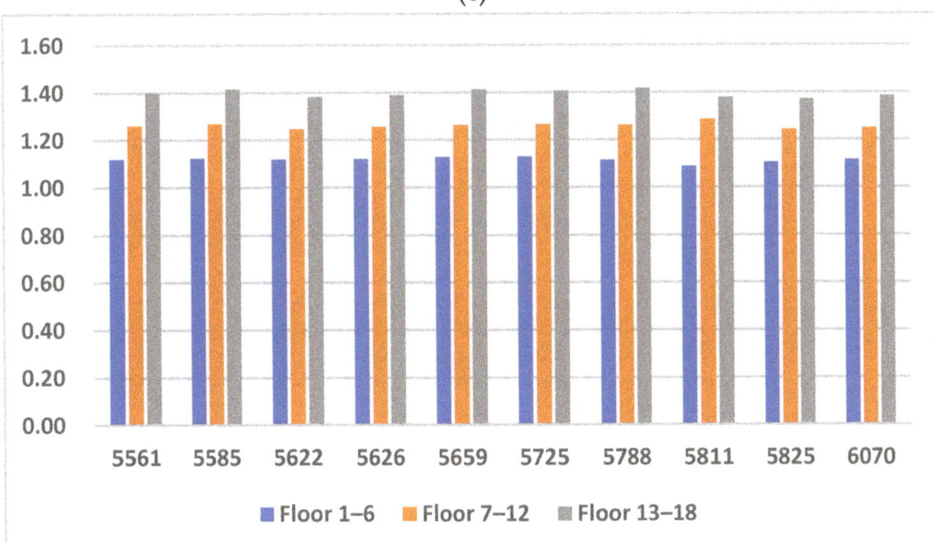

(d)

Figure 14. Layout of VFDs and results of k values in connections. (**a**) Layout plan of VFDs from C1 to C10. (**b**) Results of k values in different connections on the 11th floor. (**c**) Results of k values in connections with VFDs being installed in C4 on different floors. (**d**) Results of k values in connections with VFDs being installed in C4 on different floors.

Table 6. Results of fatigue life and k values in connections with VFDs in positions from C1 to C10.

VFD Positions	Fatigue Life	5561	5585	5622	5626	5659	5725	5788	5811	5825	6070
Without VFD	T_0 (Year)	161	353	325	543	100	95	257	3.24×10^8	186	230
C1	T_c (Year)	209	441	373	625	117	103	271	3.33×10^8	181	230
	k	1.30	1.25	1.15	1.15	1.17	1.09	1.06	1.03	0.97	0.99
C2	T_c (Year)	192	427	371	621	116	104	274	3.40×10^8	186	230
	k	1.19	1.21	1.14	1.14	1.16	1.10	1.07	1.05	1.00	1.02
C3	T_c (Year)	194	386	367	615	115	104	276	3.43×10^8	189	240
	k	1.20	1.09	1.13	1.13	1.15	1.10	1.07	1.06	1.02	1.03
C4	T_c (Year)	630	1187	202	300	369	209	494	5.34×10^8	260	340
	k	3.91	3.36	0.62	0.55	3.69	2.20	1.92	1.65	1.40	1.49
C5	T_c (Year)	188	409	363	608	114	107	274	3.56×10^8	199	250
	k	1.17	1.16	1.12	1.12	1.14	1.12	1.07	1.10	1.07	1.08
C6	T_c (Year)	190	415	374	626	116	111	270	3.69×10^8	207	260
	k	1.18	1.17	1.15	1.15	1.16	1.17	1.05	1.14	1.11	1.12
C7	T_c (Year)	311	689	605	1168	190	183	498	6.44×10^8	356	440
	k	1.93	1.95	1.86	2.15	1.90	1.93	1.94	1.99	1.91	1.91
C8	T_c (Year)	176	386	356	594	110	105	284	3.59×10^8	207	260
	k	1.09	1.09	1.09	1.09	1.10	1.11	1.10	1.11	1.11	1.12
C9	T_c (Year)	173	380	352	588	138	105	283	3.60×10^8	212	270
	k	1.07	1.08	1.08	1.08	1.38	1.10	1.10	1.11	1.14	1.16
C10	T_c (Year)	172	380	352	588	109	105	284	3.61×10^8	189	310
	k	1.07	1.08	1.08	1.08	1.09	1.11	1.11	1.11	1.02	1.35

It is found from the table and the figure that the fatigue ratio k mainly depends on two things. The first and the most important one is the relative displacement and relative velocity between the two positions which connect the two ends of the potential dampers in the model without VFDs. It is found that the k value tends to increase in the positions simultaneously with large maximum relative displacement and relative velocity. It leads to the fact that, when the dampers are installed in the positions of C4 and C7 with large relative displacement and relative velocity, the k value can even reach a rather large value. Meanwhile, when the dampers are installed in other positions, such as C2 and C3, the relative displacement is relatively small, though they have a large maximum relative velocity, which leads to a smaller k value compared with C4 and C7. It indicates that the relative displacement and relative velocity between the two positions which connect the two ends of the potential dampers determine the k value simultaneously. It is rather easy to understand, since the fatigue mitigation comes from the working of the dampers and the working efficiency is strongly related to the relative displacement and relative velocity between the two ends. Meanwhile, relative displacement and relative velocity are rather close to each other in these connections when the dampers are installed in positions other than C4 and C7, which leads to similar k values.

The second thing that needs to be mentioned is the distance from the targeted beam-to-column connection to the VFD installation positions. If the VFD is installed near to the targeted connection, a larger k value is probably obtained in these connections, such as Connection 5561 with C1 position, Connection 5561 and 5585 with C2 and C3 positions, Connection 5659 with C4 position, Connection 5788 and 5811 with C7 position, etc. Meanwhile, the installation of the VFD directly at the targeted connection usually also leads to a relatively large, though sometimes not optimum, k value, such as Connection 5561 with C2 position, Connection 5585 with C3 position, Connection 6070 with C9 position, etc. However, there is one special case where a k value smaller than 1 is found in both

Connection 5622 and 5626 when dampers are installed in the position of C4, just between the two connections, which indicates a negative fatigue damage mitigation effect. This special case is discussed in Section 4.2.

In addition to the above analysis, the layout of VFDs on different floors is also discussed. The VFDs are installed in the positions of C4 and C7 on floors from 1 to 6 (low floors), from 7 to 12 (middle floors) and from 13 to 18 (high floors). The results for C4 and C7 positions are, respectively, listed in Figure 14c,d. It is found from the figure that the k value tends to increase when the VFDs are placed on high floors in almost all the connections, although the targeted connections are all on the 10th floor, as a middle floor. It means that the VFDs should be placed on higher floors instead of lower floors or middle floors, even if they contain the targeted connection.

Based on the above analysis, it is concluded that, for the sake of a better fatigue damage mitigation effect in a targeted beam-to-column connection, the optimized VFD layout plan should be that VFDs are placed between the two connections with large relative displacement and relative velocity on higher floors to increase their working efficiency and, moreover, these two connections with VFDs should be near to the targeted connection.

4.2. Negative Mitigation Effect

It has been pointed out that Connection 5622 and Connection 5626 both show negative fatigue damage mitigation effect, where the k values are smaller than 1. To better understand this phenomenon, all the connections directly with the VFDs in the positions from C1 to C10 are illustrated in Figure 15 and the k value results of these connections are listed in Table 7. In the table, the VFDs are placed between the Connection i and Connection j. Connection i is the connection on the south side, such as 5271, 5561, 5585, etc. Connection j means the connection on the north side, such as 5286, 5317, 5333, etc.

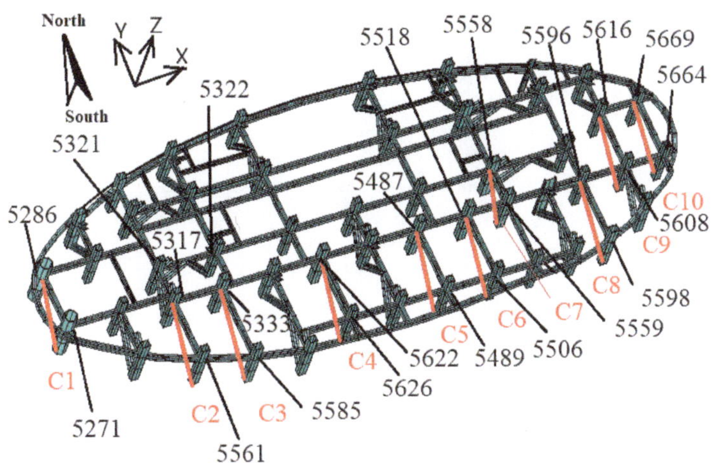

Figure 15. Layout of two end connections with VFDs in positions from C1 to C10.

It is found from the table that the connections directly equipped with VFDs in the positions of C4, C5 and C6 show a negative fatigue damage mitigation effect, while other connections all show a positive fatigue damage effect. It means that the direct equipment does not always lead to a negative mitigation effect. It is found from Figure 15 that Connection j with the damping positions of C4, C5 and C6 are all without direct lateral support of beams from the north side in the Y-axis direction, such as Connections 5622, 5487 and 5518, which means that the lateral support force exerted on these three connections in the Y-axis is generally low and thus may result in serious stress concentration within these connections when the VFD is installed. On the contrary, Connection j of most other connections are all supported by lateral beams directly from the north side in the Y-axis to provide enough lateral force and thus may avoid serious stress concentration within these connections. It is worth noting that Connection j with the position of C3 (Connection 5333), which is not directly supported by lateral beams, still shows a positive mitigation effect. This may stem from the fact that the bracing between Connection 5321 and Connection 5322 and that between Connection 5321 and Connection 5317 near to this connection provides additional lateral support and the support force is transmitted to Connection 5333, which thus may avoid serious stress concentration. On the contrary, Connections 5622, 5487 and 5518 do not have such a source of lateral support force. Therefore, it is concluded that the negative fatigue damage mitigation mainly stems from an insufficient lateral support force, so that the direct installation of VFDs may result in a negative fatigue damage mitigation effect in the connections with limited lateral support.

Table 7. Results of fatigue life and k values in the two end connections with VFDs in positions from C1 to C10.

			C1	**C2**	**C3**	**C4**	**C5**	**C6**	**C7**	**C8**	**C9**	**C10**
Model without VFD	Connection i	T_0 (Year)	136	161	353	543	319	190	258	911	210	206
	Connection j	T_0 (Year)	136	130	321	325	208	130	181	1015	230	186
Model with VFD	Connection i	T_c (Year)	181	192	386	300	203	90	270	1027	210	213
	Connection j	T_c (Year)	179	156	342	202	203	60	189	1136	270	189
	Connection i	k	1.33	1.19	1.09	0.55	0.64	0.46	1.05	1.13	1.00	1.03
	Connection j	k	1.31	1.20	1.06	0.62	0.97	0.50	1.04	1.12	1.16	1.01

5. Conclusions

(1) The VFD and the TMD systems are both effective in local fatigue damage mitigation, along with structural displacement mitigation, for both a single connection under constant amplitude cyclic loading and multi-connections in a high-rise building under stochastic wind, while the mitigation effect decreases in the low-cycle fatigue condition compared with the high-cycle fatigue condition; therefore, these two passive control systems both have promising usability in fatigue damage mitigation of welded beam-to-column connections in steel high-rise buildings.

(2) The VFD generally has a better mitigation effect than the TMD and it starts to take effect instantly with external loading and is equally effective in connections and structures with almost all ranges of natural frequencies, but it causes a phase difference in the structural responses, including stress and displacement. On the contrary, the TMD starts to take effect after a few cycles and is more effective in connections and structures with a lower natural frequency, and it does not change the phase of the structural responses. Regarding the economic effect, the VFD is more economical and easier to install for the mitigation of one or several connections, while the TMD is more economical and easier to install for the mitigation of all the connections in a high-rise building.

(3) The VFD system should be designed with a small damping exponent and a large damping coefficient in the braced installation form, while the TMD system can be designed according to the optimal parameters, so that the two systems are more optimal to both the local stress and the structural displacement, since the two indexes are found to show similar and synchronous mitigation trends and thus the VFD and the TMD can be designed based on either the local stress or the structural displacement.

(4) For the sake of a better fatigue damage mitigation effect in a targeted beam-to-column connection, the optimized VFD layout plan is that VFDs are placed between the two connections with large relative displacement and relative velocity on higher floors to increase their working efficiency; moreover, these two connections with VFDs should be near to the targeted connection.

(5) The negative fatigue damage mitigation mainly stems from an insufficient lateral support force so that the direct installation of VFDs may result in a negative fatigue damage mitigation effect in the connections with limited lateral support.

Author Contributions: Conceptualization, Z.F. and J.Z.; methodology, A.L.; software, F.Y.; validation, Z.F., J.Z. and F.Y.; formal analysis, Z.F. and J.Z.; investigation, Z.F. and J.Z.; resources, Z.F.; data curation, Z.F. and J.Z.; writing—original draft preparation, Z.F. and J.Z.; writing—review and editing, Z.F.; visualization, F.Y.; supervision, A.L.; project administration, A.L.; funding acquisition, Z.F. and A.L. All authors have read and agreed to the published version of the manuscript.

Funding: This research was supported by the National Natural Science Foundation of China (No.52008202, No.51978154 and No.52108108), the Distinguished Young Scientists of Jiangsu Province (No. BK20190013), the Natural Science Foundation of Jiangsu Province in China (No. BK20191016 and No. BK20191013) and the Beijing Postdoctoral Research Foundation (2021-zz-105).

Institutional Review Board Statement: Not applicable.

Informed Consent Statement: Not applicable.

Data Availability Statement: Due to the nature of this research, participants of this study did not agree for their data to be shared publicly, so supporting data is not available.

Conflicts of Interest: The authors declare no conflict of interest.

References

1. Plumier, A.; Agatino, M.R.; Castellani, A.; Castiglioni, C.A.; Chesi, C. Resistance of steel connections to low-cycle fatigue. In Proceedings of the 11th European Conference on Earthquake Engineering, Paris, France, 6–11 September 1998.
2. Soong, T.T.; Dargush, G.F. *Passive Energy Dissipation Systems in Structural Engineering*; John Wiley & Sons: New York, NY, USA, 1998.
3. Janbazi, R.H.; Tabeshpour, M.R. Spectral fatigue analysis of jacket platform under wave load equipped with viscous damper. *J. Mar. Sci. Technol.* **2019**, *24*, 855–870. [CrossRef]
4. Allen, C.K.; Goupee, A.J.; Viselli, A.M. A computationally-efficient frequency domain model of a floating wind turbine with hull-based tuned mass damper elements. In Proceedings of the 31st International Ocean and Polar Engineering Conference, Rhodes, Greece, 20–25 June 2021.
5. Repetto, M.P.; Solari, G. Wind-induced fatigue collapse of real slender structures. *Eng. Struct.* **2010**, *32*, 3888–3898. [CrossRef]
6. Repetto, M.P.; Solari, G. Dynamic alongwind fatigue of slender vertical structures. *Eng. Struct.* **2001**, *23*, 1622–1633. [CrossRef]
7. Bhowmik, B.; Tripura, T.; Hazra, B.; Pakrashi, V. Real time structural mod25al identification using recursive canonical correlation analysis and application towards online structural damage detection. *J. Sound Vib.* **2020**, *468*, 115101. [CrossRef]
8. Bhowmik, B.; Tripura, T.; Hazra, B.; Pakrashi, V. First-order eigen-perturbation techniques for real-time damage detection of vibrating systems: Theory and applications. *Appl. Mech. Rev.* **2019**, *71*, 060801. [CrossRef]
9. Prawin, J. Real-time reference-free breathing crack identification using ambient vibration data. *Struct. Control. Health Monit.* **2022**, *29*, e2903. [CrossRef]
10. Palmeri, A.; Ricciardelli, F. Fatigue analyses of buildings with viscoelastic dampers. *J. Wind Eng. Ind. Aerodynamics.* **2006**, *94*, 377–395. [CrossRef]
11. Golafshani, A.A.; Gholizad, A. Passive vibration control for fatigue damage mitigation in steel jacket platforms. *Int. J. Eng.* **2008**, *21*, 313–324.
12. Ambrosio, P.; Cazzulani, G.; Resta, F.; Ripamonti, F. An optimal vibration control logic for minimising fatigue damage in flexible structures. *J. Sound Vib.* **2014**, *333*, 1269–1280. [CrossRef]
13. Ripamonti, F.; Cazzulani, G.; Cinquemani, S.; Resta, F.; Torti, A. Adaptive active vibration control to improve the fatigue life of a carbon-epoxy smart structure. In *Active and Passive Smart Structures and Integrated Systems, Proceedings of the SPIE Smart Structures and Materials + Nondestructive Evaluation and Health Monitoring, San Diego, CA, USA, 8–12 March 2015*; SPIE: Bellingham, WA, USA, 2015; pp. 732–740.
14. Andersson, A.; O'Connor, A.; Karoumi, R. Passive and adaptive damping systems for vibration mitigation and increased fatigue service life of a tied arch railway bridge. *Comput.-Aided Civ. Infrastruct. Eng.* **2015**, *30*, 748–757. [CrossRef]
15. Radaj, D.; Sonsino, C.M.; Flade, D. Prediction of service fatigue strength of a welded tubular joint on the basis of the notch strain approach. *Int. J. Fatigue* **1998**, *20*, 471–480. [CrossRef]
16. Hobbacher, A. *Recommendations for Fatigue Design of Welded Joints and Components*; Springer: Dordrecht, The Netherlands, 2015.
17. Fang, Z.; Li, A.; Li, W.; Shen, S. Wind-induced fatigue analysis of high-rise steel structures using equivalent structural stress method. *Appl. Sci.* **2017**, *7*, 71. [CrossRef]
18. Fang, Z.; Li, A.; Ding, Y.; Li, W. Wind-induced fatigue assessment of welded connections in steel tall buildings using the theory of critical distances. *Eur. J. Environ. Civ. Eng.* **2020**, *24*, 1180–1205. [CrossRef]

19. Fang, Z.; Li, A.; Ding, Y.; Shen, S. Three-dimensional surface fatigue crack growth life assessment model of steel welded joints and structures. *Adv. Struct. Eng.* **2022**, *25*, 1776–1791. [CrossRef]
20. Den Hartog, J.P. *Mechanical Vibrations*; McGraw-Hill Book Company: New York, NY, USA, 1956.
21. Den Hartog, J.P. *Mechanical Vibrations*; Dover Publications: New York, NY, USA, 1985.
22. Kelly, S.G.; Kudari, S. *Mechanical Vibrations*; Tata McGraw-Hill Publishing Co., Ltd.: New Delhi, India, 2009.
23. Socie, D.F. Fatigue-life prediction using local stress-strain concepts. *Exp. Mech.* **1977**, *17*, 50–56. [CrossRef]
24. Ubertini, F.; Giuliano, F. Computer simulation of stochastic wind velocity fields for structural response analysis: Comparisons and applications. *Adv. Civ. Eng.* **2010**, *83*, 441–456. [CrossRef]

buildings

MDPI

Article

Investigation of Seismic Performance for Low-Rise RC Buildings with Different Patterns of Infill Walls

Saharat Saengyuan and Panon Latcharote *

Department of Civil and Environmental Engineering, Faculty of Engineering, Mahidol University, Nakorn Pathom 73170, Thailand
* Correspondence: panon.lat@mahidol.ac.th

Abstract: Evaluating the structural performance of low-rise RC buildings with infill walls is an essential issue in Thailand, as most infill walls were not designed for lateral load resistance. The purpose of this study was to predict the structural behavior and illustrate the effects of infill walls. Residential, commercial, and educational buildings were selected as representative buildings with different patterns of infill walls. Based on the results, infill walls contributed to considerable strength and stiffness. Most of the infill walls that affected the low-rise buildings were at the ground floor level. The behavior of the buildings that had a contribution of infill walls was found to be brittle until the infill walls collapsed, and then the buildings became ductile. Some patterns in which infill walls were placed improperly led to a torsional effect, resulting in columns in the affected areas reaching failure criteria more than those without this effect. Considering the NLRHA procedure, only infill walls on the ground floor contributed to the building being subjected to a ground motion. The fully infilled frame tended to reach the infill crack before the other patterns. For the UMRHA procedure, only the first vibration mode was adequate to predict seismic responses, such as roof displacement and top-story drift.

Keywords: pushover analysis; low-rise RC buildings; infill walls; UMRHA; seismic response

Citation: Saengyuan, S.; Latcharote, P. Investigation of Seismic Performance for Low-Rise RC Buildings with Different Patterns of Infill Walls. *Buildings* **2022**, *12*, 1351. https://doi.org/10.3390/buildings12091351

Academic Editors: Liqiang Jiang, Jihong Ye and Wei Guo

Received: 28 July 2022
Accepted: 29 August 2022
Published: 1 September 2022

1. Introduction

Seismic hazards play a critical role in the field of earthquake engineering and structural engineering in which large lateral forces are induced by earthquake shaking. Earthquakes are recognized as one of the natural hazards putting any structure at risk, especially existing low-rise RC buildings with infill walls. Most of the damage to buildings during earthquakes is within a range of light to severe damage levels. Slight damage occurs to structural members, such as beams and columns, and to nonstructural members, such as façades and ceilings, which do not contribute strength or stiffness to buildings. However, buildings with façades and ceilings can survive, and operational systems such as elevators and fire protection can still be functional. For severe levels, the overall structural performance is lost, resulting in sudden member failure and building collapse. Based on lessons learned from past earthquakes, seismic performance evaluation should be performed to investigate whether such buildings will survive future earthquakes. The effects of infill walls on RC buildings should also be examined. Several studies in the past determined that RC buildings with infill walls seemed to have both positive and negative effects on buildings.

Murty and Jain [1] indicated that the negative effects of infill walls were well known as the short column effect, soft-story effect, and out-of-plane collapse due to torsion. The beneficial effects were that infill walls can contribute to strength, stiffness, ductility, and energy dissipation capacity. Lee and Woo [2] found that infill walls showed positive behavior; for instance, an increase in earthquake inertia force was less compared to a contribution of strength by infill walls. The effect of infill walls also led to the reduction in global lateral displacement. Al-Chaar et al. [3] demonstrated that an infilled frame provided higher strength and initial stiffness than a bare frame, and these properties can be increased together, yet not linearly, with an increase in the number of bays. Asteris [4] determined that the lateral stiffness of an infilled frame can be considerably reduced due to an increase in the percentage of openings and the positions of openings moving toward the compression diagonal. The shear force on the columns was found to decrease due to the presence of infill walls. However, the shear force on the columns was higher than the shear force of the bare frame in the case of a weak-story frame. Anil and Altin [5] indicated that the arrangement of the infill as a wing wall was found to be an effective strengthening method and provided higher strength compared to the bare frame. An increase in the aspect ratio of the wing walls also increased the strength and stiffness of the corresponding specimen. However, strength, stiffness, and energy dissipation appeared to be substantially decreased due to the presence of the window opening incorporated into the frame, and a substantial reduction in strength could be observed after the infill concrete was crushed. Dolšek and Fajfar [6] demonstrated that the strength and stiffness of the corresponding frame could be considerably increased by the presence of masonry infills. Hashem et al. [7] indicated that the presence of infill walls led to an increase in frame capacity whereas the presence of openings within infill walls led to an increase in the frame lateral displacement. The presence of shear connectors in infilled frames was also found to decrease lateral displacement and increase ultimate load capacity. Niyompanitpattana and Warnitchai [8] demonstrated that an infill wall with or without an opening contributes significantly to increasing its lateral strength and stiffness, and modifying the hysteretic behavior. The overall strength and stiffness of the building were increased; however, the displacement capacity was decreased due to the presence of infill walls [9,10].

Risi et al. [11] expressed that with the comparison between the results of square infill walls and rectangular infill walls, the latter seemed to sustain higher damage than the former at approximately the same drift level. Khan et al. [12] demonstrated that infill walls played an important role in RC structures, especially for the full infill wall type, since the infill walls made structures stiffer and reduced the fundamental period and the relative displacement. Furtado et al. [13] indicated that infill walls interacting with the bounding frame during an earthquake caused various failure modes due to both in-plane and out-of-plane behavior. Out-of-plane failure was regarded as one of the most crucial failure modes of infill walls subjected to seismic loading [14,15] and was noted to occur on the lower stories of a building [15]. Chrysostomou and Asteris [16] reported that it was difficult to account for the behavior modes of infill walls since the walls were related to several parameters involving their nonlinearity and behavior. Additionally, the exhibition of infill walls was different between in-plane and out-of-plane behavior. The uncertainty regarding the characteristics of infill walls also has a dramatic impact on seismic response parameters [17].

A widely used tool to estimate the structural performance of a building is called a nonlinear static procedure, also known as pushover analysis. This tool is simple and time-saving yet effective. The results, i.e., capacity curves or pushover curves, are shown in terms of force versus displacement format or broadly recognized as base shear versus roof displacement format. Due to the benefit of a pushover curve, the global behavior of the building, such as strength and stiffness, can readily be investigated, as has been undertaken by many researchers [18–20]. Girgin and Darilmaz [21] performed pushover analysis to evaluate the seismic response of infilled frame buildings. They found that infills with no irregularity in elevation have advantages for buildings and, therefore, the global lateral displacements of buildings can be reduced. Other studies that are related to pushover analysis can be found in [22–24]. Although pushover analysis seems to provide satisfying results, this type of analysis cannot be used to simulate the actual phenomenon resulting from earthquake shaking because earthquake forces do not incrementally increase; in other words, the sign and shape of earthquake forces are randomly varied and reversed. Accordingly, nonlinear response history analysis (NLRHA) appears to be the most suitable method to compute seismic responses or seismic demands; however, NLRHA is very time-consuming and has a higher computational effort. An alternative method can be used to calculate seismic responses other than the NLRHA procedure: a simplified version of the NLRHA procedure, such as uncoupled modal response history analysis (UMRHA). This method was first developed for nonlinear systems of modal pushover analysis [25,26]. Chopra and Goel [27] performed a comparison between the NLRHA and UMRHA procedures to investigate errors in terms of seismic demands and whether the procedure was acceptable. The errors in response quantity appeared to be admissible for approximation approaches for evaluating seismic demands. In addition, the computational effort of the UMRHA procedure was significantly lower than that of the NLRHA procedure [28–30].

The purpose of this study was to evaluate the seismic performance of low-rise RC buildings with various configurations of infill walls. The types buildings studied consisted of residential, commercial, and educational buildings, which encompass most of the existing low-rise RC building types in Thailand. The infill wall patterns are classified mainly into bare frames, fully infilled frames, original frames, open ground-story frames, and open top-story frames. These representative buildings were examined using pushover analysis to investigate the global behavior of each, shown in terms of pushover curves. Local behavior in terms of the failure mechanisms of structural members, such as beams and columns, and infill walls, were also investigated. The UMRHA procedure was also performed to estimate seismic responses. The results were compared with the results from the NLRHA procedure.

2. Nonlinear Modeling Approach

Frame structures were modeled using the distributed plasticity approach, also known as fiber elements [31]. The fiber beam–column element, a mixed-type force-based element and displacement-based element, was regarded as the most precise and rigorous distributed plasticity beam–column formulation [32] and was used in this study. A great deal of research has been conducted involving the strut model together with the corresponding parameters and its constitutive law to represent the behavior of infill walls [33–37]. In this study, the nonlinear response of an infill panel was captured by modeling a four-node masonry panel element proposed by Crisafulli [38]. This masonry panel element consists of two eccentric struts in each diagonal direction carrying axial loads. The vertical separation between struts, i.e., the distance between internal and dummy nodes, is estimated as half of the contact length [38]. A shear strut was also incorporated in the proposed masonry panel to carry shear from the top to the bottom of the panel in each diagonal direction. The masonry strut hysteresis model shown in Figure 1c was employed to represent the four struts, whereas a bilinear hysteresis rule shown in Figure 1d was employed to represent the shear strut. The schematic presentation of the four-node panel element and fiber elements related to the RC cross-section [39] are shown in Figure 1a,b.

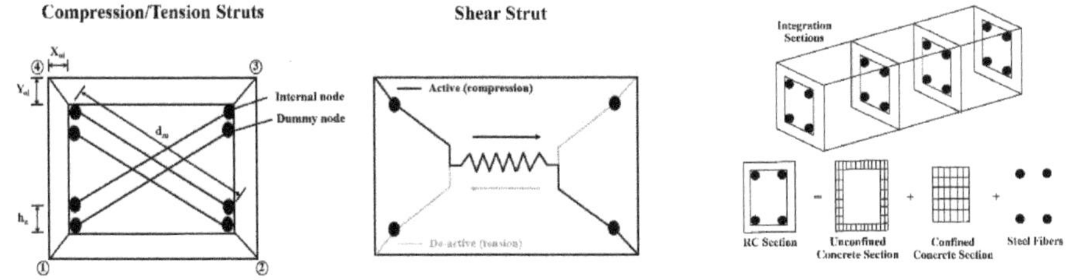

(a) (b)

Figure 1. *Cont.*

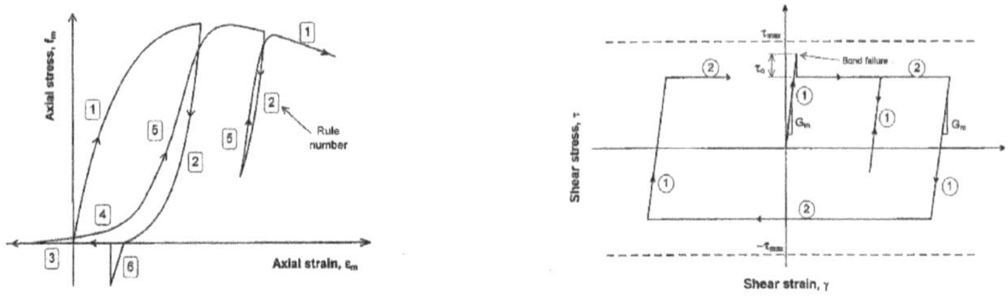

(c) (d)

Figure 1. Equivalent diagonal strut model and fiber elements: (**a**) compression or tension struts and shear strut [39]; (**b**) fiber elements associated with RC cross-section [39]; (**c**) strut curve parameters [38]; and (**d**) shear curve parameters [38].

It is essential to prove that the structural model used in this study is reliable and valid, so verification of the models must first be carried out. Three cases of experimental studies were selected from the literature for comparison with analytical models. The three experimental studies were Karayannis et al. [40], Niyompanitpattana and Warnitchai [8], and Van and Lau [41]. Selected specimens were used for the verification to compare between the experimental and analytical results consisting of a bare frame and an infilled frame from the three selected studies. This study aimed mainly at the effect of fully infilled walls, so partially infilled walls with openings, such as doors and windows, were not considered. The compared results in terms of global behavior, i.e., lateral force and lateral displacement, and local behavior such as the formation of plastic hinges and failure mechanisms, were investigated.

The structural details, such as the cross-section of the beams and columns of Karayannis et al. [40], are shown in Figure 2, while their corresponding material properties are depicted in Table 1.

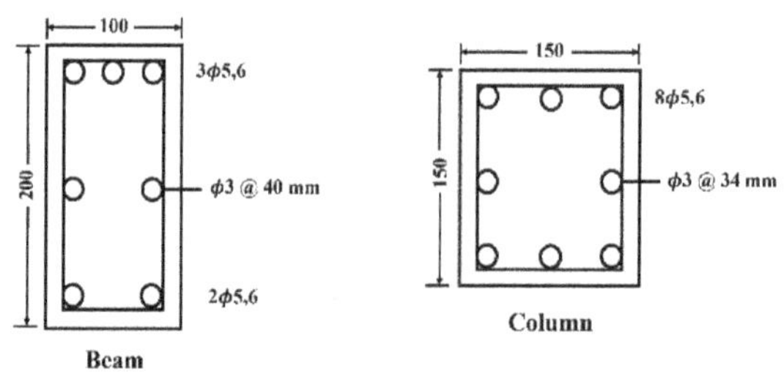

Figure 2. Cross-sections of the beam and column of Karayannis et al. [40].

Table 1. Material properties of Karayannis et al. [40].

Material Types	Material Properties	Specimen 1	Specimen 2
Concrete	f_c of column (MPa)	28.51	28.51
	f_c of beam (MPa)	28.51	28.51
	E_c (MPa)	25,096	25,096
Steel	f_y of (ϕ5,6 mm) (MPa)	390.47	390.47
	f_y of (ϕ3 mm) (MPa)	212.20	212.20
	E_s (MPa)	200,000	200,000
Infill	f_m (MPa)	-	2.630
	E_m (MPa)	-	660.66

The comparison of analytical models and experimental models [40] is shown in Figure 3. For the analytical result of the bare frame, the peak force on the positive side is equal to 33.11 kN and the lateral displacement is equal to 26.88 mm. The peak force on the negative side is equal to −36.03 kN and the lateral displacement is equal to −10.27 mm. The hysteretic loop of the experimental result is small in the inelastic range but similar in the elastic range. For local behavior regarding analytical results, since the column section is smaller than the beam section, most of the failure mechanisms are implied to occur on columns before the beam. Due to the formation of plastic hinges on the structural elements, including beams and columns, the sequence of failure mechanisms can readily be investigated. Initially, reinforcing steel yielding is found at the bottom of both columns, while nothing occurs on the beam. Subsequently, the previous failure mechanism is replaced by unconfined concrete crushing at the same locations, while reinforcing steel yielding occurs at the ends of both columns and the ends of the beam. In the final stage, unconfined concrete crushing at the bottom of both columns is replaced by confined concrete crushing, while the top of both columns is found to have unconfined concrete crushing. This analytical model seems to successfully investigate local behavior, especially for the formation of plastic hinges that occur at both ends of the beam and columns. For the analytical result of the infilled frame, the peak force on the positive side is equal to 74.79 kN, and the lateral displacement is equal to 5.92 mm. The peak force on the negative side is equal to −84.21 kN, and the lateral displacement is equal to −5.76 mm. The hysteretic loop of the analytical result is similar to the hysteretic loop of the experimental result. For local behavior regarding the analytical result, reinforcing steel yielding is found at the bottom of both columns, while nothing occurs on the beam. The diagonal cracking is also found to begin. Subsequently, unconfined concrete crushing occurs at both ends of columns, whereas reinforcing steel yielding occurs on both ends of the beam. Then, diagonal cracking is completely achieved. In the final stage, confined concrete crushing is found to occur only at the bottom of both columns. This analytical model can successfully predict local behavior in terms of the development of plastic hinges on beams and columns. The model can also be used to investigate infill cracks and infill failures.

The structural details of the cross-section of the beams and columns of Niyompanitpattana and Warnitchai [8] are shown in Figure 4, while their corresponding material properties are depicted in Table 2.

Table 2. Material properties of Niyompanitpattana and Warnitchai [8].

Material Types	Material Properties	Specimen 1	Specimen 2
Concrete	f_c of column (MPa)	17.9	23.6
	f_c of beam (MPa)	20.3	27.9
	E_c (MPa)	20,152	20,152
Steel	f_y of DB 16 (SD30) (MPa)	382.6	382.6
	f_y of RB 6 (SR24) (MPa)	373.6	373.6
	f_y of plain mild steel (ϕ3 mm) (MPa)	392.4	392.4
	E_s (MPa)	207,000	207,000
Infill	f_m (MPa)	-	5.9
	E_m (MPa)	-	887

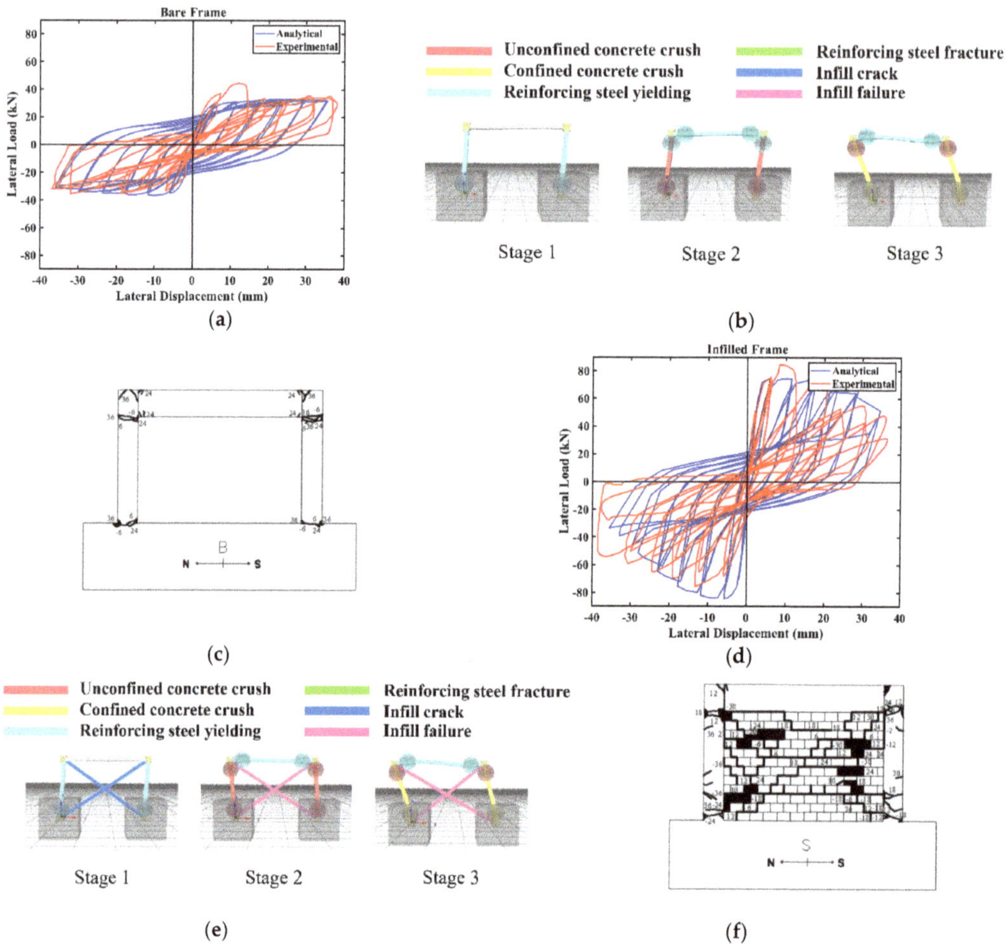

Figure 3. Comparison of analytical models in this study with experimental models [40]: (**a**) analytical and experimental results of the bare frame; (**b**) failure mechanisms of the bare frame based on the analytical model; (**c**) failure mechanisms of the bare frame based on the experimental model; (**d**) analytical and experimental results of the infilled frame; (**e**) failure mechanisms of the infilled frame based on the analytical model; (**f**) failure mechanisms of the infilled frame based on the experimental model.

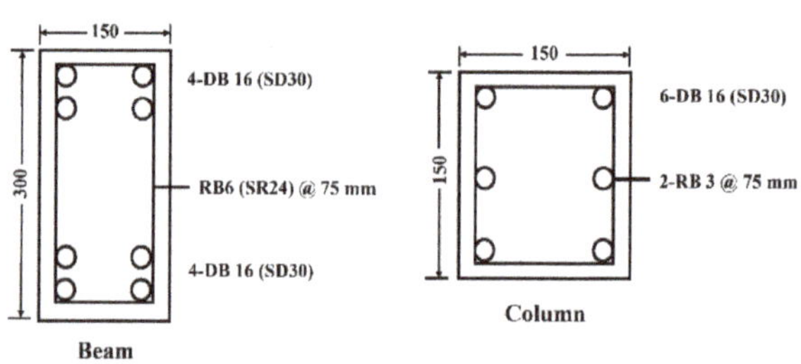

Figure 4. Cross-sections of the beam and column of Niyompanitpattana and Warnitchai [8].

The comparison of analytical models and experimental models [8] is shown in Figure 5. For the analytical result of the bare frame, the peak force on the positive side is equal to 47.8 kN, and the percentage of drift is equal to +1.25%. The peak force on the negative side is equal to −48.8 kN, and the percentage of drift is equal to −1.31%. The hysteretic loop of the analytical result is large in the inelastic range but similar in the elastic range. For local behavior, failure mechanisms of the bare frame occur only at columns. In the first stage, unconfined concrete crushing occurs at both ends of columns. Subsequently, reinforcing steel yielding occurs at the same locations and, thus, confined concrete crushing occurs in the final stage. This analytical model can successfully predict local behavior in terms of the development of plastic hinges on columns, consistent with that shown in the experimental result. In other words, deformation predominantly occurs at both ends of columns. The analytical model cannot predict the type of deformation that occurs at columns, as opposed to the experimental model. For the analytical result of the infilled frame, the peak force on the positive side is equal to 167.5 kN, and the percentage of drift is equal to +1.00%. The peak force on the negative side is equal to −168.8 kN, and the percentage of drift is equal to −0.90%. The hysteretic loop of the analytical result is similar to the hysteretic loop of the experimental result. For local behavior regarding the analytical result, failure mechanisms of the infilled frame occur only at columns and in the failure mechanisms of the bare frame. Initially, only diagonal cracking is found at the infill wall, whereas nothing occurs at the beam and columns. Subsequently, complete diagonal cracking occurs, and unconfined concrete crushing also takes place at both ends of the columns. In the final stage, reinforcing steel yielding occurs, followed by confined concrete crushing. This analytical model can predict local behavior in terms of the development of plastic hinges on columns and cracks, ranging from the initial stage to the final stage on infill walls. This model cannot predict where the location within infill walls first starts to crack, as opposed to the experimental model.

The structural details of the cross-section of the beams and columns of Van and Lau [41] are shown in Figure 6, while their corresponding material properties are depicted in Table 3.

Table 3. Material properties of Van and Lau [41].

Material Types	Material Properties	Specimen 1	Specimen 2
Concrete	f_c (MPa)	22.84	22.84
	E_c (MPa)	25,400	25,400
Steel	f_y of (ϕ2 mm) (MPa)	305.9	305.9
	f_y of (ϕ5 mm) (MPa)	581.3	581.3
	f_y of (ϕ7 mm) (MPa)	605.5	605.5
	E_s of (ϕ2 mm) (MPa)	196,800	196,800
	E_s of (ϕ5 mm) (MPa)	199,000	199,000
	E_s of (ϕ7 mm) (MPa)	198,800	198,800
Infill	f_m (MPa)	-	1.316
	E_m (MPa)	-	1316

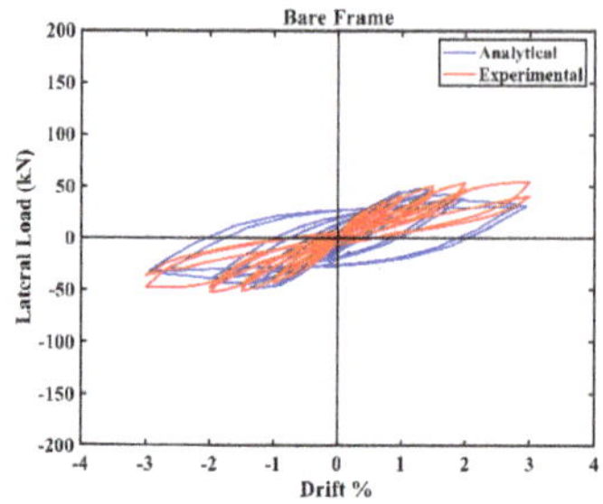

(**a**)

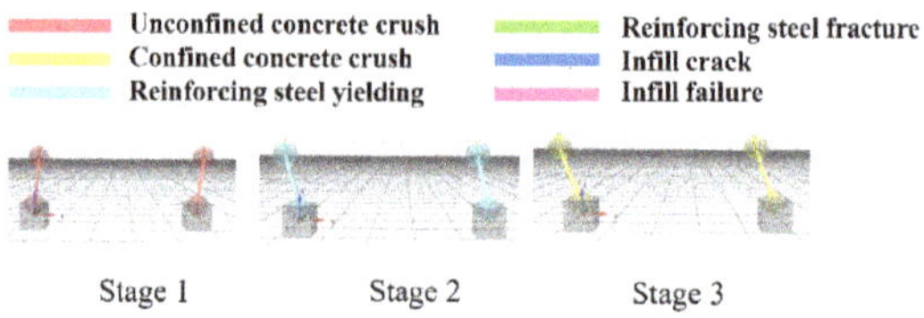

(**b**)

(**c**)

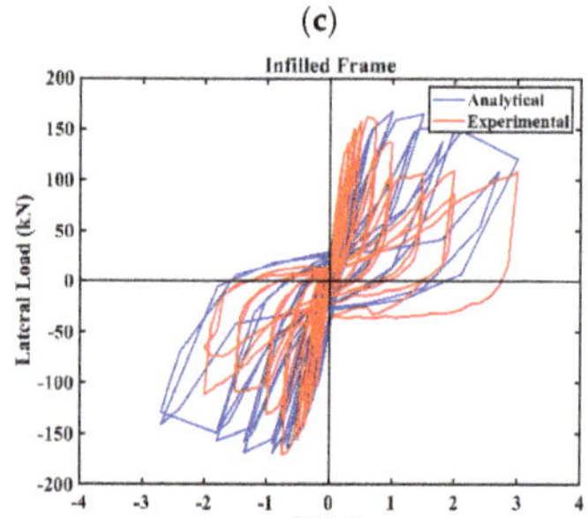

(**d**)

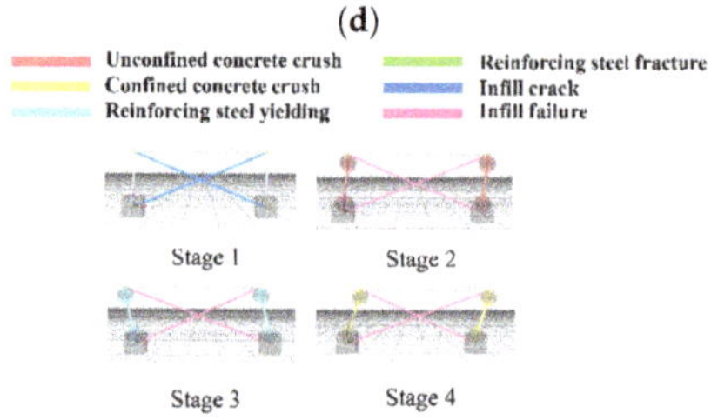

(**e**)

Figure 5. *Cont.*

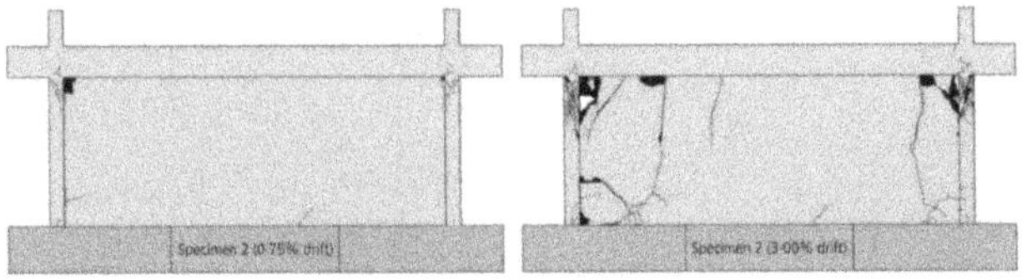

(f)

Figure 5. Comparison of analytical models in this study with experimental models [8]: (**a**) analytical and experimental results of the bare frame; (**b**) failure mechanisms of the bare frame based on the analytical model; (**c**) failure mechanisms of the bare frame based on the experimental model; (**d**) analytical and experimental results of the infilled frame; (**e**) failure mechanisms of the infilled frame based on the analytical model; (**f**) failure mechanisms of the infilled frame based on the experimental model.

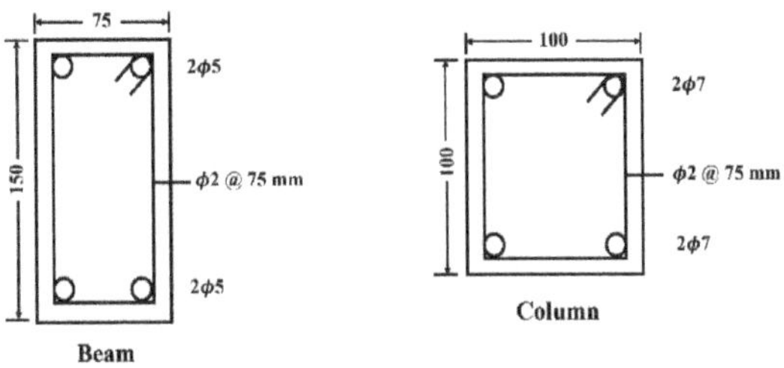

Figure 6. Cross-sections of the beam and column of Van and Lau [41].

The comparison of analytical models and experimental models [41] is shown in Figure 7. For the analytical result of the bare frame, the peak force on the positive side is equal to 14.41 kN, and the displacement is equal to 40.48 mm. The peak force on the negative side is equal to −15.18 kN, and the displacement is equal to −20.84 mm. The hysteretic loop of the analytical result is large in the inelastic range but similar in the elastic range. For local behavior regarding the analytical result, failure mechanisms of the bare frame occur at the beam before columns. Initially, reinforcing steel yielding occurs at both ends of the beam, while unconfined concrete crushing occurs at the bottom of both columns. Subsequently, the previous failure mechanism is replaced by reinforcing steel yielding at the same locations. In the final stage, confined crushing takes place only at the right-hand column, while the right-hand end of the beam changes from reinforcing steel yielding to unconfined concrete crushing. This analytical model can predict local behavior in terms of the development of plastic hinges on beams and columns. For the analytical result of the infilled frame, the peak force on the positive side is equal to 33.05 kN, and the displacement is equal to 24.67 mm. The peak force on the negative side is equal to −37.39 kN, and the displacement is equal to −13.00 mm. The hysteretic loop of the analytical result is similar to the hysteretic loop associated with the experimental result. Regarding the analytical result, for local behavior, failure mechanisms of the infilled frame occur on the beam before columns and in the failure mechanisms of the bare frame. Initially, only diagonal cracking takes place at the infill wall, whereas nothing occurs at the beam and columns. Subsequently, reinforcing steel yielding occurs at both ends of the beam, while unconfined concrete crushing occurs at the bottom of both columns. A complete diagonal crack occurs. In the final stage, reinforcing steel yielding occurs at the bottom of both columns while nothing occurs at the top of the columns. This analytical model can predict local behavior in terms of the development of plastic hinges on beams and columns. However, this model cannot indicate which type of cracks occur on such elements, for instance, flexural or shear cracks, as opposed to the experimental model.

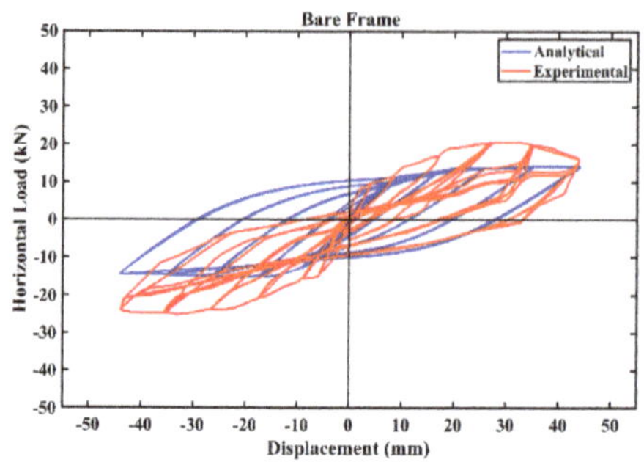

(**a**)

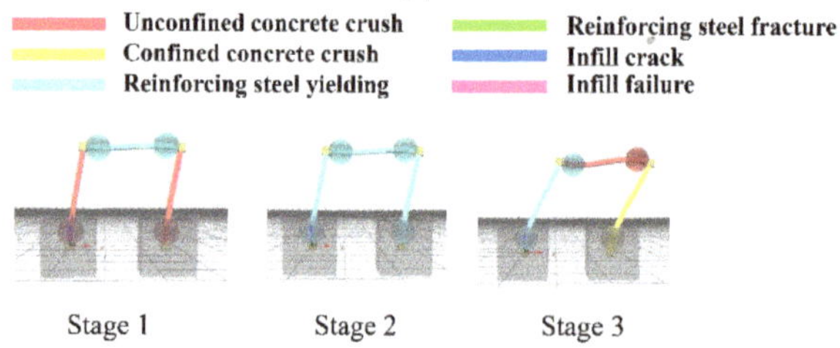

(**b**)

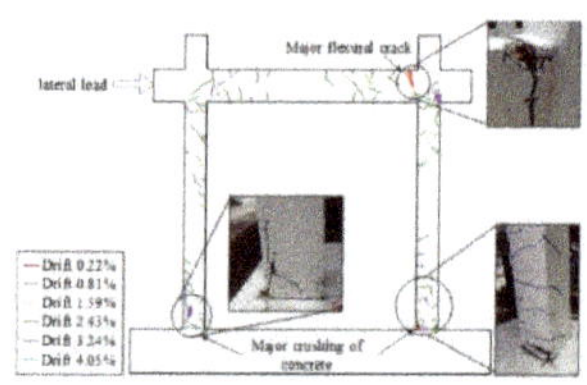

(**c**)

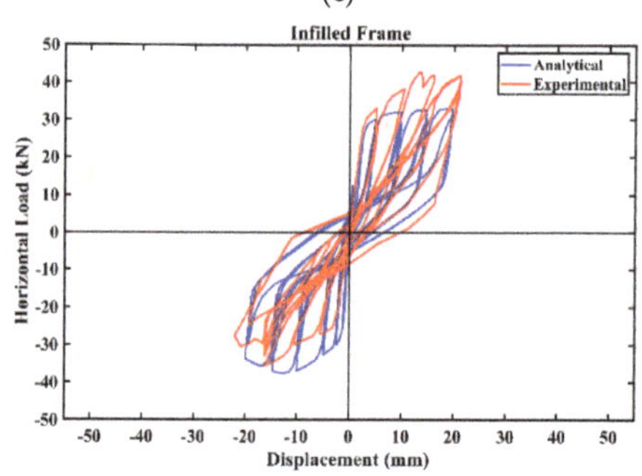

(**d**)

Figure 7. *Cont.*

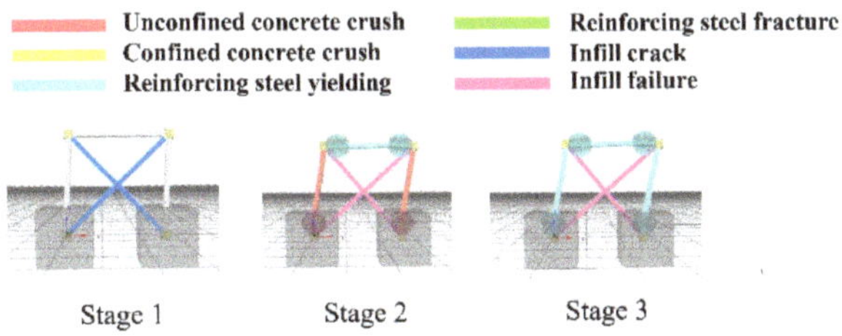

(e)

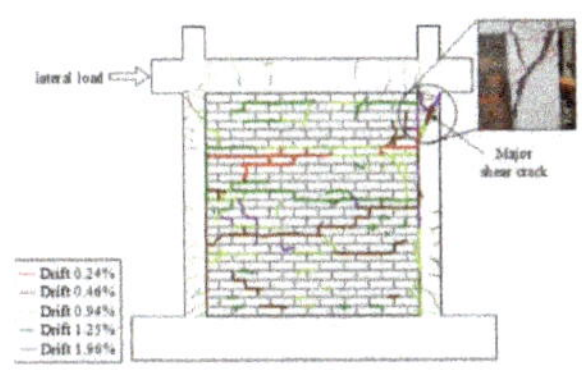

(f)

Figure 7. Comparison of analytical models in this study with experimental models [41]: (**a**) analytical and experimental results of the bare frame; (**b**) failure mechanisms of bare frame based on the analytical model; (**c**) failure mechanisms of the bare frame based on the experimental model; (**d**) analytical and experimental results of the infilled frame; (**e**) failure mechanisms of the infilled frame based on the analytical model; (**f**) failure mechanisms of the infilled frame based on the experimental model.

3. Selection of Representative Buildings

Residential, commercial, and educational buildings can easily be found in most existing low-rise buildings in Thailand. Residential buildings mostly have a single story and up to two stories, but the latter is more typical for this building type. Commercial buildings have up to four stories, but most commercial buildings can be found with two or three stories, such as stores and residences. The uniqueness of this building type is that there is always an opening in the ground story. An educational building or school building mostly has two stories constructed from a standard plan provided by the Ministry of Education. These buildings are in the seismicity region near active faults. The buildings were selected as representative buildings because their structural plan is typical. In addition, they can readily be found in some regions in Thailand that have moderate earthquakes based on the information from the earthquake observation division of Thailand. Past earthquakes have occurred in that area, leading to slight to moderate damage to buildings. Some of the buildings could be repaired and still be functional, while others collapsed. Accordingly, these types of buildings are shown to be important to consider.

There are three main types of RC moment-resisting frames: ordinary, intermediate, and special moment-resisting frames. The ordinary moment-resisting frame can easily be found in Thailand. This type does not account for the ductile behavior of the building and is not designed according to earthquake-resistant design standards. The spacing of the stirrups is very high and the hook of the stirrups is 90 degrees. The intermediate and special moment-resisting frame, however, requires at least 135 degrees on the hook of the stirrups, known as the seismic hook. Additionally, the spacing of the stirrups is very low compared to the ordinary moment-resisting frame. With this assertion, the representative buildings are regarded as intermediate moment-resisting frames because of the details of the provided reinforcing bars and stirrups, as shown in Figure 8d,e. The corresponding plans of the representative buildings are shown in Figure 8a–c for the original frame. The structural properties of the representative buildings with different infill wall patterns, as shown in Table 4, were set to be the same so that the seismic behavior of such buildings could be investigated, and the buildings could easily be compared to each other.

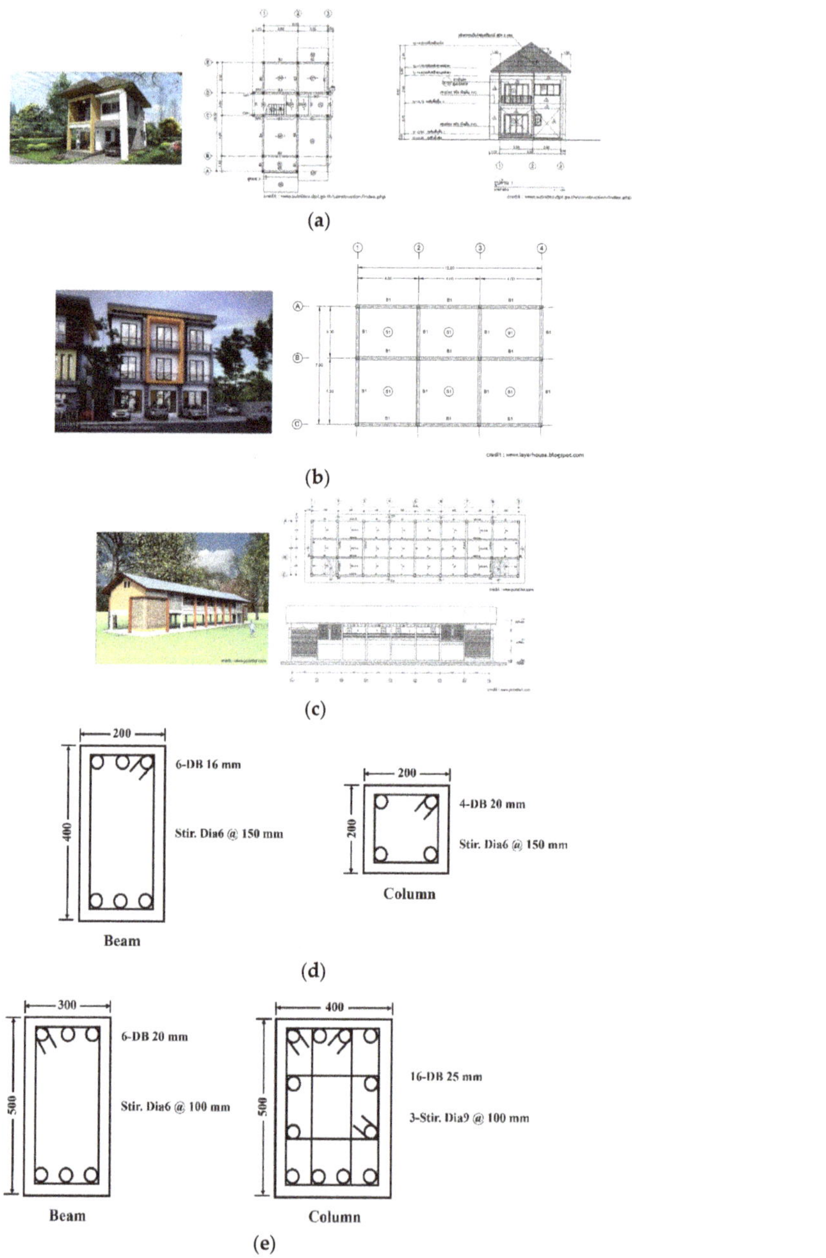

Figure 8. Structural drawing and cross-section of beams and columns for representative buildings: (**a**) residential building (RES) [42]; (**b**) commercial building (COM) [43]; (**c**) educational building (EDU) [44]; (**d**) RES and COM buildings; (**e**) EDU building.

Table 4. Structural properties of representative buildings.

Building Name	RES	COM	EDU
Type	Residential	Commercial	Educational
Height (m)	6.05	10.5	7.05
Number of stories	2	3	2
Column dimension (cm × cm)	20 × 20	20 × 20	40 × 50
Beam dimension (cm × cm)	20 × 40	20 × 40	30 × 50
Infill wall thickness (cm)	9	9	9
f_c of frame elements (MPa)	23.6	23.6	23.6
E_c of frame elements (MPa)	20,152	20,152	20,152
f_y of reinforcing steel (MPa)	382.6	382.6	382.6
f_y of stirrups (MPa)	373.6	373.6	373.6
E_s (MPa)	207,000	207,000	207,000
f_m of infill walls (MPa)	5.9	5.9	5.9
E_m of infill walls (MPa)	887	887	887

In addition to representative building types, five infill wall configuration patterns were examined. These patterns were the bare frame, original frame, fully infilled frame, open ground-story frame, and open top-story frame. The bare frame is a frame without any infill walls. The fully infilled frame is a frame that has the presence of infill walls in the exterior parts for any story. The original frame is a frame that has infill walls based on the actual structural plan without partially infilled walls, which consist of doors and windows. Only full infill panels were considered; in other words, infill walls with openings such as doors and windows were not considered. The open ground-story frame is a frame that has infill walls in the exterior parts for any floor level, except for the ground floor level. The open top-story frame is a frame that has infill walls only on the ground floor. With different patterns of infill walls, seismic performance in terms of the strength and stiffness of the building are expected to be discriminated. Especially for the bare frame and the fully infilled frame, the bare frame should provide the lowest strength and stiffness, while the fully infilled frame should provide the highest strength and stiffness. For other infill wall patterns, a different increase in strength and stiffness should be found. For the direction of the presence of infill walls, infill walls that are placed in the considered direction will affect the building rather than infill walls in the other directions. For the behavior of buildings with various configurations of infill walls, brittle and ductile behavior, or a combination of both, might obviously be observed in such buildings. The infill wall patterns of those representative buildings are shown in Figure 9.

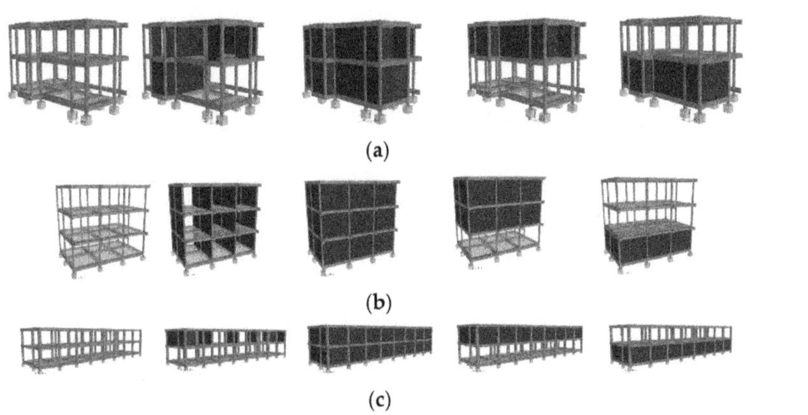

Figure 9. Various infill wall configurations of representative buildings: (**a**) RES building; (**b**) COM building; (**c**) EDU building.

4. Seismic Performance Based on Modal Pushover Analysis

In this section, the SeismoStruct [39] fiber element-based software was implemented to simulate the representative buildings and perform analyses. As described in Section 2, frame elements, i.e., beams and columns, were modeled using the fiber beam–column element, while infill walls were modeled using the proposed four-node masonry panel element [38]. A slab was modeled using a rigid diaphragm. For constitutive laws of materials, the concrete model of Mander et al. [45] was employed as a concrete material model, whereas the steel model of Menegotto and Pinto's [46] was employed as a steel material model. The former consists of a stress–strain relationship together with a simplified uniaxial concrete model, and the latter provides nonlinear behavior, which ranges from elastic to plastic behavior and is a uniaxial material with optional isotropic hardening. However, only flexural failure on RC frames, which accounts for shear failure and the P-Delta effect, was not considered in this study. Modal pushover analysis was carried out in both the x- and y-directions for all representative buildings. The displacement-controlled lateral load method was employed. In other words, such buildings were laterally pushed until reaching a predefined target displacement. The results of modal pushover analysis are known as pushover curves. These curves can be in several formats. However, the most widely used is in the base shear versus roof displacement relationship. Due to the benefits of this relationship, the global behavior of buildings can easily be investigated in terms of strength and stiffness, in addition to the local behavior of structural and nonstructural elements regarding failure mechanisms. Three types of representative buildings with various configurations of infill walls were examined.

The observed structural elements and the pushover curve of residential (RES) buildings are shown in Figure 10. According to global behavior, for the pushover curve in the x-direction, the fully infilled frame and the open top-story frame seemed to equally provide strength and stiffness. Both strength and stiffness may be implied to be largely contributed by the presence of infill walls on frame structures. Nevertheless, initially, the behavior of these frames expressed brittle behavior, and they became ductile after the collapse of infill walls. The original frame appeared to have lower strength and stiffness. In addition, this original frame behaved in the same way as the fully infilled frame and the open top-story frame. The open ground-story frame and the bare frame expressed lower strength and stiffness than others, whereas their behavior seemed to be ductile. For local behavior, in terms of structural elements, the failure mechanism of Column C1 started from unconfined concrete crushing, reinforcing steel yielding, confined concrete crushing, and reinforcing steel fracturing. However, no failure mechanism occurred on the beams. In terms of nonstructural elements for the fully infilled frame, only infill walls placed parallel to the x-direction at the ground floor were found to have both infill cracks and infill failures, while the y-direction was found to have only infill cracks. For the original frame, only infill walls placed parallel to the x-direction were found to have both infill cracks and infill failures only on the ground floor and this pattern appeared to have a torsional effect. For the open ground-story frame, there was no failure mechanism related to infill walls. For the open top-story frame, the phenomenon that occurred in this type was the same as in the fully infilled frame.

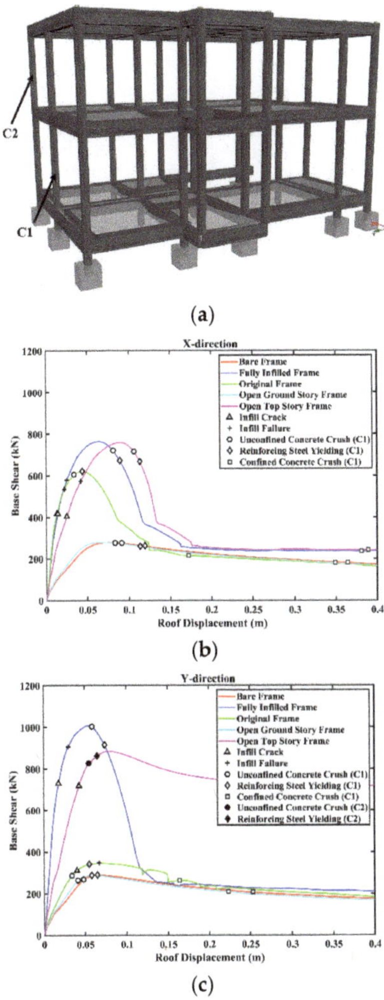

Figure 10. Observed locations and pushover curve of RES: (**a**) Columns C1 and C2; (**b**) x-direction; (**c**) y-direction.

According to global behavior, for the pushover curve in the y-direction, the fully infilled frame was found to provide the highest strength and stiffness, whereas lower strength and stiffness were provided by the open top-story frame. The original frame seemed to provide slightly higher strength and stiffness compared to the open ground-story frame and the bare frame. In addition, the behavior of the fully infilled frame in the y-direction was similar to the fully infilled frame in the x-direction, while the open top-story frame and the original frame in the y-direction seemed to be different from the x-direction because of ductile behavior. However, the bare frame and the open ground-story frame appeared to be the same as in the x-direction. The sequence of failure mechanisms regarding structural elements was similar to the sequence of failure mechanisms of the x-direction, except for Column C2. Only the case of the open top-story frame was found to have failure mechanisms of Column C2, including unconfined concrete crushing and reinforcing steel yielding. In terms of nonstructural elements, for fully infilled frames, only infill walls placed parallel to this direction at the ground floor were found to have both infill cracks and infill failures, while some belonging to the x-direction had only infill cracks. For the original frame, only infill walls placed parallel to the y-direction on the ground floor had both infill cracks and infill failures, while some in the other direction had only infill cracks. This type was also found to have a torsional effect. For the open ground-story frame, no failure mechanism was found related to infill walls. For the open top-story frame, infill walls placed parallel to the y-direction had infill cracks only on the ground floor.

The observed elements and the pushover curve of commercial (COM) buildings are shown in Figure 11. According to global behavior, for the pushover curve in the x-direction, the fully infilled frame provided the highest strength and stiffness. Initially, this frame exhibited brittle behavior, and became ductile later, after the collapse of infill walls. For other frame patterns, the sequence of strength and stiffness were as follows: the open top-story frame, the open ground-story frame, the original frame, and the bare frame. In addition, while the fully infilled frame made a large contribution due to the presence of infill walls, other types of configurations appeared to make no contributions. The other four frame patterns exhibited brittle behavior, as opposed to the fully infilled frame. For local behavior, in terms of structural elements, Columns C1 and C2 represented columns related to the ground and second floors. Only two columns were selected because no failure mechanism was found related to the beams. For Column C1, failure mechanisms started from unconfined concrete crushing, reinforcing steel yielding, and confined concrete crushing. In addition, the failure mechanism of Column C1 occurred on all frame patterns, except for the open top-story frame. However, only Column C2 had failure mechanisms including unconfined concrete crushing and reinforcing steel yielding. In terms of nonstructural elements, for the fully infilled frame, only infill walls placed parallel to the x-direction at the ground floor had both infill cracks and infill failures. For the original frame, the open ground-story frame, and the open top-story frame, no failure mechanism was found related to infill walls. However, in the case of the open ground-story frame, it seemed to have a torsional effect.

According to global behavior, for the pushover curve in the y-direction, the original frame provided the highest strength and stiffness. The reason for this was that the original frame provided more infill walls in the y-direction than other frame patterns. The fully infilled frame provided lower strength and stiffness, whereas the open ground-story frame, the open top-story frame, and the bare frame seemed to provide equal strength and stiffness. However, the failure mechanisms in terms of structural elements were similar to the frame patterns of the x-direction. For the fully infilled frame, only infill walls placed parallel to this direction on the ground floor were found to have both infill cracks and infill failures. For the original frame, infill walls placed parallel to the y-direction on the ground floor had both infill cracks and infill failures, and this pattern appeared to have a torsional effect. For the open ground-story frame, no failure mechanism was found related to infill walls. However, this pattern appeared to have a torsional effect. For the open top-story frame, this pattern had failure mechanisms related to infill walls similar to failure mechanisms of the open ground-story frame, except for the torsional effect.

The observed elements and pushover curve of educational (EDU) buildings are shown in Figure 12. According to global behavior, for the pushover curve in the x-direction, the fully infilled frame and the open top-story frame seemed equal to provide higher strength and stiffness than other frame patterns. Initially, the behavior of this frame exhibited brittle behavior and became ductile after the collapse of infill walls. The open ground-story frame, the original frame, and the bare frame expressed strength and stiffness at the same level. In addition, the behavior of these was ductile, as opposed to the fully infilled frame and the open top-story frame. For local behavior in terms of structural elements, Column C1 had failure mechanisms that started from unconfined concrete crushing, reinforcing steel yielding, confined concrete crushing, and reinforcing steel fracturing. In terms of nonstructural elements, for the fully infilled frame, infill walls placed parallel to this direction at the ground floor had both infill cracks and infill failures, while some of the infill walls regarding the y-direction had only infill cracks. In addition, some of the infill walls placed parallel to the x-direction on the second floor had infill cracks. For the original frame, only infill walls placed parallel to the x-direction had infill cracks. For the open ground-story frame, no failure mechanism was found related to infill walls. For the open top-story frame, the phenomenon that occurred on this type was similar to the fully infilled frame, regardless of the failure mechanism on the second floor.

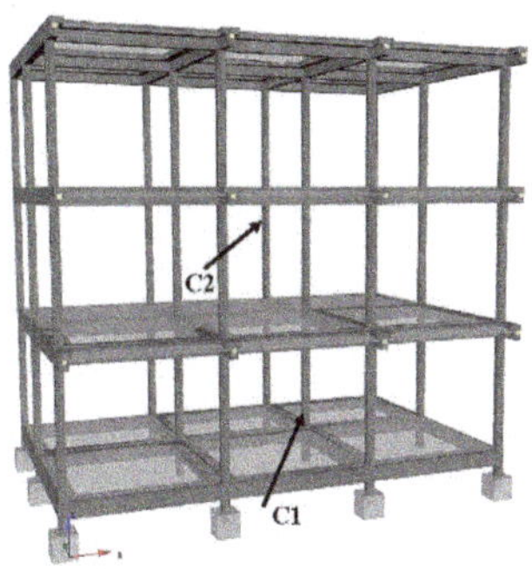

(**a**)

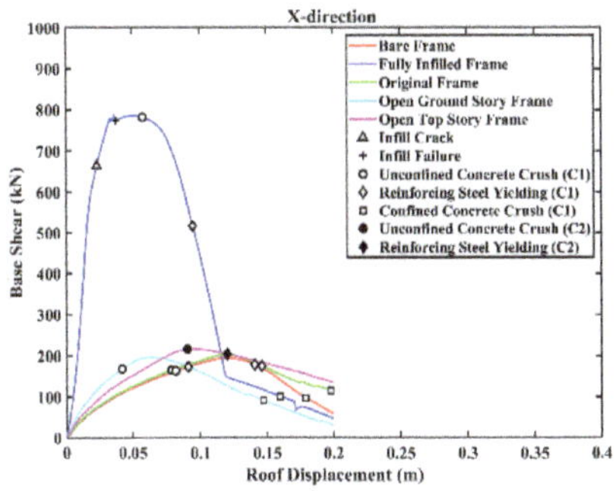

(**b**)

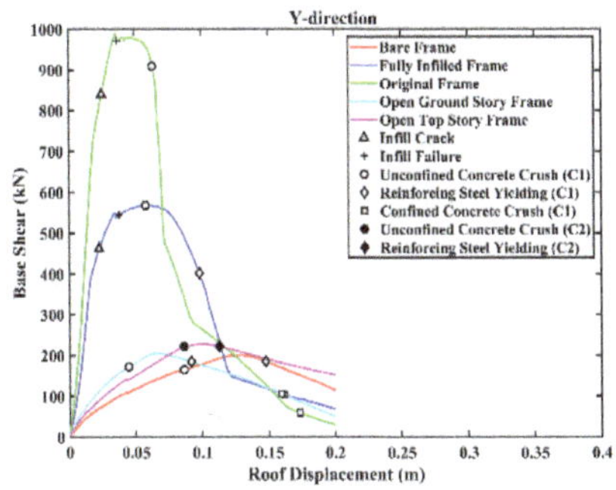

(**c**)

Figure 11. Observed locations and pushover curve of COM: (**a**) Columns C1 and C2; (**b**) x-direction; (**c**) y-direction.

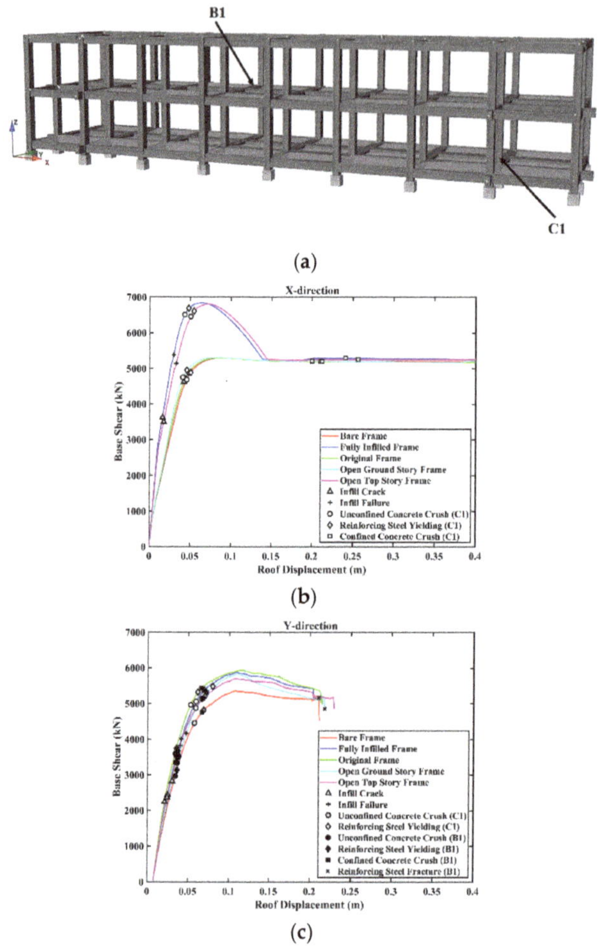

Figure 12. Observed locations and pushover curve of EDU: (**a**) Column C1 and beam B1; (**b**) x-direction; (**c**) y-direction.

According to global behavior, for the pushover curve in the y-direction, the original frame provided higher strength and stiffness than other frame patterns, followed by the fully infilled frame, the open ground-story frame, the open top-story frame, and the bare frame. For local behavior in terms of structural elements of all frame patterns, regarding Column C1, only unconfined concrete crushing and reinforcing steel yielding occurred, while failure mechanisms occurred on Beam B1, consisting of unconfined concrete crushing, reinforcing steel yielding, confined concrete crushing, and reinforcing steel fracture. For the fully infilled frame, only infill walls placed parallel to the y-direction on either the ground floor or the second floor had both infill crack and infill failures. For the original frame, infill walls placed parallel to the y-direction had both infill cracks and infill failures. For the open ground-story frame and the open top-story frame, only infill walls placed parallel to the y-direction had both infill cracks and infill failures.

In conclusion, the comparison of three types of representative buildings resulted in some findings. For bare frames and open ground-story frames, the behavior was similar for most of the aforementioned building types, as they showed ductility. This phenomenon occurred as bare frames had a soft story effect. In addition, for open ground-story frames, there was no contribution of infill walls. For fully infilled frames, buildings behaved in a brittle manner first; after the collapse of infill walls, the buildings became ductile again. For the original frames and open top-story frames, buildings behaved differently for some building types because some buildings had infill walls while some did not. This result indicates that the shape of buildings and the direction of the presence of infill walls were the major factors producing their effects on buildings. The advantages and disadvantages of each infill wall pattern are summarized in Table 5.

Table 5. The advantages and disadvantages of different infill wall patterns.

Patterns of Infill Walls	Advantages	Disadvantages
Original frame	- If infill walls are placed in the considered direction, strength and stiffness are higher than that of the bare frame, yet brittle then ductile again.	- If infill walls are not placed in the considered direction, strength and stiffness are equal to that of the bare frame, yet ductile throughout. - A torsional effect makes columns in an affected area undergo more failure mechanisms than columns where the position has less effect of torsion, including reinforcing steel yielding, unconfined concrete crushing, and confined concrete crushing. - Only infill walls on the ground floor affect the buildings.
Fully infilled frame	- The highest strength and stiffness. Infill walls placed in both x- and y-directions affect the buildings.	- Brittle behavior and then ductile behavior.
Open ground-story frame	- Ductile behavior throughout.	- There is no contribution from infill walls. The lowest strength and stiffness. - A torsional effect makes columns in an affected area undergo more failure mechanisms than columns where the position has less effect of torsion, including reinforcing steel yielding, unconfined concrete crushing, and confined concrete crushing.
Open top-story frame	- Strength and stiffness are as high as the fully infilled frame or higher than the bare frame. - Infill walls placed in the x-direction and some placed in the y-direction affect the buildings. - Columns on the ground floor do not undergo any failure mechanisms.	- Brittle behavior and then ductile behavior. - Columns on floors other than the ground floor undergo failure mechanisms, including reinforcing steel yielding, unconfined concrete crushing, and confined concrete crushing.

5. Comparison of Seismic Responses between UMRHA and NLRHA

For the UMRHA procedure, only RES buildings with various configurations of infill walls were selected because residential buildings can easily be found in Thailand. The RES building is characterized as 6.50 m × 10.70 m in plan and 6.05 m in elevation, including 1.00 m of ground column height, and 3.15 and 2.9 m of columns on the first and second floors, respectively. This procedure uses the concept of a sum of inelastic SDOF systems in each vibration mode, as shown in Figure 13. The first two vibration modes in the x-direction of the RES bare frame are shown in Figure 14A and the vibration periods are 0.44 s and 0.15 s In other words, only the first and second vibration modes in the x-direction were accounted for.

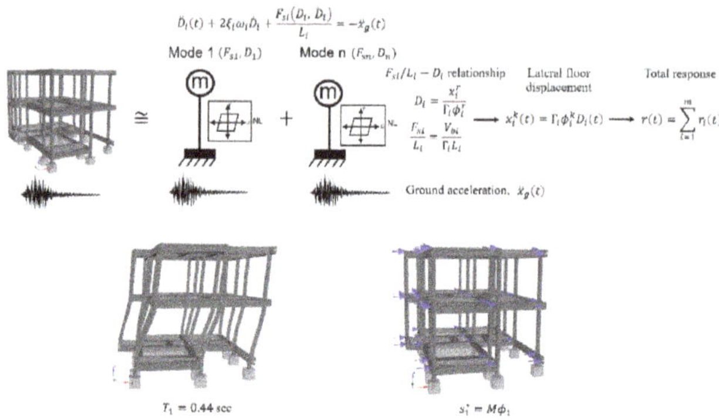

Figure 13. Schematic presentation of the UMRHA procedure.

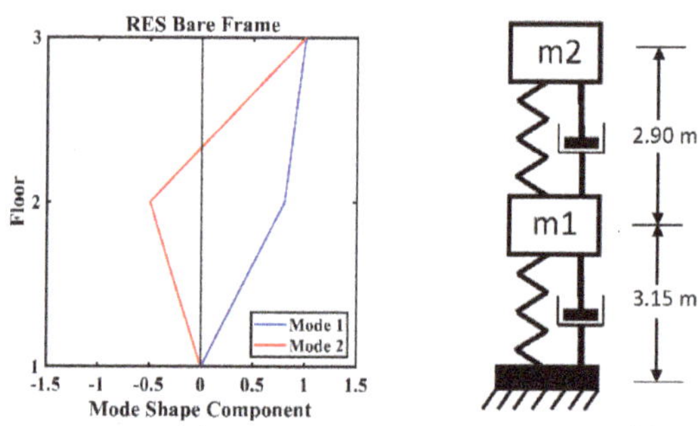

(a) (b)

Figure 14. Mode shape and simplified concept of building RES: (a) first two vibration modes in the x-direction of the RES bare frame; and (b) 2 DOFs of RES.

In this study, the properties of RES buildings, such as masses and mode shapes, were obtained from a 3D full model together with eigenvalue analysis. The mass on each floor was computed by a sum of nodal masses on the considered floor. Mode shapes were first obtained for all nodes, and then a single value of mode shape on each floor was selected as the representative value because of the symmetrical shape of the buildings. The UMRHA procedure regarding RES buildings accounted for 2 degrees of freedom (DOFs) where the level of ground beams was considered at the base, neglecting the DOF on this floor. Only the second floor and roof floor were considered, so the DOFs were reduced from 3 DOFs to 2 DOFs, as shown in Figure 14b. Due to the reduction in DOFs, damage to the ground columns was neglected. The masses and mode shapes used in the UMRHA procedure were the values that were converted from the 3D model to the 2D model by means of the previous explanation. Figure 15a to Figure 15c show the comparison between the modal coordinates of the hysteretic curves and the response of idealized inelastic systems to three ground motions, i.e., Imperial Valley, Kobe, and Landers. The first mode response was the inelastic response and was higher than the second mode response. The latter seemed to be lower or in the linear range compared to the first mode. Seismic responses, regardless of their types, were contributed largely by the first mode to the total response because the second mode contributed slight responses or had no contribution to the total response. Since the building RES is a low-rise building, the effects from the higher modes to the response appeared to be insignificant, as opposed to high-rise buildings or tall buildings. With this assertion, the hysteretic model was only used in the first vibration mode, while the linear model was used in the second vibration mode instead. This was found to be the simplified method. Concurrent with the UMRHA procedure, 3D full models of building

RES were used to implement the NLRHA procedure. The following is a brief explanation of the UMRHA procedure. Figure 16 expresses the idealized curves that were transformed into the $F_{si}/L_i - D_i$ relationship to obtain the nonlinear force deformation function so that a standard governing equation of motion regarding the UMRHA procedure could be solved. The properties of the modal inelastic SDOF systems of building RES used as input parameters to perform the UMRHA procedure are shown in Table 6.

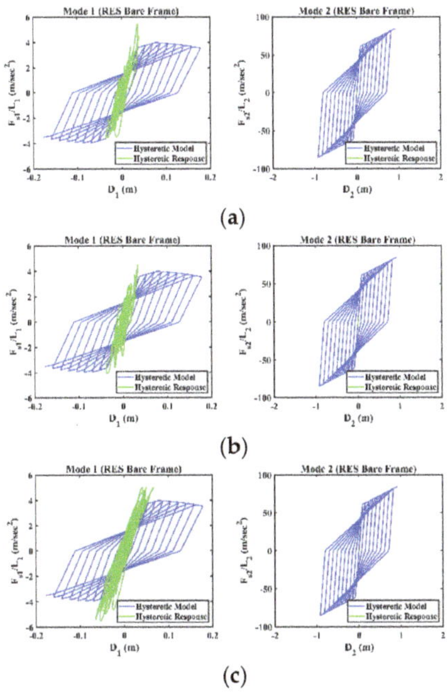

Figure 15. The hysteretic models of the RES bare frame versus the hysteretic response of idealized inelastic SDOF systems to three ground motions: (**a**) Imperial Valley; (**b**) Kobe; (**c**) Landers.

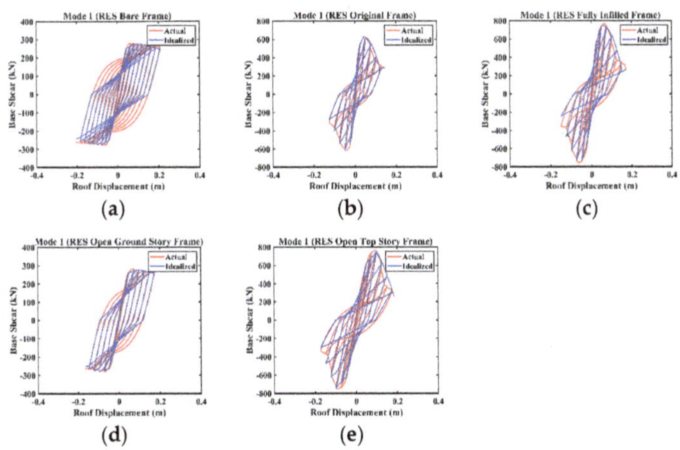

Figure 16. Actual cyclic pushover curves versus idealized nonlinear SDOF system curves of building RES: (**a**) bare frame; (**b**) original frame; (**c**) fully infilled frame; (**d**) open ground-story frame; (**e**) open top-story frame.

Table 6. Properties of the first modal inelastic SDOF systems of RES.

Types	Bare Frame	Original Frame	Fully Infilled Frame	Open Ground-Story Frame	Open Top-Story Frame
Γ_i	1.159	1.239	1.220	1.123	1.300
F_{siy}/L_i (m/s^2)	3.62	6.74	6.83	3.33	8.58
D_{iy} (m)	0.038	0.016	0.021	0.044	0.039

Ground motions were conveniently selected from the Pacific Earthquake Engineering Research Center (PEER) ground motion database [47]. The ten ground motions were selected as representative ground motions with a moment magnitude from 5.6 to 7.51, while the peak ground acceleration of the selected ground motions ranged from 0.059 to 0.28 g. The distance from the recording site to the epicenter of the selected ground motions was less than 50 km to represent the possible earthquake event near an active fault that will probably occur in northern Thailand. Most of the earthquakes were regarded as having strong ground motion. The earthquakes were also one component of ground motion and were performed in the x-direction for both UMRHA and NLRHA procedures. Figure 17 and Table 7 show the response spectrum and the details of the selected ground motions. With the Ruaumoko-2D computer program [48], the standard governing equation of motion for inelastic SDOF systems can be solved conveniently. When the force-based response or displacement-based response is obtained, the seismic responses, for instance, roof displacement, story drift, base shear, and base moment, can be computed. The total response can be achieved through a sum of individual responses in each vibration mode. Finally, the comparison between the seismic responses of the NLRHA procedure and the UMRHA procedure must be examined to investigate the results from the UMRHA procedure and determine whether the responses were accurate and valid.

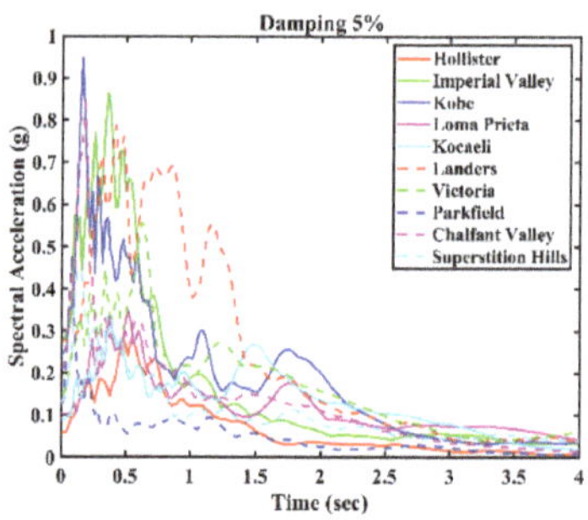

Figure 17. The response spectrum of the selected ground motions.

Figures 18 and 19 show the time history of the seismic responses of the RES open ground-story frame and the RES open top-story frame to the three selected ground motions, while Tables 8–12 show peak values regarding roof displacement and top-story drift regarding RES buildings to the ten selected ground motions. The seismic responses include the time history of roof displacement and top-story drift. These responses were compared between the UMRHA procedure and the NLRHA procedure. Based on the results, the presence of infill walls in buildings can significantly reduce floor displacement and story drift. With the UMRHA procedure together with the proposed assumption, seismic responses appeared to be underestimated. The seismic responses in terms of time history did not match well between the UMRHA and NLRHA procedures, possibly because of a lack of consistency regarding tuning hysteretic models. As shown in Figures 15 and 16, the tuning idealized curve matched well in the large deformation range, whereas the curve did not match well in the small deformation range. Due to the previous statement, seismic responses cannot be accurately predicted and can lead to underestimated results. For failure mechanisms in terms of infill walls, infill cracks were first found on the fully infilled frame, whereas infill failure was finally found to occur on the original frame. This result indicated that the configuration of infill walls as the original frame expressed more durability than other patterns when they were subjected to strong ground motion.

Table 7. List of the selected ground motions.

No	Record Seq. #	Event	Year	M_w	Mechanism	R_{rup} (km)	V_{s30} (m/s)	$D_{5\text{-}95}$ (s)	PGA (g)
1	26	Hollister-01	1961	5.6	Strike-slip	19.56	198.77	18.7	0.059
2	162	Imperial Valley-06	1979	6.53	Strike-slip	10.45	231.23	14.8	0.28
3	1107	Kobe, Japan	1995	6.9	Strike-slip	22.5	312	13.2	0.24
4	731	Loma Prieta	1989	6.93	Reverse Oblique	41.88	391.91	16.5	0.069
5	1166	Kocaeli, Turkey	1999	7.51	strike-slip	30.73	476.62	19.5	0.091
6	864	Landers	1992	7.28	Strike-slip	11.03	379.32	27.1	0.27
7	266	Victoria, Mexico	1980	6.33	Strike-slip	18.96	242.05	19	0.15
8	31	Parkfield	1966	6.19	Strike-slip	12.9	256.82	13.1	0.09
9	548	Chalfant Valley-02	1986	6.19	Strike-slip	21.92	370.94	16.6	0.21
10	718	Superstition Hills-01	1987	6.22	Strike-slip	17.59	179	15.2	0.13

Table 8. Peak values regarding roof displacement and top-story drift of RES bare frame.

No	Roof Displacement (m)		Top-Story Drift (m)	
	UMRHA	NLRHA	UMRHA	NLRHA
1	0.0246	0.0463	0.0052	0.0076
2	0.0446	0.0683	0.0096	0.0133
3	0.0448	0.0517	0.0102	0.0112
4	0.0288	0.0589	0.0060	0.0100
5	0.0222	0.0440	0.0047	0.0072
6	0.0781	0.1278	0.0161	0.0198
7	0.0656	0.0876	0.0137	0.0142
8	0.0099	0.0139	0.0023	0.0028
9	0.0363	0.0708	0.0084	0.0141
10	0.0239	0.0299	0.0054	0.0049

Table 9. Peak values regarding roof displacement and top-story drift of RES original frame to the selected ground motions.

No	Roof Displacement (m)		Top-Story Drift (m)	
	UMRHA	NLRHA	UMRHA	NLRHA
1	0.0057	0.0132	0.0017	0.0038
2	0.0175	0.0210	0.0054	0.0060
3	0.0161	0.0259	0.0053	0.0087
4	0.0064	0.0215	0.0019	0.0066
5	0.0070	0.0169	0.0021	0.0054
6	0.0207	0.0306	0.0061	0.0089
7	0.0088	0.0139	0.0028	0.0045
8	0.0024	0.0031	0.0008	0.0010
9	0.0085	0.0162	0.0031	0.0054
10	0.0074	0.0182	0.0025	0.0064

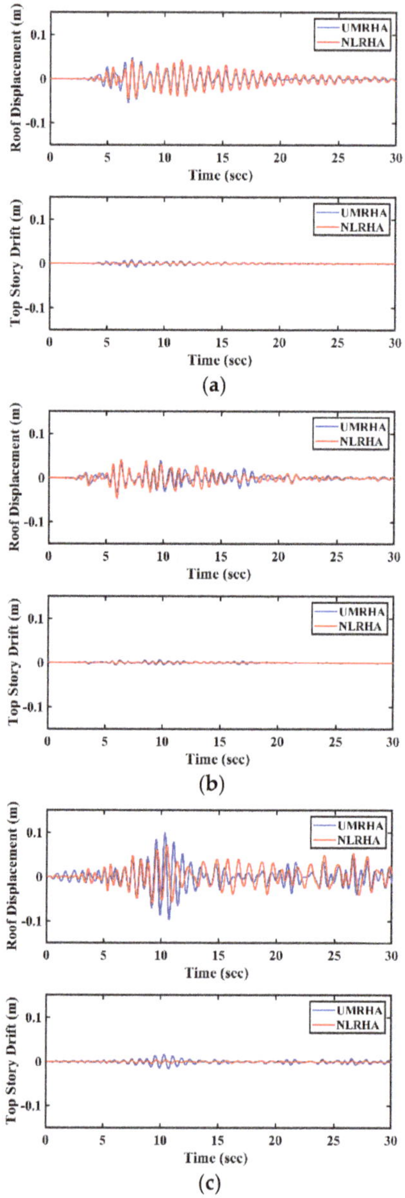

Figure 18. Comparison of roof displacement and top-story drift of RES open ground-story frame subjected to the selected ground motions: (**a**) Imperial Valley; (**b**) Kobe; (**c**) Landers.

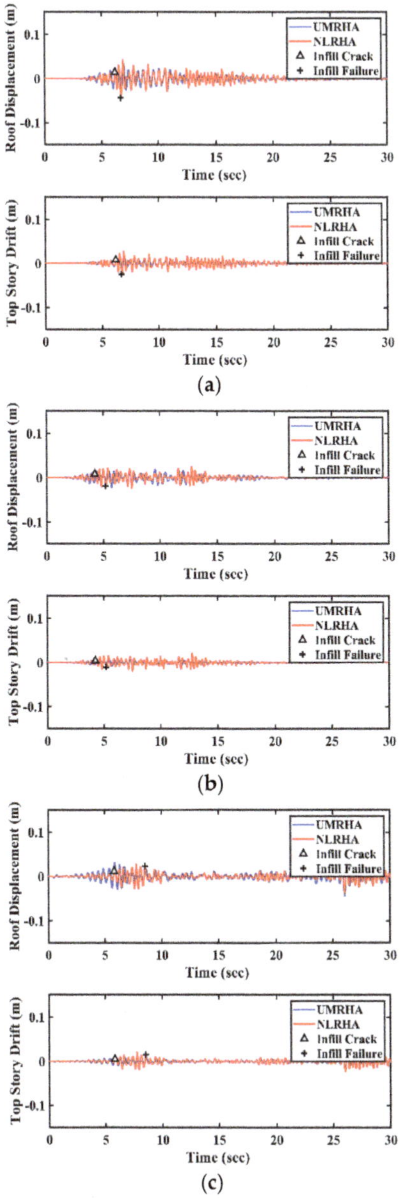

Figure 19. Comparison of roof displacement and top-story drift of RES open top-story frame subjected to the selected ground motions: (**a**) Imperial Valley; (**b**) Kobe; (**c**) Landers.

Table 10. Peak values regarding roof displacement and top-story drift of RES fully infilled frame to the selected ground motions.

No	Roof Displacement (m)		Top-Story Drift (m)	
	UMRHA	NLRHA	UMRHA	NLRHA
1	0.0063	0.0122	0.0018	0.0041
2	0.0307	0.0468	0.0087	0.0158
3	0.0207	0.0337	0.0059	0.0173
4	0.0097	0.0190	0.0027	0.0083
5	0.0097	0.0118	0.0027	0.0036
6	0.0212	0.0367	0.0060	0.0094
7	0.0158	0.0155	0.0045	0.0059
8	0.0029	0.0030	0.0010	0.0010
9	0.0120	0.0151	0.0035	0.0069
10	0.0117	0.0132	0.0034	0.0046

Table 11. Peak values regarding roof displacement and top-story drift of RES open ground-story frame to the selected ground motions.

No	Roof Displacement (m)		Top-Story Drift (m)	
	UMRHA	NLRHA	UMRHA	NLRHA
1	0.0339	0.0316	0.0056	0.0027
2	0.0553	0.0448	0.0092	0.0035
3	0.0400	0.0466	0.0067	0.0038
4	0.0248	0.0336	0.0041	0.0033
5	0.0238	0.0191	0.0040	0.0018
6	0.1002	0.0716	0.0167	0.0041
7	0.0468	0.0549	0.0078	0.0040
8	0.0107	0.0075	0.0018	0.0009
9	0.0314	0.0414	0.0053	0.0039
10	0.0283	0.0199	0.0048	0.0018

Table 12. Peak values regarding roof displacement and top-story drift of RES open top-story frame to the selected ground motions.

No	Roof Displacement (m)		Top-Story Drift (m)	
	UMRHA	NLRHA	UMRHA	NLRHA
1	0.0115	0.0175	0.0044	0.0103
2	0.0330	0.0445	0.0125	0.0288
3	0.0254	0.0258	0.0108	0.0216
4	0.0147	0.0106	0.0055	0.0070
5	0.0162	0.0140	0.0060	0.0093
6	0.0453	0.0411	0.0165	0.0289
7	0.0227	0.0221	0.0087	0.0144
8	0.0062	0.0024	0.0026	0.0013
9	0.0169	0.0165	0.0077	0.0132
10	0.0163	0.0200	0.0067	0.0156

In conclusion, only one vibration mode seems to be adequate to estimate seismic responses such as floor displacement and story drift. A slight discrepancy in responses between the two procedures can be observed in the framed building affected by infill walls, as opposed to the framed buildings without the effects of infill walls. Other hysteretic models should be considered, especially for buildings that have contributions from infill walls, which may lead to more accurate results in terms of evaluating seismic responses.

6. Conclusions

For the buildings affected by infill walls, the behavior of buildings is first brittle; then, after the collapse of infill walls occurs, the behavior will become ductile again. However, for buildings without the effect of infill walls, the behavior is ductile. For bare frames and open ground-story frames, the behavior is similar for most of the aforementioned building types, as they behaved in a ductile manner. This phenomenon takes place as bare frames have a soft story effect. The presence of infill walls in some patterns significantly contributes strength and stiffness to buildings as high as the fully infilled frame or even more. However, some may have no contribution from infill walls and lead to a torsional effect. Due to the torsional effect on buildings, columns in an affected area undergo more failure mechanisms than columns where the position has less effect of torsion, including reinforcing steel yielding, unconfined concrete crushing, and confined concrete crushing. Some infill wall patterns, such as the open top-story frame, lead columns on the ground floor to have no failure mechanisms, whereas those on the other floors have failure mechanisms, including reinforcing steel yielding, unconfined concrete crushing, and confined concrete crushing. Considering the position of infill walls on a framed building, most of the infill walls that affect the framed buildings are found on the ground floor level. Nevertheless, a few walls on floors other than the ground floor level are also found to affect the framed buildings. In terms of directions, infill walls in both the x- and y-directions affect the buildings when the infill walls of the considered direction all collapsed before those in the other direction. For infill walls, only those on the ground floor contribute to the building when they are subjected to a ground motion in the NLRHA procedure. The fully infilled frame tends to experience infill cracking before the other patterns. The original frame, however, is the last type to experience infill failure, implying that the original frame is more durable than other patterns when it is subject to earthquake shaking.

In the UMRHA procedure, the reduction from 3 to 2 DOFs is used, considering the ground beam level as the support level, and the first and second DOFs are regarded as the second floor and roof floor, respectively. In contrast, the 3D full model is carried out according to the structural plan for the NLRHA procedure. Based on the results, only the first vibration mode is found to be adequate to estimate seismic responses. The use of a linear model in the second vibration mode is found to be an alternative to evaluating seismic responses, provided that the response contributed from the second vibration mode is less than or in the elastic range.

Author Contributions: Conceptualization, P.L.; Investigation, S.S.; Methodology, P.L.; Software, S.S.; Validation, S.S.; Visualization, S.S.; Writing—original draft, S.S.; Writing—review & editing, P.L. All authors have read and agreed to the published version of the manuscript.

Funding: This research project was supported by Mahidol University (Basic Research Fund: fiscal year 2021). The APC was funded by Faculty of Engineering, Mahidol University.

Institutional Review Board Statement: Not applicable.

Informed Consent Statement: Not applicable.

Data Availability Statement: MDPI Research Data Policies.

Conflicts of Interest: The authors declare no conflict of interest.

References

1. Murty, C.V.R.; Jain, S.K. Beneficial Influence of Masonry Infill Walls on Seismic Performance of RC Frame Buildings. In Proceedings of the 12th World Conference on Earthquake Engineering, Auckland, New Zealand, 30 January–4 February 2000.
2. Lee, H.-S.; Woo, S.-W. Effect of masonry infills on seismic performance of a 3-storey R/C frame with non-seismic detailing. *Earthq. Eng. Struct. Dyn.* **2002**, *31*, 353–378. [CrossRef]
3. Al-Chaar, G.; Issa, M.; Sweeney, S. Behavior of Masonry-Infilled Nonductile Reinforced Concrete Frames. *J. Struct. Eng.* **2002**, *128*, 1055–1063. [CrossRef]
4. Asteris, P.G. Lateral Stiffness of Brick Masonry Infilled Plane Frames. *J. Struct. Eng.* **2003**, *129*, 1071–1079. [CrossRef]
5. Anil, O.; Altin, S. An experimental study on reinforced concrete partially infilled frames. *Eng. Struct.* **2007**, *29*, 449–460. [CrossRef]

6. Dolšek, M.; Fajfar, P. The effect of masonry infills on the seismic response of a four-storey reinforced concrete frame—A deterministic assessment. *Eng. Struct.* **2008**, *30*, 1991–2001. [CrossRef]
7. Hashem, Y.M.; Mahmoud, A.A.; Adam, M.; Shanour, A.S. Behavior of Reinforced Concrete Infilled Frames under Cyclic Loading. *Ain Shams J. Civ. Eng.* **2010**, *2*, 169–182.
8. Niyompanitpattana, S.; Warnitchai, P. Effects of masonry infill walls with openings on seismic behaviour of long-span GLD RC frames. *J. Inst. Civ. Eng.* **2017**, *69*, 1–21. [CrossRef]
9. Cavaleria, L.; Trapani, F.D.; Asteris, P.G.; Sarhosis, V. Influence of column shear failure on pushover based assessment of masonry infilled reinforced concrete framed structures: A case study. *Soil Dyn. Earthq. Eng.* **2017**, *100*, 98–112. [CrossRef]
10. Fiore, A.; Porco, F.; Raffaele, D.; Uva, G. About the influence of the infill panels over the collapse mechanisms actived under pushover analyses: Two case studies. *Soil Dyn. Earthq. Eng.* **2012**, *39*, 11–22. [CrossRef]
11. Risi, M.T.D.; Domenico, M.D.; Ricci, P.; Verderame, G.M.; Manfredi, G. Experimental investigation on the influence of the aspect ratio on the inplane/out-of-plane interaction for masonry infills in RC frames. *Eng. Struct.* **2019**, *189*, 523–540. [CrossRef]
12. Khan, N.A.; Tahir, M.F.; Nuti, C.; Briseghella, B.; Bergami, A.V. Influence of Brick Masonry Infill Walls on Seismic Response of RC Structures. *Tech. J.* **2019**, *24*, 15–23.
13. Furtado, A.; Rodrigues, H.; Arêde, A.; Varum, H. Experimental evaluation of out-of-plane capacity of masonry infill walls. *Eng. Struct.* **2016**, *111*, 48–63. [CrossRef]
14. Mazza, F. In-plane–out-of-plane non-linear model of masonry infills in the seismic analysis of r.c.-framed buildings. *Earthq. Eng. Struct. Dyn.* **2019**, *48*, 432–453. [CrossRef]
15. Pasca, M.; Liberatore, L.; Masiani, R. Reliability of analytical models for the prediction of out-of-plane capacity of masonry infills. *Struct. Eng. Mech.* **2017**, *64*, 765–781.
16. Chrysostomou, C.Z.; Asteris, P.G. On the in-plane properties and capacities of infilled frames. *Eng. Struct.* **2012**, *41*, 385–402. [CrossRef]
17. Celarec, D.; Ricci, P.; Dolšek, M. The sensitivity of seismic response parameters to the uncertain modelling variables of masonry-infilled reinforced concrete frames. *Eng. Struct.* **2012**, *35*, 165–177. [CrossRef]
18. Inel, M.; Ozmen, H.B. Effects of plastic hinge properties in nonlinear analysis of reinforced concrete buildings. *Eng. Struct.* **2006**, *28*, 1494–1502. [CrossRef]
19. Kadid, A.; Boumrkik, A. Pushover Analysis of Reinforced Concrete Frame Structures. *Asian J. Civ. Eng.* **2008**, *9*, 75–83.
20. Daniel, D.M.; John, S.T. Pushover Analysis of RC Building. *Int. J. Sci. Eng. Res.* **2016**, *7*, 88–92.
21. Girgin, K.; Darılmaz, K. Seismic Response of Infilled Framed Buildings Using Pushover Analysis. *Bull. Istanb. Tech. Univ.* **2007**, *54*, 5.
22. Hakim, R.A.; Alama, M.S.; Ashour, S.A. Seismic Assessment of RC Building According to ATC 40, FEMA 356 and FEMA 440. *Arab. J. Sci. Eng.* **2014**, *39*, 761–7699. [CrossRef]
23. Choudhary, N.; Wadia, M. Pushover Analysis of R.C. Frame Building with Shear Wall. *IOSR J. Mech. Civ. Eng.* **2014**, *11*, 9–13. [CrossRef]
24. Jalaeefar, A.; Zargar, A. Effect of infill walls on behavior of reinforced concrete special moment frames under seismic sequences. *Structures* **2020**, *28*, 766–773. [CrossRef]
25. Chopra, A.K. *Dynamics of Structures: Theory and Applications to Earthquake Engineering*, 4th ed.; Prentice-Hall: Hoboken, NJ, USA, 2012.
26. Chopra, A.K.; Goel, R.K. A modal pushover analysis procedure for estimating seismic demands for buildings. *Earthq. Eng. Struct. Dyn.* **2002**, *31*, 561–582. [CrossRef]
27. Chopra, A.K.; Goel, R.K. A modal pushover analysis procedure to estimate seismic demands for unsymmetric-plan buildings. *Earthq. Eng. Struct. Dyn.* **2004**, *33*, 903–927. [CrossRef]
28. Munir, A.; Warnitchai, P. The cause of unproportionately large higher mode contributions in the inelastic seismic responses of high-rise core-wall buildings. *Earthq. Eng. Struct. Dyn.* **2012**, *41*, 2195–2214. [CrossRef]
29. Najam, F.A.; Warnitchai, P. A modified response spectrum analysis procedure to determine nonlinear seismic demands of high-rise buildings with shear walls. In Proceedings of the 16th World Conference on Earthquake, 16WCEE 2017, Santiago, Chile, 9–13 January 2017.
30. Mehmood, T.; Warnitchai, P.; Suwansaya, P. Seismic Evaluation of Tall Buildings Using a Simplified but Accurate Analysis Procedure. *J. Earthq. Eng.* **2018**, *22*, 356–381. [CrossRef]
31. Fragiadakis, M.; Papadrakakis, M. Modeling, Analysis and Reliability of Seismically Excited Structures: Computational Issues. *Int. J. Comput. Methods* **2008**, *5*, 483–511. [CrossRef]
32. Spacone, E.; Filippou, F.C.; Taucer, F.F. Fibre beam-column model for non-linear analysis of R/C frames: Part I. formulation. *Earthq. Eng. Struct. Dyn.* **1996**, *25*, 711–725. [CrossRef]
33. Cavaleri, L.; Trapani, F.D. Cyclic response of masonry infilled RC frames: Experimental results and simplified modeling. *Soil Dyn. Earthq. Eng.* **2014**, *65*, 224–242. [CrossRef]
34. Crisafulli, F.J.; Carr, A.J.; Park, R. Analytical Modelling of Infilled Frame Structures—A General Review. *Bull. New Zealand Soc. Earthq. Eng.* **2000**, *33*, 30–47. [CrossRef]
35. Crisafulli, F.J.; Carr, A.J. Proposed Macro-model for the Analysis of Infilled Frame Structures. *Bull. New Zealand Soc. Earthq. Eng.* **2007**, *40*, 69–77. [CrossRef]

36. Smyrou, E.; Blandon-Uribes, C.; Antoniou, S.; Pinho, R.; Crowley, H. Implementation and Verification of A Masonry Panel Model for Nonlinear Pseudo-Dynamic Analysis of Infilled RC Frames. In Proceedings of the First European Conference on Earthquake Engineering and Seismology, Geneva, Switzerland, 3–8 September 2006.
37. Srechai, J.; Leelataviwat, S.; Wararuksajja, W.; Limkatanyu, S. Multi-strut and empirical formula-based macro modeling for masonry infilled RC frames. *Eng. Struct.* **2022**, *266*, 114559. [CrossRef]
38. Crisafulli, F.J. *Seismic Behaviour of Reinforced Concrete Structures with Masonry Infills*; University of Canterbury: Christchurch, New Zealand, 1997.
39. Seismosoft. SeismoStruct 2021—A Computer Program for Static and Dynamic Nonlinear Analysis of Framed Structures. Available online: https://seismosoft.com/ (accessed on 6 April 2021).
40. Karayannis, C.G.; Kakaletsis, D.J.; Favvata, M.J. Behavior of bare and masonry infilled R/C frames under cyclic loading: Experiments and analysis. In *Earthquake Resistant Engineering Structures V*; WIT Press: Southampton, UK, 2005; pp. 429–438.
41. Van, T.C.; Lau, T.L. Experimental Evaluation of Reinforced Concrete Frames with Unreinforced Masonry Infills under Monotonic and Cyclic Loadings. *Int. J. Civ. Eng.* **2020**, *19*, 401–419. [CrossRef]
42. Available online: https://office.dpt.go.th/construction (accessed on 6 April 2021).
43. Available online: www.layerhouse.blogspot.com (accessed on 6 April 2021).
44. Available online: https://www.yotathai.com/ (accessed on 6 April 2021).
45. Mander, J.B.; Priestley, M.J.N.; Park, R. Theoretical stress-strain model for confined concrete. *J. Struct. Eng.* **1988**, *114*, 1804–1826. [CrossRef]
46. Menegotto, M.; Pinto, P.E. Method of analysis for cyclically loaded R.C. plane frames including changes in geometry and non-elastic behaviour of elements under combined normal force and bending. In Proceedings of the Symposium on the Resistance and Ultimate Deformability of Structures Acted on by Well Defined Repeated Loads; International Association for Bridge and Structural Engineering: Zurich, Switzerland, 1973; pp. 15–22.
47. Pacific Earthquake Engineering Research Center (PEER) PEER Ground Motion Database. Available online: https://ngawest2.berkeley.edu/ (accessed on 6 April 2021).
48. Carr, A.J. *Ruaumoko—The Maori God of Volcanoes and Earthquakes*; University of Canterbury, Inelastic Analysis Finite Element Program: Christchurch, New Zealand, 2005.

buildings

Article

Seismic Performance of Flat Steel Plate Shear Walls with Atmospheric Corrosion

Xiaoming Ma [1,2,3], Yi Hu [4], Xinyuan Cheng [5,*], Liqiang Jiang [5,*], Yun Li [6] and Hong Zheng [3]

1 Key Laboratory of Environmental Change and Natural Disasters of Ministry of Education, Beijing Normal University, Beijing 100875, China
2 Engineering Research Center for Disaster Prevention and Mitigation of Southeast Coastal Engineering Structures (JDGC03), Fujian University, Putian 351100, China
3 School of Civil Engineering, Chang'an University, Xi'an 710061, China
4 School of Civil Engineering, Central South University of Forestry and Technology, Changsha 410004, China
5 School of Civil Engineering, Central South University, Changsha 410075, China
6 Hunan Tieyuan Civil Engineering Testing Co., Ltd., Changsha 410075, China
* Correspondence: xinyuancc@csu.edu.cn (X.C.); jianglq2019@csu.edu.cn (L.J.)

Abstract: The seismic performance of four different kinds of steel plate shear walls (SPSWs) before and after corrosion are investigated in this paper, including flat steel plate shear walls (FSPSWs), SPSWs with vertical slots in the middle (VSSPSWs), SPSWs with orthogonal stiffeners (OSPSWs) and SPSWs with silts on both sides (BSPSWs). A numerical model that can be validated by existing quasi-static cyclic tests is developed by ABAQUS 6.13. The seismic performance of SPSWs under cyclic loading and atmospheric corrosion is investigated. The results show that (1) compared with FSPSWs, the ultimate shearing strength, initial stiffness, energy dissipation, and ductility of OSPSWs are significantly improved under cyclic loads when both VSSPSWs and BSPSWs are reduced in ultimate shearing strength and energy dissipation. (2) The performance of the FSPSW is most affected by atmospheric corrosion but setting stiffeners can significantly improve the hysteretic performance after corrosion. Meanwhile, the effect of the VSSPSW is better than that of the BSPSW in decreasing the ultimate shearing strength, although both decrease more tardily than that of the FSPSW. (3) The relationships between the ultimate shearing strength and corrosion time of SPSWs are fitted into four equations in this paper, which can be used in practical situations.

Keywords: steel plate shear wall (SPSW); atmospheric corrosion; hysteretic performance; finite element method (FEM)

Citation: Ma, X.; Hu, Y.; Cheng, X.; Jiang, L.; Li, Y.; Zheng, H. Seismic Performance of Flat Steel Plate Shear Walls with Atmospheric Corrosion. *Buildings* **2022**, *12*, 1881. https://doi.org/10.3390/buildings12111881

Academic Editor: Rita Bento

Received: 27 September 2022
Accepted: 31 October 2022
Published: 4 November 2022

1. Introduction

Steel structure buildings have the advantages of better seismic performance, shorter construction periods, and environmental protection and can be widely used for China's urban construction in the future. With the advantages of high initial stiffness, excellent ductility, and uncomplicated connection, the steel plate shear wall (SPSW) is an ideal lateral force-resisting system [1–5]. Meanwhile, the SPSW can carry loads using the post-buckling strength and dissipate energy by the plastic development of steel under earthquake loads, so there exist outstanding mechanical properties and energy dissipation capacity on SPSW [6,7]. However, the SPSW easily buckles and has poor out-of-plane stiffness when subjected to earthquake loads, and due to connecting with the boundary column, the tension field can be formed in the SPSW, which can decrease the ultimate shearing resistance of the SPSW [8,9].

In order to solve the above problems of the SPSW, many experimental and theoretical investigations were conducted. Ahmadi et al. [10] conducted experimental and numerical investigations on the seismic performance of steel slit shear walls, and the results showed this lateral force-resisting system had a good seismic performance. Compared to the non-stiffened flat steel plate, the stiffener can effectively improve the seismic performance of the SPSW [11]. Haddad et al. [12] conducted an experimental investigation on the seismic performance of stiffened and unstiffened SPSWs; the results showed that the SPSW with stiffeners could considerably improve the ultimate shearing resistance, ductility, and energy dissipation capacity.

All this time, the corrosion problem of steel structures has attracted much attention. There have been lots of studies on the corrosion mechanism and mechanical property degradation of corroded steel; the results showed that corrosion could lead to the rapid aging of steel, reduction in strength, and stress concentration, which finally decreases the ultimate shearing strength of structure [13–26], so it is significant to investigate the performance of corroded steel structures. Woloszyk et al. [27–29] applied a random field approach to the modeling of corroded steel surfaces and confirmed that the irregularity of corrosion for steel surfaces is one of the main reasons for the decrease in mechanical properties. Zhao et al. [30–32] conducted a numerical investigation of pitting steel and systematically explored how factors such as the shape, depth, and arrangement of pitting holes can reduce the tensile performance of steel.

In addition to material properties, some experimental and numerical investigations have also been conducted on the properties of corroded components. Zheng et al. [33,34] implemented an accelerated corrosion test in a chloride environment and obtained a mutual verification between his finite element (FE) model and the experimental results. Wang et al. [35,36] investigated the effect of corroded steel frame joints on seismic performance through experimental and numerical investigations. Xu and Zhang et al. [37–39] established FE numerical models of corroded steel columns in simulated corrosion and cyclic loading tests. The above results indicated that the seismic performance of the steel frame and components was seriously damaged by corrosion. As the corrosion degree increased, the deformation capacity, energy dissipation capacity, and mechanical properties of the steel frame and components were significantly reduced.

Although scholars have conducted a large number of experimental and numerical studies for SPSWs, the comparative research on the seismic performance of different types of flat steel plates is still insufficient. Meanwhile, although many experiments and numerical analyses on corroded steel components have been finished, there is still little research on the seismic performance of SPSWs after corrosion [40–42]. Therefore, in order to investigate the seismic performance of different kinds of SPSWs before and after corrosion, four different types of SPSWs are proposed in this paper, and numerical analyses of four kinds of SPSWs are performed by using ABAQUS based on the test-validated modeling method. In addition, the seismic performance of specimens at different corrosion levels is investigated. Finally, the fitted formulae of ultimate shearing resistance for corroded SPSWs are proposed. The analyses results of seismic performance and fitted formulae can be applied to practical engineering design.

2. FEM of SPSWs

2.1. Model Design

Four different kinds of SPSWs are studied in this paper, including flat steel plate shear walls (FSPSWs), SPSWs with orthogonal stiffeners (OSPSWs), SPSWs with vertical slots in the middle (VSSPSWs), and SPSWs with silts on both sides (BSPSWs), as presented in Figure 1. According to the Chinese codes GB50017-2017 [43], JCJ/T380-2015 [44], and GB50011-2010 [45], the dimension details of specimens are designed as shown in Figure 1. Meanwhile, HW 400 $\times$ 400 $\times$ 20 $\times$ 24, HN 350 $\times$ 250 $\times$ 14 $\times$ 20, and HN 300 $\times$ 200 $\times$ 12 $\times$ 20 are used as the boundary column, boundary beams (top beam and bottom beam), and middle beam, respectively. The flat steel plate, which has a 6 mm thickness, is only welded to the boundary beam, and the welded connection is adopted between the beams and boundary columns. The section dimension of the stiffener of specimen OSPSW is $-3000 \times 50 \times 6$, and stiffeners are arranged on two surfaces of flat steel plate. Q235 steel is used for the infilled flat steel plate, and Q355 steel is selected for all other components.

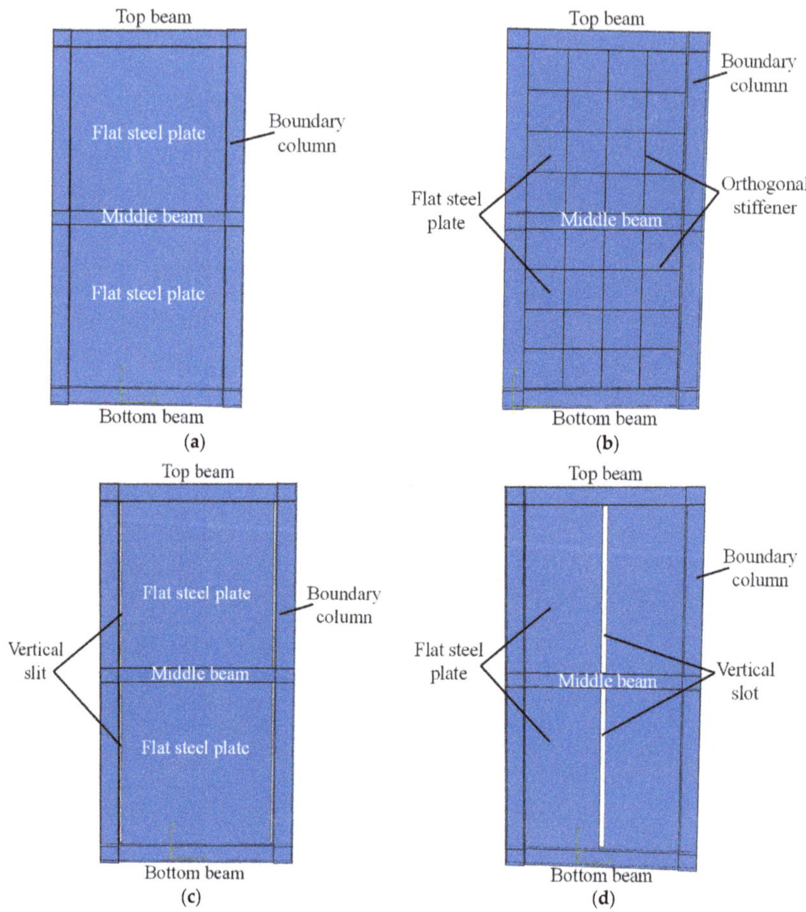

Figure 1. Diagrams of specimens: (**a**) FSPSW; (**b**) OSPSW; (**c**) BSPSW; (**d**) VSSPSW.

The detailed dimensions and profiles of the four kinds of SPSW specimens are presented in Figure 2. The boundary conditions of FEM models for specimens are shown in Figure 3, and the boundary condition of flat steel plates is that the upper and lower sides are only connected with the frame beam. The constitutive relation of the material is shown in Figure 4, and σ_y is the yield strength, σ_u is the ultimate strength, σ_{st} is 85% of σ_u. Modeling with S4R four-node shell elements, the "Merge" command is used to simulate the connection between the beam and frame column, and the "Tie" command is applied to simulate the infilled steel plate welded to the beam. The element size of the infilled steel plate is 30 mm × 30 mm, and the element size of the boundary frame is 35 mm × 35 mm. The vertical load, such as gravity, is simulated with an axial force of 2500 kN, and the horizontal load is applied at a coupling point consisting of the top of the boundary column until the specimen is damaged.

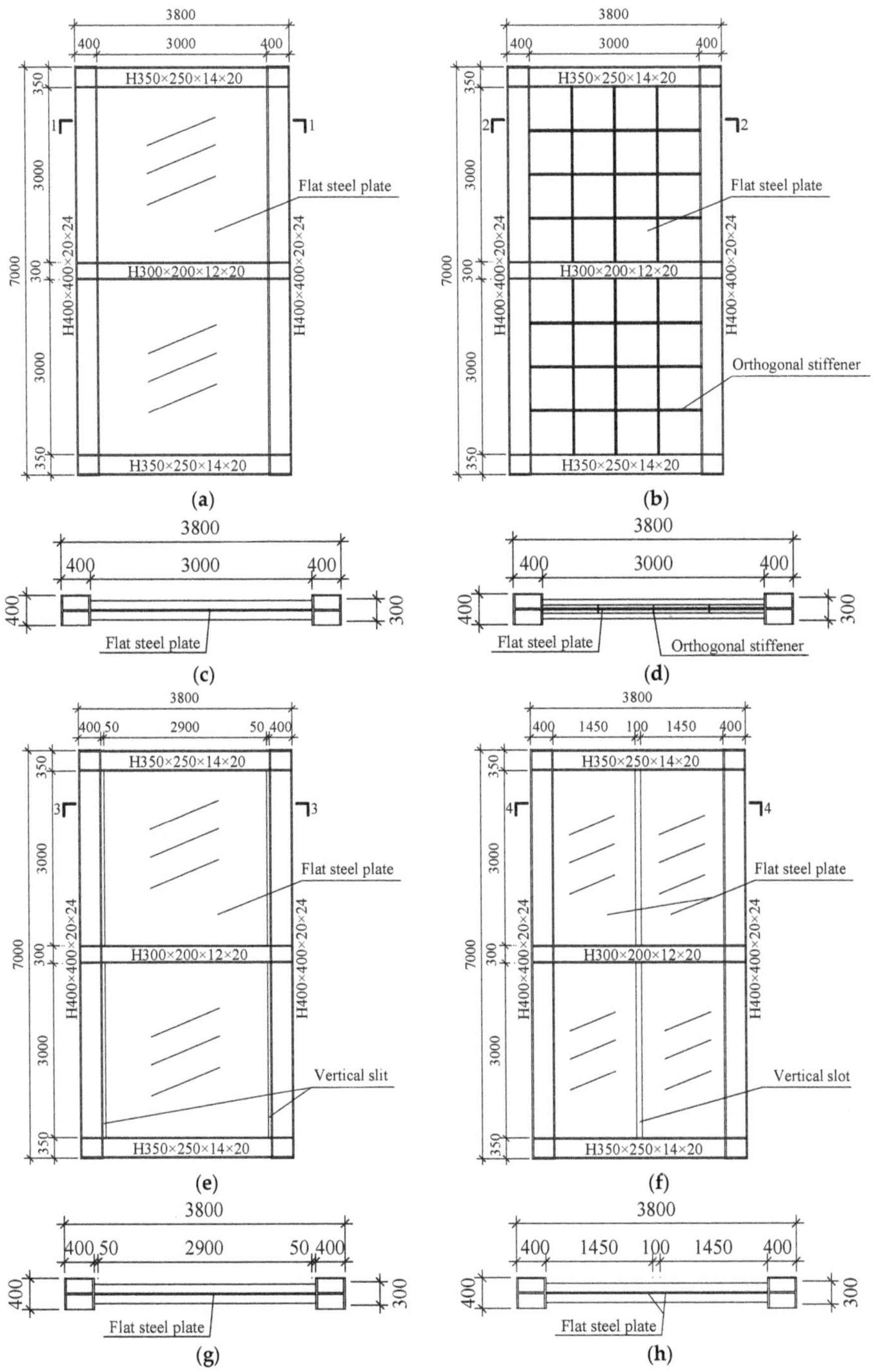

Figure 2. Dimension details of specimens: (**a**) FSPSW; (**b**) OSPSW; (**c**) Section 1-1 of FSPSW; (**d**) Section 2-2 of OSPSW; (**e**) BSPSW; (**f**) VSSPSW; (**g**) Section 3-3 of BSPSW; (**h**) Section 4-4 of VSSPSW.

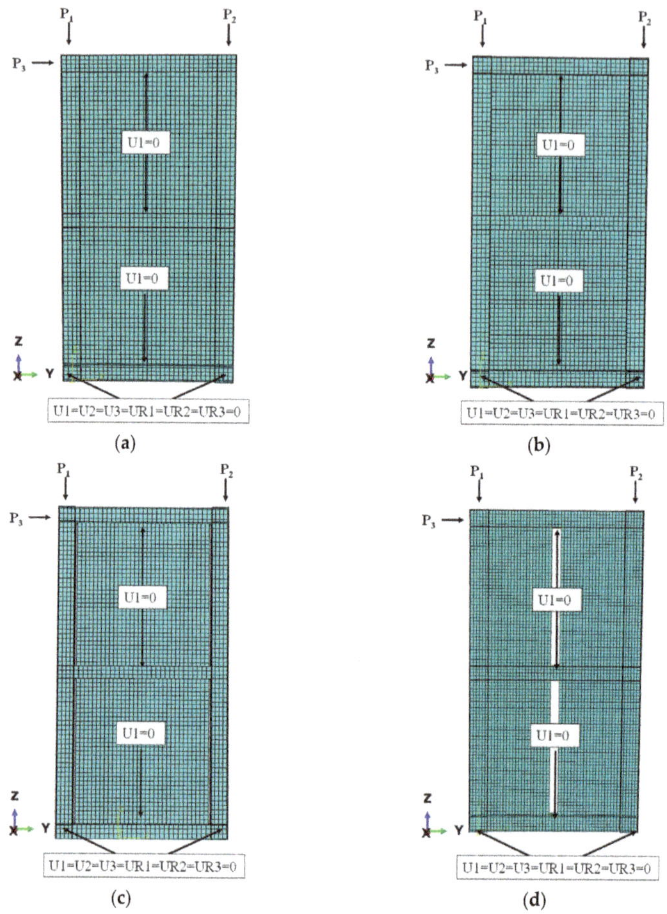

Figure 3. The boundary conditions of FEM models for SPSWs: (**a**) FSPSW; (**b**) OSPSW; (**c**) VSSPSW; (**d**) BSPSW.

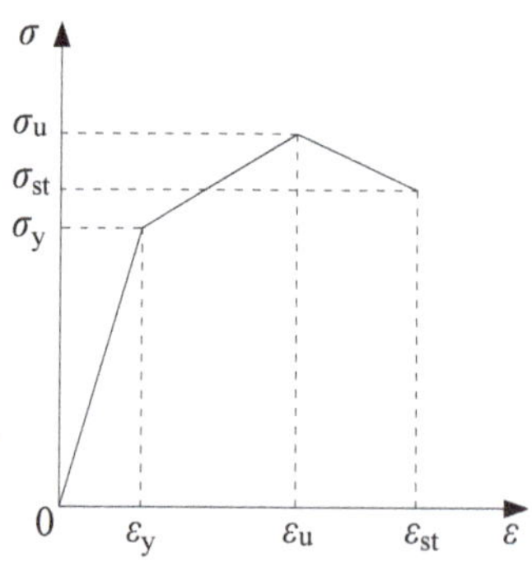

Figure 4. The constitutive relation of steel in the FEM model.

2.2. Model Validation

Due to the lack of tests on flat steel plate shear walls, test specimen S-4 from Ref. [46] is selected to compare the test results with the FEM results. The dimensions of each component are shown in Table 1, the material properties are shown in Table 2, and the test specimens are shown in Figure 5.

Table 1. Dimension details of test components [46].

Component	Sectional Dimension/mm	Steel Grade
Upper and lower beams Middle beam Column	HN 250 × 200 × 12 × 14 HN 175 × 175 × 8 × 10 HW 200 × 200 × 8 × 12	Q355
Infilled vertical infilled corrugated steel plate	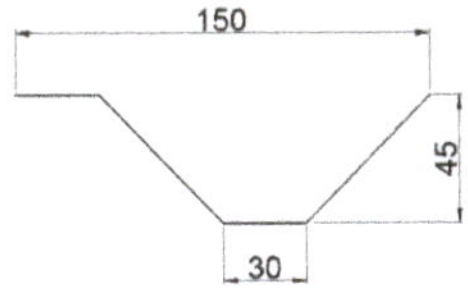	Q235

Table 2. Material properties [46].

Component	E/GPa	f_y/MPa	f_u/MPa	Elongation%	f_u/f_y	Steel Grade
Infilled steel plate	176.0	301.43	397	34.0	1.32	Q235
Middle beam web	228.0	388.83	518	27.0	1.33	Q355
Middle beam flange	219.0	370.27	495	29.2	1.34	Q355
Upper beam flange	203.0	330.90	529	32.9	1.60	Q355
Upper beam web	205.7	360.17	507	32.1	1.41	Q355
Column web	216.7	360.93	533	33.4	1.48	Q355
Column flange	207.3	363.43	540	34.6	1.49	Q355

Note: *E* represents the elastic modulus; f_y and f_u represent the yield stress and ultimate stress, respectively.

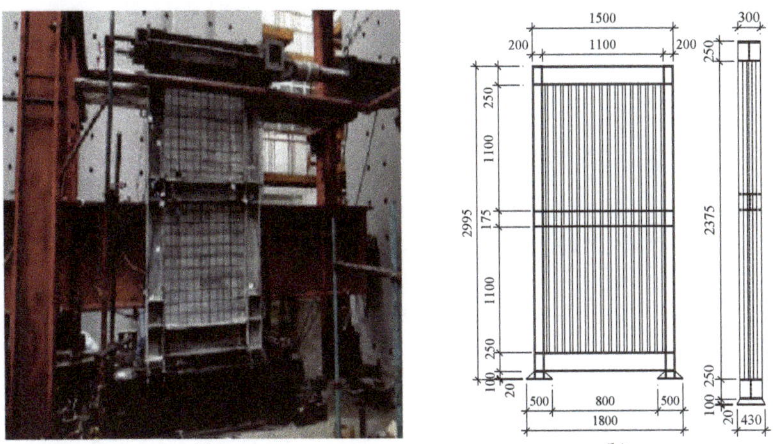

(a) (b)

Figure 5. Test specimen S-4 [46]: (**a**) Test site equipment; (**b**) Dimensions of the test specimen.

According to the literature [46] and the Chinese standards of JGJ101-96 [47], the lateral load was applied by using the combination of the load control method and displacement control method, as shown in Figure 6. Before the test specimen yielded, the lateral load was applied by using the load control method, the spacing between each loading step was 100 kN and repeated once at each loading step; when the test specimen yielded, the lateral load was applied by using the displacement control method, and the story drift ratio θ was used as load displacement, the spacing of θ between each load step was 0.25%, and the spacing of θ between each load step increased to 0.5% after of the θ reached 1.0%, with two cycles for each loading step.

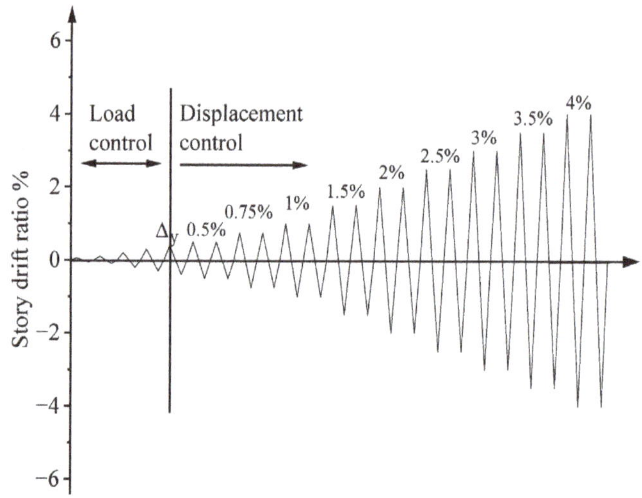

Figure 6. Cyclic loading of S−4 [46].

The comparisons of load-displacement curves derived from the FEM and test are shown in Figure 7, and the characteristic values are shown in Table 3. There existed strong consistency between the two hysteretic curves, which were obtained from the FEM and test, respectively, and there was little difference in the initial stiffness of the FEM model and test at the elastic stage. The general trend of the skeleton curves was primarily in line with the ultimate shearing resistance. Compared to the test data, the errors of the lateral stiffness, load, and displacement at the yield point, peak load, and corresponding displacement from the FEM model were 4%, 15%, 11%, 16%, 16%, and 8%, respectively, so the FEM results are by and largely coincident with the test.

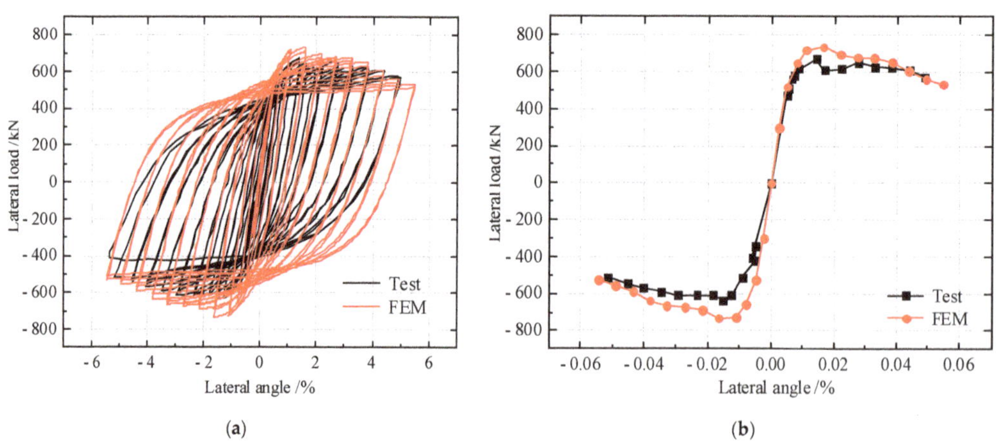

Figure 7. The contrast of load-angle curves of S−4: (**a**) Hysteretic curves; (**b**) Envelope curves.

Table 3. Characteristic parameters of specimens.

Results	K_0/(kN/mm)	P_y/kN	Δ_y/mm	P_m/kN	Δ_m/mm
FEM	25.81	758.39	22.35	876.05	42.70
Test	24.81	654.67	25.23	750.39	39.55
FEM/Test	1.04	1.15	0.89	1.16	1.08

Note: K_0 represents the initial lateral stiffness; P_y and P_m represent the yield load and ultimate load, respectively; Δ_y represents the yield displacement; Δ_m represents the displacement at P_m.

The comparisons of the invalidation phenomenon of specimen S−4 from the test and FEM are shown in Figure 8. There was an awfully noticeable out−of−plane deformation on the whole steel plates, parts of them were torn, and severe buckling occurred at the boundary column flanges and webs. By contrast, it can be concluded that the model accurately forecasts the buckling effects of specimen S−4. Therefore, it is observed that the modeling method in Section 2.1 is practicable, and subsequent studies can also be modeled using this method.

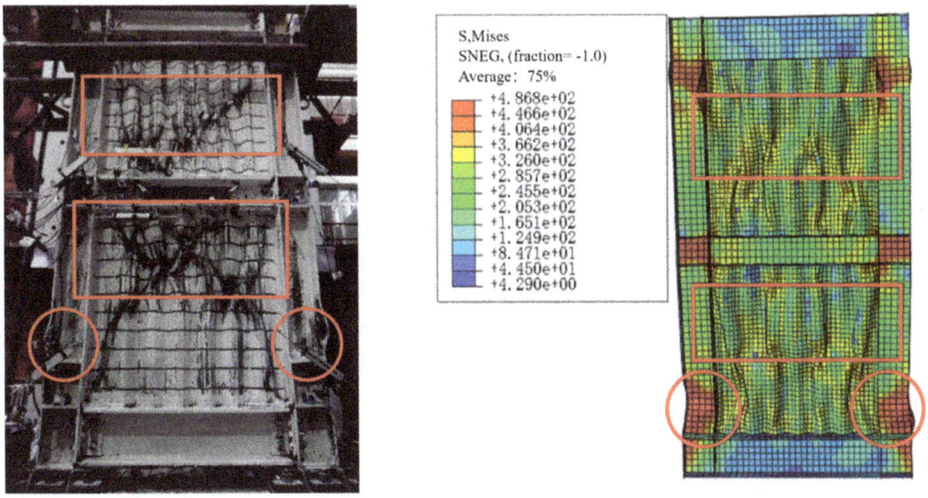

(a) (b)

Figure 8. Comparison of the invalidation phenomenon: (**a**) Test phenomenon; (**b**) FEM result.

3. Comparative Study of SPSW Seismic Performance

3.1. Comparison of the Load−Displacement Curves

The FEM and numerical analysis of the four kinds of SPSWs under cyclic loads are conducted through a modeling method similar to that in Section 2. Figure 9 depicts the load-displacement curves of the specimens, and Table 4 shows their characteristic results. After yielding, the hysteretic loop of the OSPSW is the fullest among them, the pinching of the hysteretic loop of BSPSW is the most obvious, and the initial stiffness of the FSPSW decreases more slowly after arriving at the peak load, while the trends of initial stiffness degradation for the VSSPSW, OSPSW, and BSPSW are mainly consistent. Compared with the FSPSW, the initial stiffness, yield load, peak load, and ductility of the OSPSW increased by 86.3%, 25.9%, 29.3%, and 34%, respectively, while the four corresponding values of the BSPSW decreased by 15.1%, 31.4%, 28.8%, and 48.2%, respectively. The initial stiffness and ductility of the VSSPSW are 35.6% and 22.2% lower than those of the BSPSW, respectively, but the yield and peak loads are 27.9% and 19.6% higher, respectively.

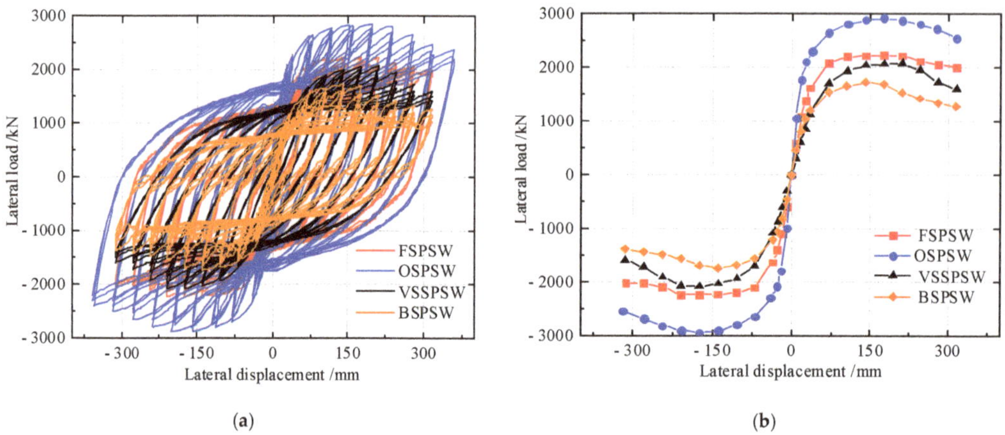

Figure 9. Load-displacement curves of the FSPSW, OSPSW, VSSPSW, and BSPSW: (**a**) Hysteretic curves; (**b**) Envelope curves.

Table 4. Characteristic results of the FSPSW, OSPSW, VSSPSW, and BSPSW.

Specimen	K_0/(kN/mm)	P_y/kN	Δ_y/mm	P_m/kN	Δ_m/mm	μ
FSPSW	54.64	1781.69	47.09	2225.77	174.91	6.67
OSPSW	101.77	2243.52	35.23	2877.79	171.82	8.94
VSSPSW	35.01	1734.91	74,63	2067.81	209.95	3.68
BSPSW	47.46	1356.42	51.07	1728.70	140.00	4.50

Note: μ is the ductility factor.

The results show that the OSPSW has a larger initial stiffness, ultimate shear-resistant capacity, and better stretchability, while the initial stiffness, yield load, peak load, and ductility of the VSSPSW and BSPSW are lower than those of the FSPSW. Therefore, setting stiffeners can effectively improve the hysteretic performance of SPSWs. Slotting results in a reduction in the hysteretic performance of the SPSW to a certain extent, while the hysteretic property of the VSSPSW is superior to that of the BSPSW.

3.2. Comparison of Energy Dissipation Capacity

How much energy a structure or component absorbs after plastic deformation during an earthquake is its energy dissipation capacity, which is evaluated by the energy dissipation coefficient h_e and calculated according to Equation (1).

$$h_e = \frac{S_{ABC} + S_{CDA}}{S_{OBE} + S_{ODF}}, \tag{1}$$

where S_{ABC} and S_{CDA} represent the areas surrounded by the hysteretic curve and the x-axis, respectively, and S_{OBF} and S_{ODE} represent the areas of OBE and ODF, respectively, as shown in Figure 10.

The energy dissipation values at each loading step for the specimens are shown in Figure 11. Compared with FSPSW, VSSPSW, and BSPSW, the total energy dissipation of OSPSW increases by 38%, 89.3%, and 138.6%, respectively. At each loading step, the energy dissipation value and rate of the OSPSW are larger than those of the FSPSW, VSPSW, and BSPSW. The energy dissipation value of each loading step of the FSPSW is larger than that of the VSSPSW and BSPSW, and the energy dissipation rate of the FSPSW slightly decreases after the loading step reaches 6δ. Meanwhile, before reaching 4δ, the energy dissipation value of the BSPSW is greater than that of the VSSPSW. With increasing loading steps, the energy dissipation rate of the BSPSW becomes slower, and the energy dissipation value is less than that of the VSSPSW. Therefore, setting reinforcement can enhance the energy dissipation capacity of SPSWs, and slotting will decrease the energy dissipation capacity of SPSWs. It is noted that slotting in the middle part has a better energy dissipation capacity than slotting on both sides.

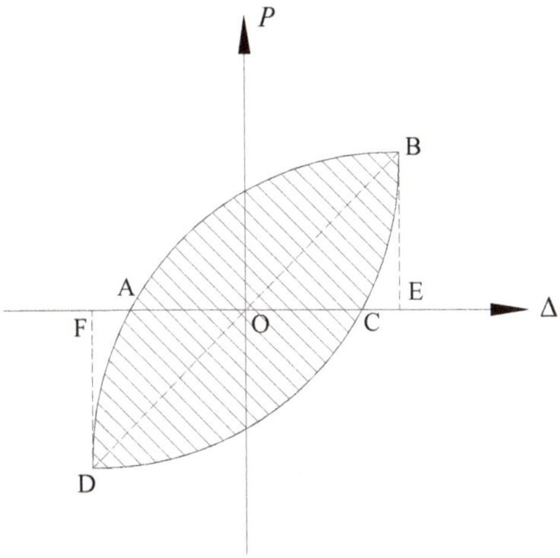

Figure 10. Calculation diagram of the energy dissipation coefficient.

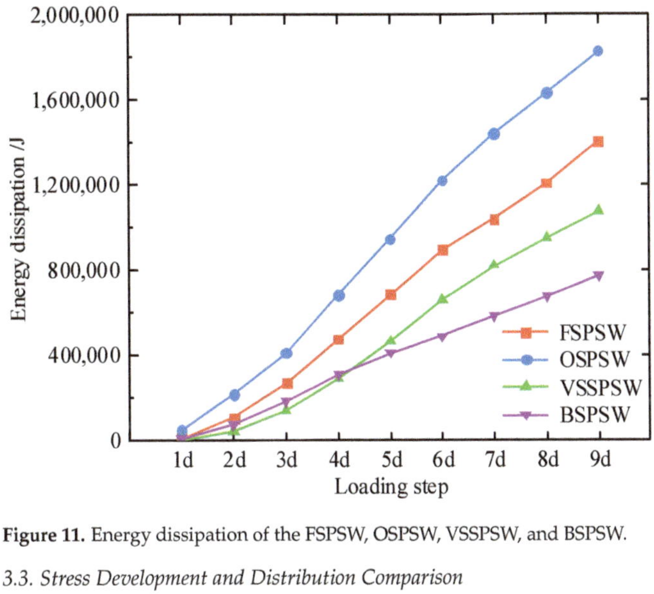

Figure 11. Energy dissipation of the FSPSW, OSPSW, VSSPSW, and BSPSW.

3.3. Stress Development and Distribution Comparison

The stress distributions of the SPSWs at failure are shown in Figure 12. For the FSPSW, the infilled steel plate buckles slightly, and severe buckling occurs at the top and bottom parts of the boundary column and connections between the columns and middle beam. For the OSPSW, because of the stiffeners, the buckling on the flat steel plate is less than that of the FSPSW, and slight buckling deformation occurs at the top and bottom parts of the columns. The plastic hinge occurs successively at both ends of the middle beam and column. For the VSSPSW, due to slotting at the middle part, the distribution of stress on the flat steel plate is well proportioned, slight buckling deformation occurs at the top and bottom of the columns, and there is no full section yielding. For the BSPSW, the variation in stresses for the steel plate is diagonally distributed along the tensile band. The peak stress still arises at the bottom of the column and the beam-column connection. This result is consistent with the failure mode of this result and FSPSW, and the bottom of the boundary column exhibits severe buckling. Therefore, setting stiffeners and slotting in the middle part can inhibit the generation of the plastic hinge at the column end and inhibit the buckling of the flat steel plate, in which slotting in the middle part has a better effect.

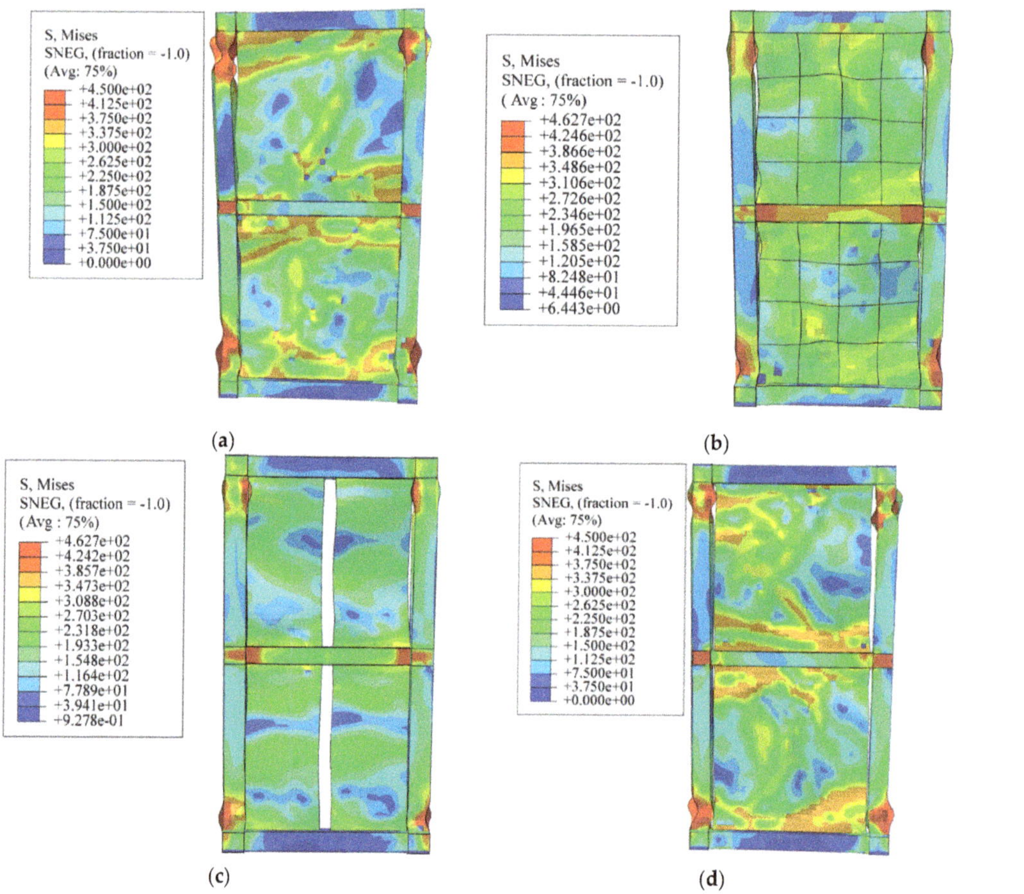

Figure 12. Stress distribution of damaged FEM specimens: (**a**) FSPSW; (**b**) OSPSW; (**c**) VSSPSW; (**d**) BSPSW.

3.4. Comparison of Implementation Costs of SPSWs

The usage amount of steel specimens is presented in Table 5. Compared to FSPSW, the usage amount of steel of OSPSW increases by 19.9%, and the usage amount of steel of VSSPSW and BSPSW decreases by 3.4%. It can be seen from Section 3.1 that compared to FSPSW, the K_0, P_m, and μ of OSPSW increase 86.3%, 29.3%, and 34%, and the three corresponding values of VSSPSW decrease 56.1%, 7.6%, and 81.3%, respectively; and the three corresponding values of BSPSW decrease 15.1%, 28.8%, and 48.2%. It is revealed that the OSPSW has better seismic performance. Meanwhile, it can be seen from Section 3.1 that compared with FSPSW, VSSPSW, and BSPSW, the total energy dissipation of OSPSW increases by 38%, 89.3%, and 138.6%, respectively. The above analysis results show that although the specimen OSPSW has a little more usage amount of steel than other specimens, the values of K_0, P_m, and μ of OSPSW are much larger. Therefore, the specimen OSPSW is suggested to be used for practical engineering.

Table 5. Characteristic results and usage amount of steel of specimens.

Specimen	Usage Amount of Steel/*t*	K_0/(kN/mm)	P_m/kN	μ
FSPSW	0.848	54.64	2225.77	6.67
OSPSW	1.017	101.77	2877.79	8.94
VSSPSW	0.82	35.01	2067.81	3.68
BSPSW	0.82	47.46	1728.70	4.50

4. Comparative Research on the Seismic Performance of SPSWs under Atmospheric Corrosion

4.1. Material Properties after Corrosion

The results of the accelerated corrosion test from Ref. [33] are selected in this paper, and the fitting formula of Ref. [33] is picked to compute the yield strength of steel after corrosion, as shown in Equation (2).

$$\frac{f_y}{f_{y0}} = 1 - 0.902 \times \frac{1 - E/E_0}{0.897}, \quad (2)$$

where f_y represents the yield strength of corroded steel; *E* represents the elastic modulus of corroded steel; f_y and E_0 represent the yield strength and elastic modulus of steel that is not corroded, respectively, and f_{y0} = 235 MPa or 355 MPa, E_0 = 206,000 MPa.

In this paper, the corrosion rate of the test is calculated to be 1208 $\mu m \cdot a^{-1}$, and the atmospheric corrosion rate in the Beijing area is 11.7 $\mu m \cdot a^{-1}$ [48]. The actual corrosion time is obtained through these two rates. Table 6 shows the switch from the simulated corrosion test time to the actual corrosion time.

Table 6. The actual corrosion time and elastic modulus of corrosion steel.

Corrosion Time/d	20	60	120
Actual corrosion time/a	5.657	16.972	33.945
E/MPa [33]	204,798	194,170	178,805
Modulus loss ratio η/%	0.58%	5.74%	13.2%

Note: $\eta = 1 - E/E_0$, and E_0 is 206,000 Mpa.

4.2. Comparison of the Hysteretic Performance of Eroded SPSWs

Hysteretic curves and envelope curves of the eroded test piece for three kinds of corrosion days is shown in Figure 13a–f, with characteristic results shown in Table 7. At different corrosion levels, the hysteretic curve of the OSPSW is the fullest, with the largest initial stiffness, peak load, and ductility coefficient, but the stiffness degradation of the OSPSW is the most obvious at 60 and 120 days of corrosion. The BSPSW, which has the minimum hysteretic loop area, is basically consistent with the FSPSW in the trend of stiffness degradation and ultimate shearing resistance decrease. VSPSWs have the lowest stiffness and ductility among them, but the energy dissipation capacity of VSPSWs is second only to that of OSPSWs.

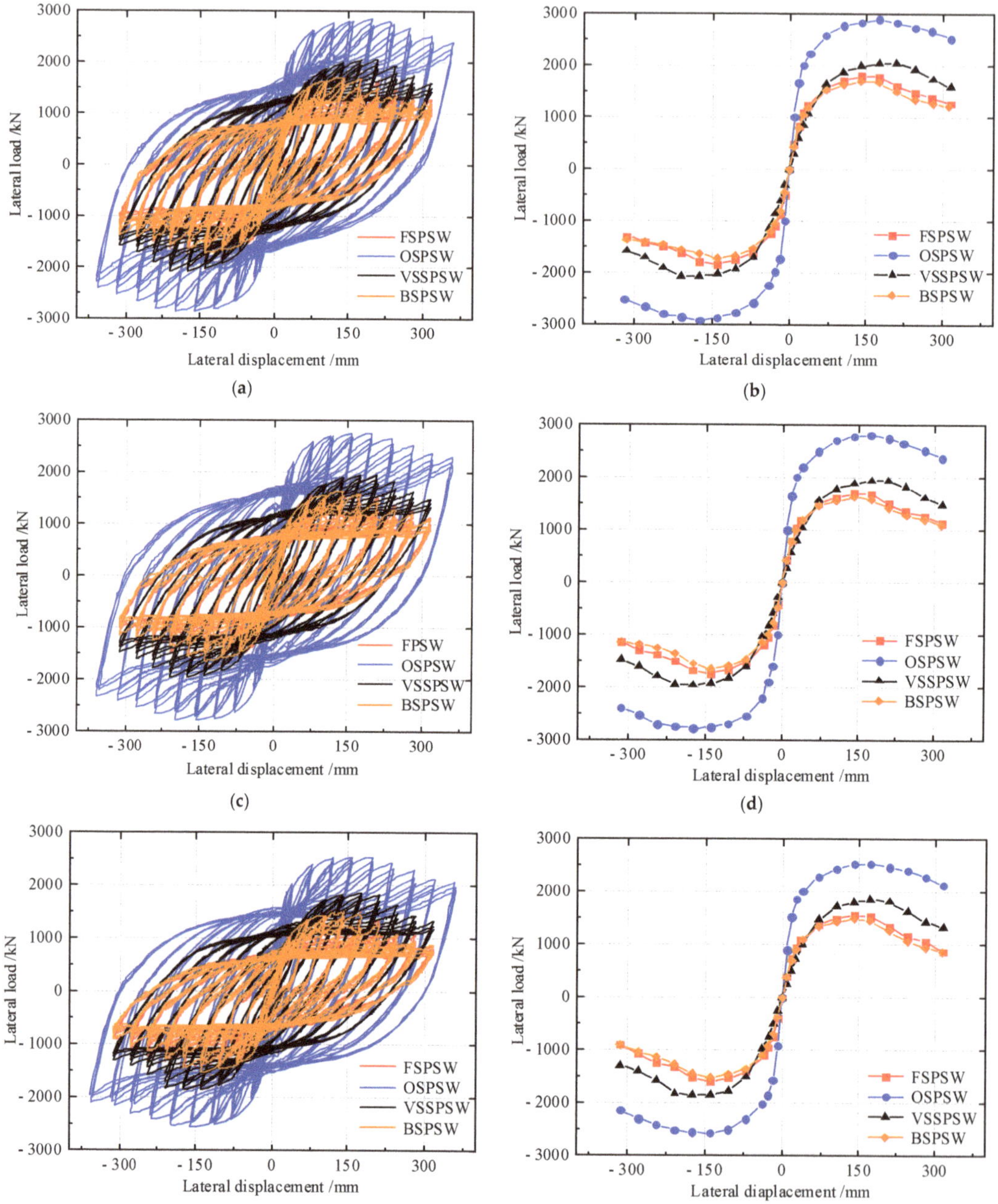

Figure 13. Load-displacement curves of corrosion SPSWs: (**a**) hysteretic curves of corrosion for 20 d; (**b**) envelope curves of corrosion for 20 d; (**c**) hysteretic curves of corrosion for 60 d; (**d**) envelope curves of corrosion for 60 d; (**e**) hysteretic curves of corrosion for 120 d; (**f**) envelope curves of corrosion for 120 d.

Table 7. Characteristic results of FSPSW, OSPSW, VSSPSW, and BSPSW at diverse corrosion days.

Corrosion Time/d	Specimen	K_0/(kN/mm)	P_y/kN	Δ_y/mm	P_m/kN	Δ_m/mm	μ
20	FSPSW	43.13	1425.15	53.68	1806.20	139.93	4.33
	OSPSW	98.90	2223.85	36.36	2847.81	171.84	8.66
	VSSPSW	32.35	1702.09	77.20	2055.89	209.36	3.61
	BSPSW	40.87	1352.97	52.13	1707.28	138.99	4.24
60	FSPSW	40.69	1348.88	53.35	1708.60	139.91	4.16
	OSPSW	95.53	2170.32	36.98	2768.22	171.23	8.39
	VSSPSW	31.40	1607.10	76.58	1940.59	209.07	3.57
	BSPSW	38.93	1293.38	51.41	1630.94	139.94	4.23
120	FSPSW	37.29	1241.67	53.44	1572.02	139.91	3.95
	OSPSW	88.65	2062.38	36.83	2575.38	171.05	8.25
	VSSPSW	28.72	1531.44	79.46	1848.06	171.68	3.20
	BSPSW	35.79	1181.56	50.13	1487.89	139.99	4.18

When the experimental corrosion time of SPSWs is 20 days, in contrast to FSPW, the characteristic results of the OSPSW are increased dramatically; more specifically, the initial stiffness and ductility coefficient is improved by 129.3% and 100.0%, respectively. Meanwhile, the yield and peak load increased by 56.0% and 57.7%, respectively. The initial stiffness and ductility coefficient of the VSSPSW decrease by 25.0% and 16.6%, respectively, whereas the yield load and the peak load increase by 19.4% and 13.8%, respectively. The ductility coefficient is reduced by only the initial 2.1% less than the other initial stiffness, and the yield load and peak load are decreased by 5.2%, 5.1%, and 5.5%, respectively.

When the experimental corrosion time of SPSWs is 60 days, in contrast to FSPW, the characteristic results of the OSPSW are obviously increased. The initial stiffness and ductility coefficient improved by 134.8% and 101.7%. Simultaneously, the yield load and peak load increased by 60.9% and 62.0%, respectively. The initial stiffness and ductility coefficients of the VSSPSW decrease by 22.8% and 14.2%, respectively, while the yield and peak load increase by 19.1% and 13.6%, respectively. The BSPSW decreases by 4.3%, 4.1%, and 4.5% in the initial stiffness, yield load, and peak load, respectively, while the ductility coefficient increases by 1.7%.

When the experimental corrosion time of SPSWs is 120 days, in contrast to FSPW, the characteristic results of the OSPSW are increased. The initial stiffness and ductility coefficient improved by 137.7% and 108.9%, respectively, while the yield load and peak load rise by 66.1% and 63.8%, respectively. The initial stiffness and ductility coefficients of the VSSPSW decrease by 23.0% and 19.0%, respectively, while the yield and peak load increase by 23.3% and 17.6%, respectively. The BSPSW decreases by 4.0%, 4.8%, and 5.4% in the initial stiffness, yield load, and peak load, respectively, while the ductility coefficient increases by 5.8%.

Consequently, the initial stiffness, ductility, yield load, and energy dissipation capacity of SPSWs are improved effectively through setting stiffeners, which also can inhibit the stiffness degradation of SPSWs under different corrosion levels. With increasing corrosion time, the initial stiffness and ductility coefficient of the FSPSW decrease faster than those of the VSSPSW and BSPSW. Thus, the slotting of the flat steel plate can effectively inhibit the hysteretic performance decrease after corrosion.

4.3. The Hysteretic Performance of SPSWs under Different Corrosion Days

Load-displacement curves of SPSWs under different corrosion days are shown in Figure 14. Pinching occurs in the hysteretic loop of the FSPSW with the passage of corrosion time, and the ultimate shearing resistance decreases rapidly in the early stage of corrosion. Compared to FSPSW without corrosion, when the corrosion time reaches 20 days, 60 days, and 120 days, the initial stiffness of the FSPSW falls by 21.1%, 25.5%, and 31.8%, respectively; the yield load decreases by 20.0%, 24.3%, and 30.3%, respectively; the peak load decreases by 18.9%, 23.2%, and 29.4%, respectively; and the ductility coefficient decreases by 35.1%, 37.6%, and 40.8%, respectively.

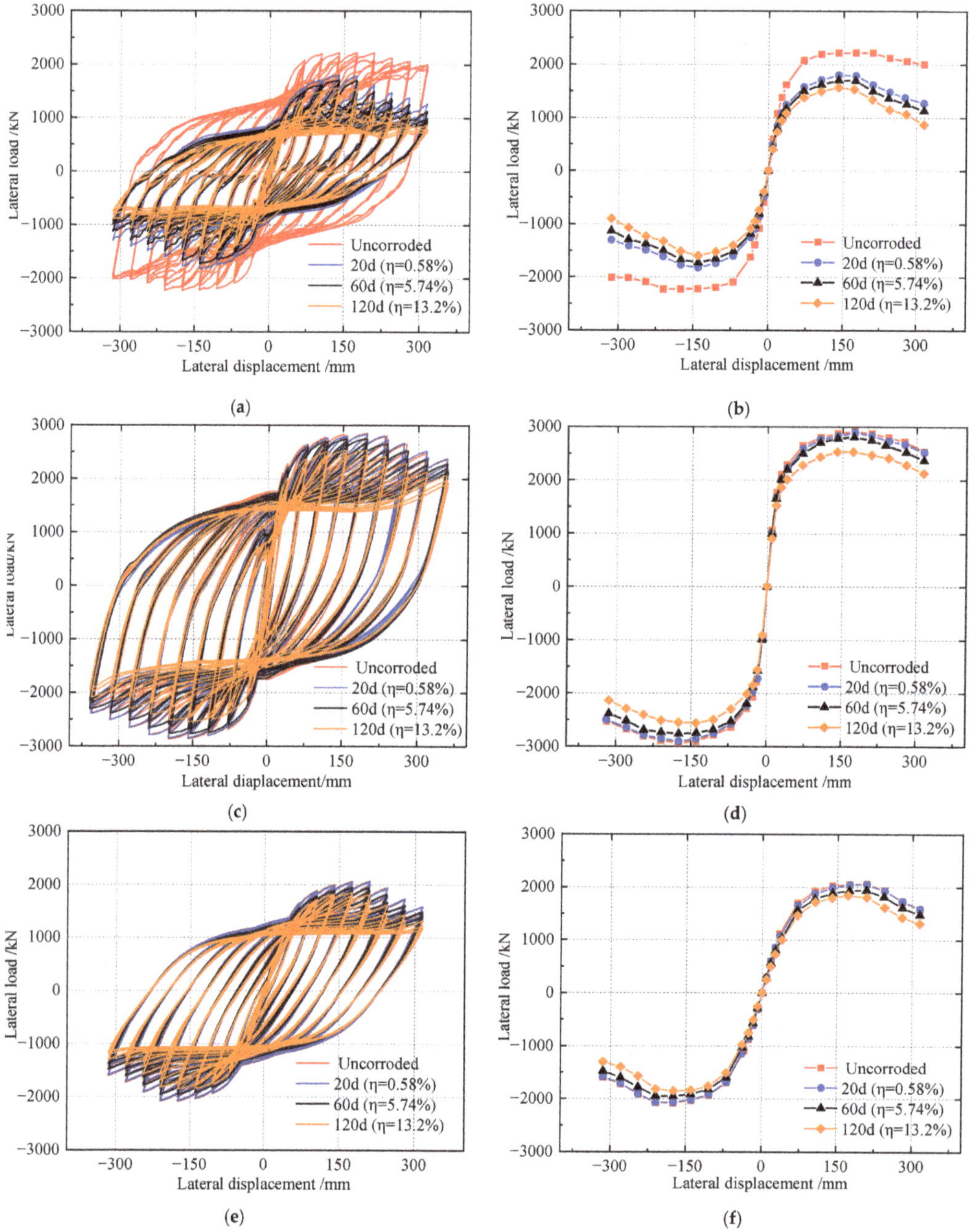

Figure 14. *Cont.*

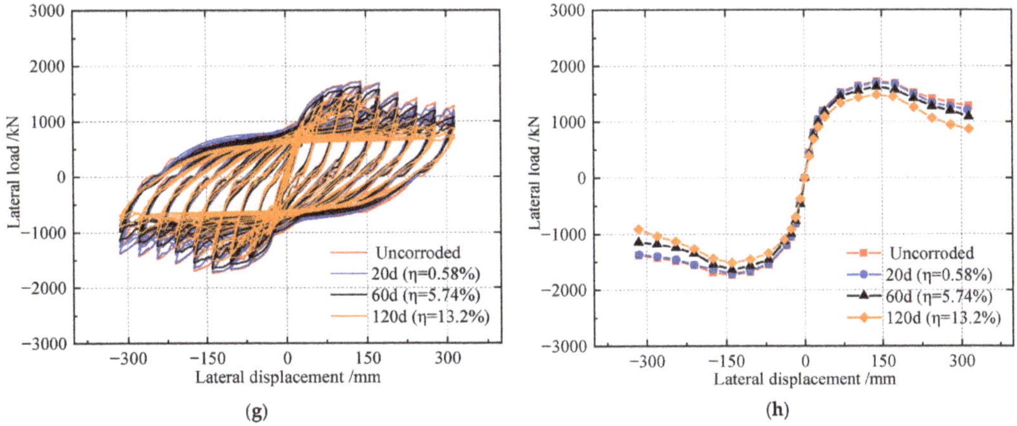

Figure 14. Comparison of SPSWs under different corrosion days: (**a**) Hysteretic curves of FSPSW; (**b**) Envelop curves of FSPSW; (**c**) Hysteretic curves of OSPSW; (**d**) Envelop curves of OSPSW; (**e**) Hysteretic curves of VSSPSW; (**f**) Envelop curves of VSSPSW; (**g**) Hysteretic curves of BSPSW; (**h**) Envelop curves of BSPSW.

As the corrosion time increases, no pinching occurs in the hysteretic loop of the OSPSW, but the ultimate shearing resistance decreases significantly when the corrosion time is 120 d, indicating that the OSPSW has good hysteretic performance after corrosion. Compared to OSPSW without corrosion, when the corrosion time reaches 20 days, 60 days, and 120 days, the stiffness of the corroded OSPSW reduces by 2.8%, 6.1%, and 12.9%, respectively; the yield load decreases by 0.8%, 3.3%, and 8.1%, respectively; the peak load decreases by 1.0%, 3.8%, and 10.5%, respectively, bringing ductility coefficients down 3.1%, 6.2%, and 7.7%, respectively.

As the corrosion time increases, the pinching occurs in the hysteretic loop of the VSSPSW, and the overall trend of the envelope curves is consistent in principle. Compared with the non-corrosion, the initial stiffness of the VSSPSW decreases by 7.6%, 10.3%, and 18.0%, respectively, when the accelerated corrosion time reaches 20 days, 60 days, and 120 days; the yield load decreases by 1.9%, 7.4%, and 11.7%, respectively; the peak load decreases by 0.6%, 6.1%, and 10.6%, respectively; and the ductility factor decreases by 1.9%, 3.0%, and 13.0%, respectively.

As the corrosion time increases, the pinching occurs in the hysteretic loop of the BSPSW similarly, and the overall tendency of the envelope curves is broadly the same. Compared with the BSPSW without corrosion, the initial stiffness of corroded BSPSW decreases by 13.9%, 18.0%, and 24.6%, respectively, after 20 days, 60 days, and 120 days; yield load decreases by 0.3%, 4.6%, and 12.9%, respectively; peak load decreases by 1.2%, 5.7%, and 13.9%, respectively; and ductility factor decreases by 5.8%, 6.0%, and 7.1%, respectively.

Consequently, due to the corrosion, the ultimate shearing resistance, initial stiffness, ductility, and energy dissipation capability of specimens are reduced. For the FSPSW, when uncorroded, the ultimate shearing resistance decreases, and the stiffness degradation is gentle. After corrosion, the decline in the envelope curve is significant, and furthermore, the initial stiffness, yield load, and ultimate load decrease more severely than those of the other three specimens. At the same time, the area of the FSPSW hysteretic loop is greatly reduced after corrosion, so corrosion greatly affects the seismic performance of the FSPSW. For the OSPSW, the initial stiffness, yield load, and ductility coefficient decrease slightly before 60 days of corrosion. The performance indexes decrease after corrosion for 120 days, but the reduction is the gentlest among the four specimens. The hysteretic loop is still full, indicating that the OSPSW has good seismic performance. For corroded VSSPSWs and BSPSWs, the decline in the indexes is gentler than that of FSPW, indicating that slotting can effectively inhibit the decline in SPSW performance after corrosion. At the same time, the hysteretic curve of the corroded VSSPSW is more abundant than that of the corroded BSPSW, with greater ultimate shearing resistance and better seismic performance.

4.4. Comparison of Stress Distribution of SPSWs before and after Corrosion

The stress distributions after the corrosion of the specimens are shown in Figure 15. As the corrosion time increases, the plastic area at the boundary column end of the FSPSW gradually increases. After corrosion for 120 days, serious buckling deformation occurs in the flat steel plate. Comparative analyses show that for the corroded SPSW, setting stiffeners and slotting can restrain the generation of plastic hinges at both ends of the boundary column, inhibit the buckling of flat steel plate, and improve the ductility after corrosion. Among them, the middle vertical slot effect is better.

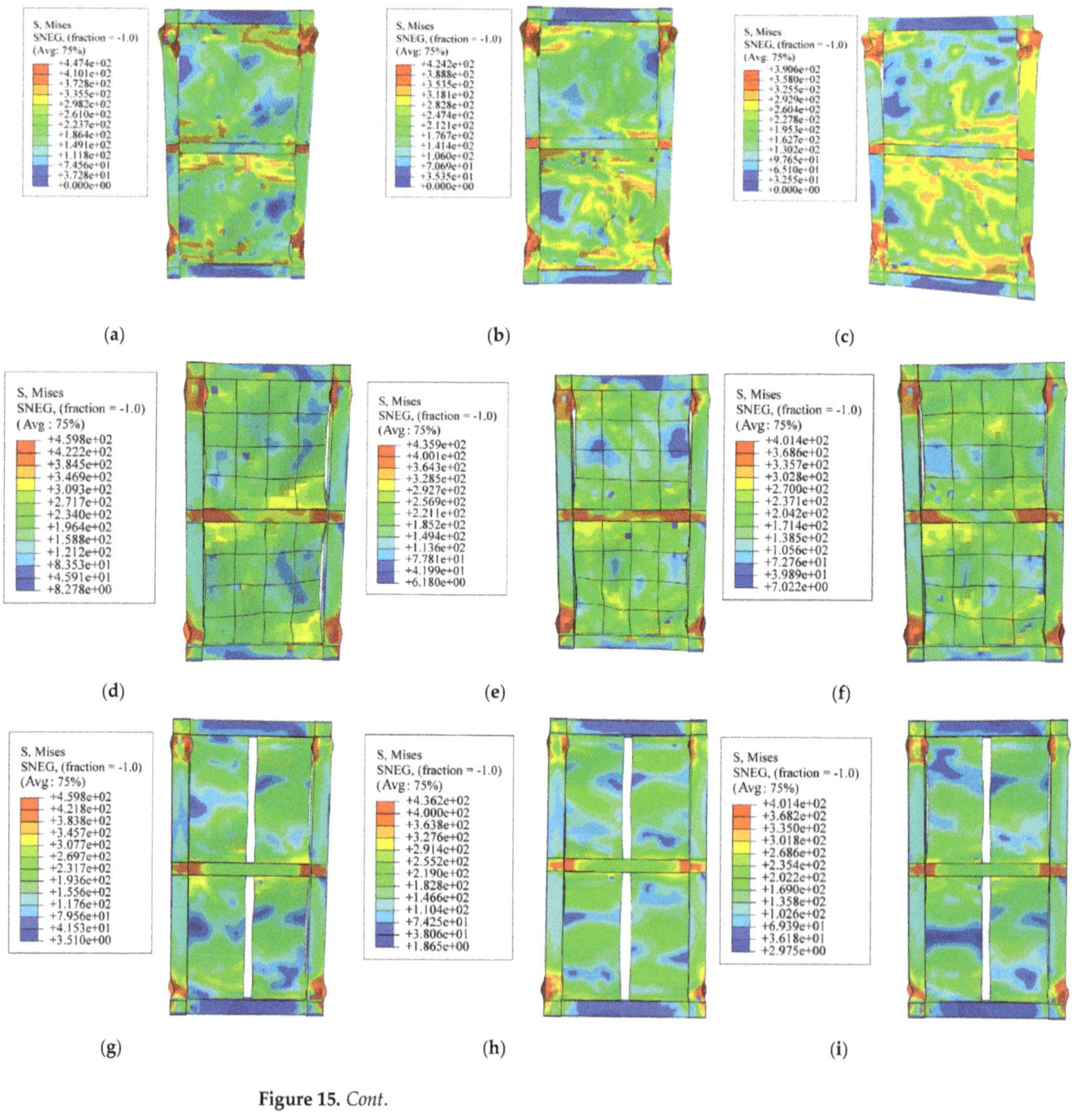

Figure 15. *Cont.*

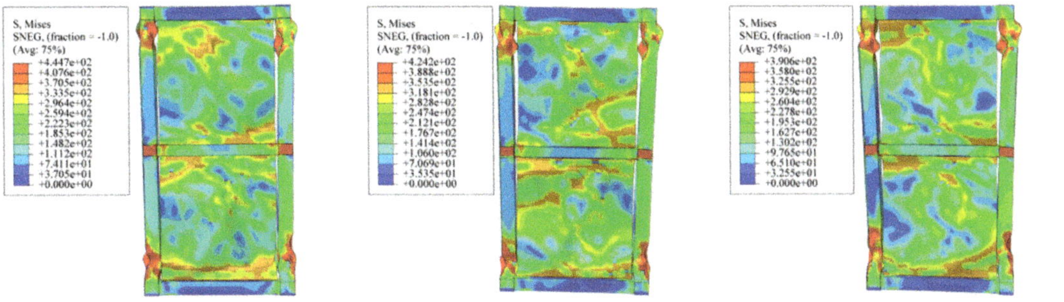

(j) (k) (l)

Figure 15. Stress distribution of FEM results for different corrosion times: (**a**) corroded FSPSW for 20 d; (**b**) corroded FSPSW for 60 d; (**c**) corroded FSPSW for 120 d; (**d**) corroded OSPSW for 20 d; (**e**) corroded OSPSW for 60 d; (**f**) corroded OSPSW for 120 d; (**g**) corroded VSSPSW for 20 d; (**h**) corroded VSSPSW for 60 d; (**i**) corroded VSSPSW for 120 d; (**j**) corroded BSPSW for 20 d; (**k**) corroded BSPSW for 60 d; (**l**) corroded BSPSW for 120 d.

4.5. Fitting the Ultimate Shearing Resistance

After converting 0, 20, and 120 days, which are accelerated corrosion test times to 5.657, 16.972, and 33.945 years in actual outdoor corrosion times, respectively, the ultimate shearing resistance of SPSWs with changing etching time is shown in Figure 16. Equations (3)–(6) are fitting formulae for the ultimate shearing resistance of the SPSWs with an exponential function.

$$y = 602.178e^{-x/5.362} + 1618.560, \tag{3}$$

$$y = -144.203e^{x/30.030} + 3021.959, \tag{4}$$

$$y = -29.545e^{x/16.630} + 2068.212, \tag{5}$$

$$y = -205.511e^{x/43.391} + 1936.920, \tag{6}$$

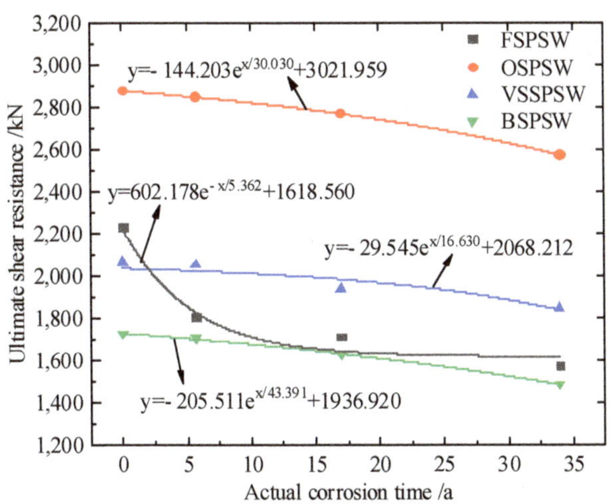

Figure 16. Fitting formulae of ultimate shearing resistance of SPSWs.

From the fitting curve, it is observed that with increasing corrosion time, the ultimate shearing resistance of the FSPSW decreases remarkably, and it mainly occurs when it starts to corrode. Stiffeners or slotting of the flat steel plate can effectively inhibit the decline of ultimate shearing resistance after corrosion.

5. Conclusions

This paper proposes four kinds of SPSWs and studies the seismic performance of the four specimens before and after corrosion through FE models. The research results are summarized as follows:

(1) The FEM model that can be validated by quasi-static cyclic tests of specimen S-4 in Ref. [46] is developed. The comparative results indicate that the errors between the FEM results and test data in lateral stiffness, yield load, yield displacement, ultimate load, and ultimate displacement are 4%, 15%, 11%, 16%, and 8%, respectively, so the results of the FEM model can validate well the test results.

(2) Under cyclic loading, compared to other specimens, the ultimate shearing resistance, initial stiffness, and ductility of OSPSW are higher, and the total energy dissipation value of the OSPSW is 38% higher than that of FSPSW, while that of the VSSPSW and BSPSW are less 23.5% and 44.8%, respectively. The stiffener can also retard the formation of a plastic hinge at the column end and inhibit the buckling of the flat steel plate. It is revealed that although the OSPSW has a little more usage amount of steel, setting stiffeners can make the SPSW has a better seismic performance.

(3) Under atmospheric corrosion, the seismic performance of OSPSW is affected least among them. Compared to other specimens, the ultimate shearing resistance, ductility, and energy dissipation capacity of the OSPSW are largest under different corrosion levels. With corrosion time increasing, the stiffness of OSPSW degrades most slowly; stiffeners can also inhibit the occurrence of plastic hinges and out-of-plane buckling. It is indicated that setting stiffeners can make corroded SPSWs have better seismic performance.

(4) Under atmospheric corrosion, the seismic performance of FSPSW is affected most among them, but its peak load and energy dissipation capacity are still higher than those of the BSPSW. Although slotting reduces the energy dissipation capacity and ultimate shearing resistance of SPSWs, it can effectively inhibit the reduction in the seismic performance after corrosion, and slotting at the middle part is better than at both sides.

(5) In order to truly reflect the variation of ultimate shearing resistance with corrosion time, the formulae of ultimate shearing resistance for the four specimens are fitted, which can be used for practical engineering.

Author Contributions: X.M.: Software, Investigation, Data curation, Writing—original draft preparation, Writing—review and editing; Y.H.: Conceptualization, Methodology, Formal analysis, Project administration, Writing—review and editing; X.C.: Software, Validation, Data curation, Writing—original draft preparation; L.J.: Conceptualization, Methodology, Project administration, Visualization, Funding acquisition; Y.L.: Visualization, Funding acquisition; H.Z.: Supervision. All authors have read and agreed to the published version of the manuscript.

Funding: This research was funded by open scientific funding in the Key Laboratory of Environmental Change and Natural Disasters of the Ministry of Education, Beijing Normal University (No. 01. Funder: Liqiang Jiang); open scientific funding in Fujian University Engineering Research Center for Disaster Prevention and Mitigation of Southeast Coastal Engineering Structures in China (No. 2019003. Funder: Liqiang Jiang); Natural Science Foundation of Hunan Province, China (No. 2021JJ41083. Funder: Liqiang Jiang); the Scientific Research Project of Hunan Education Department (No. 20B598. Funder: Liqiang Jiang); the Scientific Research Fund of Institute of Engineering Mechanics, China Earthquake Administration (No. 2020D12. Funder: Liqiang Jiang); Open Funding of Hunan Tieyuan Civil Engineering Testing Co., Ltd. (No. HNTY2021K05. Funder: Liqiang Jiang).

Data Availability Statement: The raw data required to reproduce these findings cannot be shared at this time due to time limitations. The processed data required to reproduce these findings cannot be shared at this time due to time limitations.

Conflicts of Interest: The authors declare no conflict of interest.

References

1. Yang, R. Study on Buckling Analysis and Hysteretic Behavior of Composite Shear Walls with Pseudo Corrugated Cross-ribbed Flat Steel Plates. Master's Thesis, Chang'an University, Xi'an, China, 2021.
2. Tahaii, S.M.; Hamidi, H.; Vaseghi Amiri, J. Inelastic Seismic Demand of Steel-Plate Shear Wall Structures: Emphasis on the PTD Effect. *Int. J. Civ. Eng.* **2022**, *20*, 1145–1163. [CrossRef]
3. Jiang, L.; Zheng, H.; Hu, Y. Experimental seismic performance of steel-and composite steel-panel wall strengthened steel frames. *Arch. Civ. Mech. Eng.* **2017**, *17*, 520–534. [CrossRef]
4. Hu, Y.; Zhao, J.; Jiang, L. Seismic risk assessment of steel frames equipped with steel panel wall. *Struct. Des. Tall Spec. Build.* **2017**, *26*, e1368. [CrossRef]
5. Paslar, N.; Farzampour, A.; Hatami, F. Infill plate interconnection effects on the structural behavior of steel plate shear walls. *Thin-Walled Struct.* **2020**, *149*, 106621. [CrossRef]
6. Azandariani, M.; Gholhaki, M.; Kafi, M. Experimental and numerical investigation of low-yield-strength (LYS) steel plate shear walls under cyclic loading. *Eng. Struct.* **2020**, *203*, 109866. [CrossRef]
7. Tan, J.; Su, M.; Wang, Y.; Wang, K.; Cao, Y.; Li, P. Experimental study on cyclic shear performance of steel plate shear wall with different buckling restraints. *Structures* **2022**, *35*, 469–482. [CrossRef]
8. Ye, J.; Jiang, L. Collapse mechanism analysis of a steel moment frame based on structural vulnerability theory. *Arch. Civ. Mech. Eng.* **2018**, *18*, 833–843.
9. Lv, Y.; Li, L.; Wu, D.; Chen, Y.; Li, Z.; Chouw, N. Shear–displacement diagram of steel plate shear walls with precompression from adjacent frame columns. *Struct. Des. Tall Spec. Build.* **2019**, *28*, e1585. [CrossRef]
10. Ahmadi, Z.; Aghakouchak, A.A.; Mirghaderi, S.R. Steel slit shear walls with an efficient geometry. *Thin-Walled Struct.* **2021**, *159*, 107296. [CrossRef]
11. Suo, Y.; Fan, S.; Li, C.; Zeng, S.; Liu, C. Parametric analysis on hysteresis performance and restoring force model of LYP steel plate shear wall with two-side connections. *Int. J. Steel Struct.* **2020**, *20*, 1960–1978. [CrossRef]
12. Haddad, O.; Sulong, N.H.R.; Ibrahim, Z. Cyclic performance of stiffened steel plate shear walls with various configurations of stiffeners. *J. Vibroeng.* **2018**, *20*, 459–476. [CrossRef]
13. Sultana, S.; Wang, Y.; Sobey, A.J.; Wharton, J.A.; Shenoi, R.A. Influence of corrosion on the ultimate compressive strength of steel plates and stiffened panels. *Thin-Walled Struct.* **2015**, *96*, 95–104. [CrossRef]
14. Jamil, I.; Bano, H.; Castano, J.G.; Mahmood, A. Characterization of atmospheric corrosion near the coastal areas of Arabian sea. *Mater. Corros.* **2018**, *69*, 898–907. [CrossRef]
15. Nguyen, M.N.; Wang, X.; Leicester, R.H. An assessment of climate change effects on atmospheric corrosion rates of steel structures. *Corros. Eng. Sci. Techn.* **2013**, *48*, 359–369. [CrossRef]
16. Tidblad, J. Atmospheric corrosion of metals in 2010–2039 and 2070–2099. *Atmos. Environ.* **2012**, *55*, 1–6. [CrossRef]
17. Tanjung, I.; Nasution, A.R.; Harahap, A.G.; Fonna, S.; Ariffin, A.K.; Huzni, S. Atmospheric Corrosion Analysis on Low Carbon Steel Plate Profile and Elbow in Medan Belawan District. *Key Eng. Mater.* **2021**, *892*, 142–149.
18. Seechurn, Y.; Surnam, B.Y.R.; Wharton, J.A. Marine atmospheric corrosion of carbon steel in the tropical microclimate of Port Louis. *Mater. Corros.* **2022**, *72*, 1474–1489. [CrossRef]
19. Nuraini, L.; Prifiharni, S.; Priyotomo, G.; Sundjono; Gunawan, H.; Purawiardi, I. Atmospheric corrosion performance of different steels in early exposure in the coastal area region West Java, Indonesia. In *AIP Conference Proceedings*; AIP Publishing LLC: Melville, NY, USA, 2018.
20. Kumar, V.; Sharma, N.; Tiwari, S.K.; Kango, S. Atmospheric corrosion of materials and their effects on mechanical properties: A brief review. *Mater. Today* **2021**, *44*, 4677–4681. [CrossRef]
21. Xu, Y.; Huang, Y.; Cai, F.; Lu, D.; Wang, X. Study on corrosion behavior and mechanism of AISI 4135 steel in marine environments based on field exposure experiment. *Sci. Total Environ.* **2022**, *830*, 154864. [CrossRef]
22. Liu, H.; Gu, T.; Asif, M.; Zhang, G.; Liu, H. The corrosion behavior and mechanism of carbon steel induced by extracellular polymeric substances of iron-oxidizing bacteria. *Corros. Sci.* **2017**, *114*, 102–111. [CrossRef]
23. Tang, F.; Chen, G.; Brow, R.K. Chloride-induced corrosion mechanism and rate of enamel-and epoxy-coated deformed steel bars embedded in mortar. *Cement Concrete Res.* **2016**, *82*, 58–73. [CrossRef]
24. Qin, S.; Cui, W. Effect of corrosion models on the time-dependent reliability of steel plated elements. *Mar. Struct.* **2003**, *16*, 15–34. [CrossRef]
25. Li, L.; Li, C.; Mahmoodian, M.; Shi, W. Corrosion induced degradation of fatigue strength of steel in service for 128 years. *Structures* **2020**, *23*, 415–424. [CrossRef]
26. Lu, Y.; Wang, R.; Han, Q.; Yu, X.; Yu, Z. Experimental investigation on the corrosion and corrosion fatigue behavior of butt weld with G20Mn5QT cast steel and Q355D steel under dry–wet cycle. *Eng. Fail Anal.* **2022**, *134*, 105977. [CrossRef]
27. Woloszyk, K.; Garbatov, Y. Random field modelling of mechanical behaviour of corroded thin steel plate specimens. *Eng. Struct.* **2020**, *212*, 110544. [CrossRef]
28. Wang, Y.; Zhou, X.; Wang, H.; Kong, D.; Xu, S. Stochastic constitutive model of structural steel based on random field of corrosion depth. *Case Studies Constr. Mater.* **2022**, *16*, e00972. [CrossRef]
29. Woloszyk, K.; Garbatov, Y. An enhanced method in predicting tensile behaviour of corroded thick steel plate specimens by using random field approach. *Ocean Eng.* **2020**, *213*, 107803. [CrossRef]

30. Zhao, Z.; Zhang, H.; Xian, L.; Liu, H. Tensile strength of Q345 steel with random pitting corrosion based on numerical analysis. *Thin-Walled Struct.* **2020**, *148*, 106579. [CrossRef]
31. Sheng, J.; Xia, J. Effect of simulated pitting corrosion on the tensile properties of steel. *Constr. Build. Mater.* **2017**, *131*, 90–100. [CrossRef]
32. Guo, Q.; Zhao, Y.; Xing, Y.; Jiao, J.; Fu, B.; Wang, Y. Experimental and numerical analysis of mechanical behaviors of long-term atmospheric corroded Q235 steel. *Structures* **2022**, *39*, 115–131. [CrossRef]
33. Zhang, X.; Zheng, S.; Zhao, X. Experimental and numerical study on seismic performance of corroded steel frames in chloride environment. *J. Constr. Steel Res.* **2020**, *171*, 106164. [CrossRef]
34. Zheng, S.; Zhang, X.; Zhao, X.; Liu, Y. Experimental and restoring force model research on the seismic behavior of corroded steel frame beams in offshore atmospheric environment. *Eng. Mech.* **2018**, *35*, 98–106, 115.
35. Wang, H.; Xu, S.; Li, A.; Kang, K. Experimental and numerical investigation on seismic performance of corroded welded steel connections. *Eng. Struct.* **2018**, *174*, 10–25. [CrossRef]
36. Xu, Q.; Shang, X.; Zheng, S. Cyclic Behavior of Steel Frame Joints in the Offshore Atmospheric Environment. *Int. J. Steel Struct.* **2020**, *20*, 870–884. [CrossRef]
37. Xu, S.; Zhang, Z.; Qin, G. Study on the seismic performance of corroded H-shaped steel columns. *Eng. Struct.* **2019**, *191*, 39–61. [CrossRef]
38. Karagah, H.; Shi, C.; Dawood, M.; Belarbi, A. Experimental investigation of short steel columns with localized corrosion. *Thin-Walled Struct.* **2015**, *87*, 191–199. [CrossRef]
39. Zheng, S.; Zhang, X.; Zhao, X. Experimental investigation on seismic performance of corroded steel columns in offshore atmospheric environment. *Struct. Des. Tall Spec. Build.* **2019**, *28*, e1580. [CrossRef]
40. Beben, D. Backfill corrosivity around corrugated steel plate culverts. *J. Perform. Constr. Facil.* **2015**, *29*, 04014159. [CrossRef]
41. Wei, X.; Li, G.; Xiao, L.; Zhou, L.; He, K.; Han, B. Shear strength reduction of trapezoidal corrugated steel plates with artificial corrosion pits. *J. Constr. Steel Res.* **2021**, *180*, 106583.
42. Kudus, S.A. Modal analysis of corrugated plate by finite element analysis. *Int. Integr. Eng.* **2020**, *12*, 252–258.
43. *GB 50017-2017*; Standard for Design of Steel Structures. Chinese Architecture and Industry Press: Beijing, China, 2017.
44. *JCJ/T 380-2015*; Technical Specification for Steel Plate Shear Wall. Chinese Architecture and Industry Press: Beijing, China, 2016.
45. *GB 50011-2010*; Code for Seismic Design of Buildings. Chinese Architecture and Industry Press: Beijing, China, 2016.
46. Hao, B. Cyclic Test Vertically Corrugated Steel Plate Shear Walls with Various Openings. Master's Thesis, Tianjin University, Tianjin, China, 2018.
47. *JGJ101-96*; Specificating of Testing Method for Earthquake Resistant Building. Chinese Architecture and Industry Press: Beijing, China, 1997.
48. Ma, X.; Hu, Y.; Jiang, L.; Jiang, L.; Nie, G.; Zheng, H. Study on the Seismic Performance of Stiffened Corrugated Steel Plate Shear Walls with Atmospheric Corrosion. *Materials* **2022**, *15*, 4920. [CrossRef] [PubMed]

buildings

MDPI

Article

Study on Seismic Response and Parameter Influence in a Transformer–Bushing with Inerter Isolation System

Ruoyu Zhang [1], Meigen Cao [2,*] and Jizhong Huang [3]

1 School of Mechanics and Engineering Science, Shanghai University, Shanghai 200444, China; 21820359@shu.edu.cn
2 School of Civil Engineering, North China University of Technology, Beijing 100144, China
3 Institute for Conservation of Cultural Heritage, Shanghai University, Shanghai 200444, China; hjizhong@163.com
* Correspondence: caomeigen@ncut.edu.cn

Abstract: In this paper, a mechanical model of a transformer–bushing with an inerter isolation system (IIS) is established. An IIS is composed of an inerter element, a damping element, and a spring element connected in parallel between the same two terminals. Vibration control equations and frequency response functions are also established. The influence of parameters on IIS, including inerter–mass ratio, damping ratio, and frequency ratio, was studied. In the extremum condition that represents the most efficient parameter set of inerter–mass ratio and damping ratio for relative displacement response ratio, an optimal design method was developed by exploiting a performance demand. Finally, the seismic response of the transformer–bushing with IIS was carried out to verify the isolation performance of IIS. The research shows that the equivalent mass coefficient and damping coefficient of IIS can be amplified by an inerter element and the inerter–mass ratio and damping ratio are reduced simultaneously under the conditions of meeting the performance demand after parameter optimization. Meanwhile, the parameter optimization design method proved to be effective for meeting the target demand of the relative displacement response of the bushing and tank, while base shear force and isolation displacement were reduced simultaneously. Based on the results from a response history analysis under ground motion records, IISs can significantly suppress the resonance response of a structure and the continuous vibration response in the stable state. The peak displacement can be reduced by 50% compared with a traditional isolation system.

Keywords: transformer–bushing system; inerter element; isolation system; stochastic response; parameter optimization; seismic response

Citation: Zhang, R.; Cao, M.; Huang, J. Study on Seismic Response and Parameter Influence in a Transformer–Bushing with Inerter Isolation System. *Buildings* **2022**, *12*, 530. https://doi.org/10.3390/buildings12050530

Academic Editor: Jia-Bao Yan

Received: 26 March 2022
Accepted: 15 April 2022
Published: 22 April 2022

1. Introduction

Lifeline structures have high vulnerability under earthquakes [1], such as transformer–bushing systems [2]. The main reason is that the natural vibration frequency of the structure is close to the predominant frequency of external excitation [3], resulting in the amplification of vibration response [4]; therefore, vibration mitigation and control in the structure is particularly important. Many researchers have noticed that isolation technology can effectively reduce the base shear force at the bottom of the transformer and reduce the displacement response of tank, bushing and auxiliary facilities [5], so as to protect the transformer from various disasters in strong earthquakes. Fujita et al. [6,7] carried out a test analysis of laminated rubber isolation bearings for transformers and other large gravity equipment, focusing on the development of isolation devices and verifying vibration control effect, without considering the interaction between transformers and bushings. Thu Pham [8] carried out a shaking table test of transformer–bushing. During the test, a vibration isolation device composed of steel cable was installed between the transformer frame and bushing, and a viscous damping system was set under the transformer frame to transformer oil. The test showed that the vibration isolation device has the vibration isolation effect of significantly reducing the acceleration response at the top of the bushing. Murota et al. [9–11] designed three types of isolation systems at the bottom of the transformer model: BS isolation, the friction pendulum system (FPS), and the segmented combined high–damping rubber isolation system (SHRB). To verify the effectiveness of the above three types of isolation systems, shaking table tests were carried out and the numerical models of base isolation system and SHRB isolation system were established, after which theoretical analysis and research were carried out. However, the displacement of the isolation layer of the traditional isolation measures was large, and the vibration control effect of the upper bushing with large height was not satisfactory. Hence, it is necessary to develop an isolation system that can control the vibration response of bushing and isolation layer simultaneously. An inerter isolation system (IIS) for seismic response mitigation of transformer–bushings is proposed in this paper.

In the last ten years, the electromechanical similarity theory [12] has provided a theoretical basis for the proposal of an inerter element, and the vibration mitigation and isolation technology based on inerter element has been developed. Compared with the traditional tuned mass damper (TMD), a damper with an inerter element can directly and effectively control inertial force at two terminals. Moreover, the inerter element can effectively enlarge the small actual mass through methods such as ball screw to convert the translational motion into rotary motion. In 2001, Smith [13,14] put forward the concept of an inerter element based on electromechanical similarity theory, described the basic forms of a ball screw inerter element and a rack and pinion inerter element; later designing a hydraulic inerter element in 2013 [15] that has a simpler structure and greater robustness. Subsequently, shock absorbers based on different inerter element connection forms were developed, such as the tuned viscous mass damper (TVMD) with a mass element in parallel with damping element, and the tuned inerter damper (TID) with a stiffness element in parallel with damping element. At the same time, the design method of inerter systems has also been studied. Ikago et al. [16] derived a simple formula for TVMD optimization design based on fixed–point theory. Pan et al. [17] studied the parameters of single–degree–of–freedom structures with different inerter systems considering the natural damping of the original structure and the output cost control of the damper, made up for deficiencies in design method based on the fixed–point theory, and proposed the SPIS–II inerter damping system design method [18]. Hwang et al. [19] proposed a ball screw inerter system connected with a toggle brace. Through theoretical analysis and numerical calculation, it is shown that the system can be effectively used in a structure even when the drift is very small. Zhang et al. [20,21] applied the inerter damper system to high–rise structures such as chimneys and wind power towers, conducted theoretical analysis and parameter influence analysis and proved the effectiveness of the inerter damping system in high–rise structures. Gao et al. [22] put forward an optimum design procedure of VID based on the output feedback control theory for controlling specific cable mode vibrations. While some Japanese scholars have used the inerter damping support in practical engineering [23], most of the research on the inerter damping system is still in the stage of theoretical analysis and numerical simulation. Only the simplified mechanical model is used for the damping analysis of various structures, and only a few scholars have proposed the connection mode and design method for inerter systems applied in building structures [24]. Xie et al. [25,26] put forward a cable–bracing inerter system (CBIS), which is composed of cable and an inerter energy–dissipation system fixed at the bottom of the interlayer of a structure. Parameter analysis and optimization design were carried out and showed that it is easy to install and can effectively control structural displacement. Wang et al. [27] put forward a new tuned inerter negative–stiffness damper (TINSD) for seismic protection of structures, which is more effective than the TID, TVMD, and INSD in reducing the dynamic response of structures. Zhang et al. [28,29] have proposed a hybrid isolation system by employing the recently developed inerter element for seismic response mitigation of a storage tank, in terms of sloshing height, base shear force, and isolation displacement.

In this paper, an inerter isolation system (IIS) for transformer–bushings is proposed. The IIS is composed of an inerter element, a damping element, and a spring element connected in parallel between the same two terminals. First, Section 2 introduces the basic principle of IIS; the equivalent mass and equivalent damping of the isolation layer are improved. In Section 3, the motion control equations and frequency response functions of transformer–bushings with IIS are established. In Section 4, the parameter analysis and parameter optimization is carried out to obtain the minimum inerter–mass ratio and damping ratio of IIS under the condition of meeting the performance demand. Finally, the dynamic response analyses of transformer–bushings with IIS under the ground motion are carried out in Section 5 to verify the vibration control effect of IIS; the mechanism of vibration mitigation of inerter element in the IIS is explained in terms of displacement mitigation and energy dissipation in detail. The research in this paper can provide reference for the design of efficient and lightweight isolation vibration mitigation schemes of transformer–bushing systems based on an inerter system.

2. Theoretical Analysis of Transformer–Bushing with Inerter Isolation Systems

2.1. Mechanical Model of Inerter Element and IIS

Compared with the mass unit, the inertia unit can increase inertia by rotating. The two ends of the unit have different accelerations, and its output is also directly proportional to the relative acceleration at both ends, which can be expressed as:

$$f_I = m_d(a_2 - a_1) \tag{1}$$

where, f_I is the output force of the inerter unit, a_1 and a_2 are the accelerations at both ends, and Figure 1a is the mechanical model of the inerter element. The inerter element is the same as the mass element and cannot dissipate energy by itself. It is generally used in combination with the damper. For example, it forms an inerter isolation system (IIS) in parallel with the damper and spring at the bottom of the structure. IIS obtains greater damping and equivalent mass through the amplification of the inerter and does not increase the apparent volume and mass of the system itself, so as to realize the lightweight of the isolation control system. The mechanical model is shown in Figure 1b.

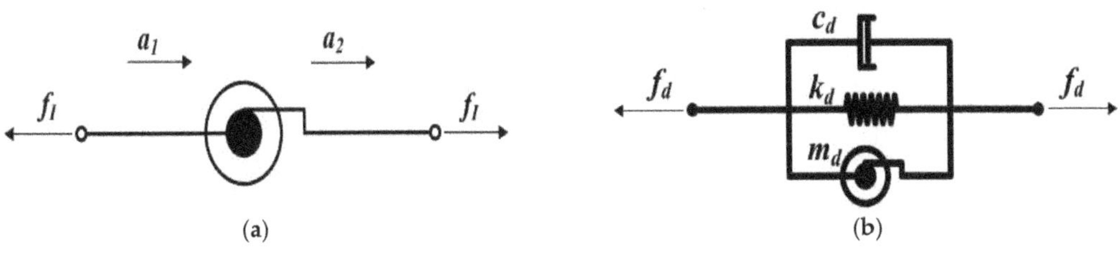

Figure 1. Mechanical model: (**a**) an inerter element; (**b**) IIS.

2.2. Mechanical Model of Transformer–Bushing System with IIS

The transformer–bushing system is generally composed of the main body of the transformer and the upper bushings. The main body of the transformer is composed of a tank, conservator, radiator, stiffener, and other components, and the upper bushings system is composed of an elevated seat, porcelain bushings or composite bushings, grading ring, and other components. The center of gravity of the tank of the transformer is lower, and the mass is much larger than that of the bushings. Therefore, when simplifying the transformer–bushing system, the lower tank, conservator radiator, and other structures can be simplified as a single mass point without considering the deformation of the wall of the tank, which is considered to undergo rigid body motion during earthquakes; the bushing is flexible, the stiffness is smaller, and the center of gravity is high, so the relative displacement between the bushing and the tank should be considered. Figure 2 shows the transformer–bushing system and isolation layer, where h_t and h_b are the height of the center of gravity of the tank and the center of gravity of the bushing from the ground, respectively.

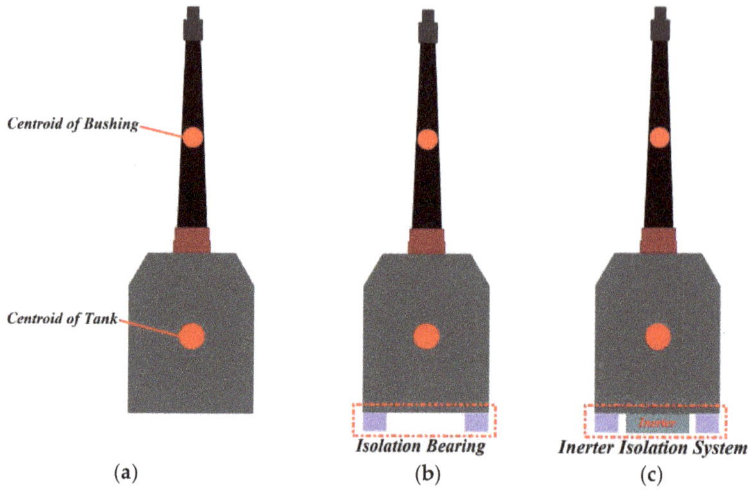

Figure 2. Transformer–bushing system and isolation layer: (**a**) original; (**b**) IB; (**c**) IIS.

For traditional isolation systems [30,31], the isolation bearings, e.g., elastomeric [32] or wire rope isolators [33], are installed at the bottom of the transformer–bushing system. The mechanical model of the bearing can be simplified as the parallel connection of the stiffness element and damping element, which is called the isolation bearing (IB) in this paper. For the inerter isolation system, an additional inerter is added on the isolation bearings. Figure 3 shows the mechanical model of the transformer–bushing system with IIS and IB installed. Where m_t, m_b, and b are the mass of the tank, bushing, and the equivalent mass of inerter, respectively, c_{eq} and c_b are the damping coefficient of isolation bearing and bushing respectively, and k_{eq} and k_b are the stiffness of isolation bearing and bushing, respectively.

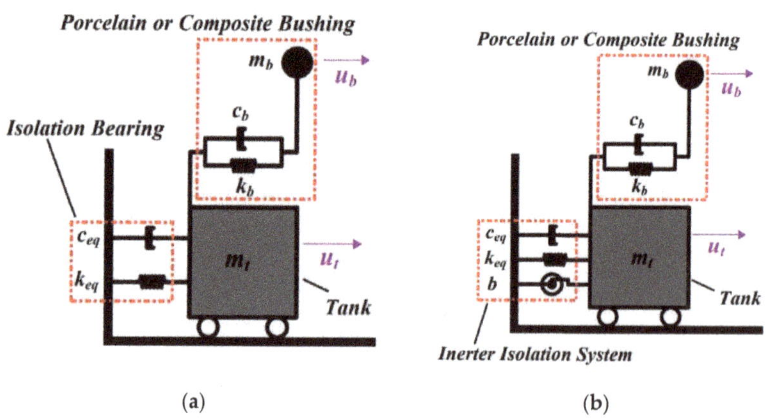

Figure 3. Mechanical model of transformer–bushing system: (**a**) IB; (**b**) IIS.2.3. Motion Control Equation of Transformer–Bushing System with IIS.

According to the mechanical model shown in Figure 3, if the axial displacements of the isolation base are neglected at the same time, the motion equations of two particles of the transformer–bushing system are established:

$$M\ddot{X} + C\dot{X} + KX = -M_0 a_g \tag{2}$$

where: $\ddot{X}, \dot{X}, X$ is the acceleration, velocity, and displacement vector of each mass of the transformer–bushing system, and a_g is the acceleration under seismic action; M, C, and K are the mass, damping, and stiffness matrix of the isolation system respectively, and M_0 is the mass matrix of the transformer–bushing system without isolation.

When considering the isolation bearings installed, as in the case of traditional base-isolated buildings [34] or rigid blocks [35], the M, C, K and motion vectors can be expressed as:

$$M = M_0 = \begin{bmatrix} m_t & \\ & m_b \end{bmatrix}, C = \begin{bmatrix} c_{eq} + c_b & -c_b \\ -c_b & c_b \end{bmatrix}, K = \begin{bmatrix} k_{eq} + k_b & -k_b \\ -k_b & k_b \end{bmatrix} \tag{3}$$

$$X = \begin{Bmatrix} u_{t,IB} \\ u_{b,IB} \end{Bmatrix}, \dot{X} = \begin{Bmatrix} \dot{u}_{t,IB} \\ \dot{u}_{b,IB} \end{Bmatrix}, \ddot{X} = \begin{Bmatrix} \ddot{u}_{t,IB} \\ \ddot{u}_{b,IB} \end{Bmatrix} \tag{4}$$

when considering the inerter isolation system installed, where M_0, C and K are the same as the isolation bearings system, the mass matrix M and the vectors displacement, velocity and acceleration can be expressed as:

$$M = \begin{bmatrix} m_t + b & \\ & m_b \end{bmatrix}, X = \begin{Bmatrix} u_{t,IIS} \\ u_{b,IIS} \end{Bmatrix}, \dot{X} = \begin{Bmatrix} \dot{u}_{t,IIS} \\ \dot{u}_{b,IIS} \end{Bmatrix}, \ddot{X} = \begin{Bmatrix} \ddot{u}_{t,IIS} \\ \ddot{u}_{b,IIS} \end{Bmatrix} \tag{5}$$

where: u_t and u_b are displacement vectors of tank and bushing, IB and IIS are inerter isolation system and isolation bearing.

For the convenience of parameter analysis, the following dimensionless parameters are defined. The parameters of isolation layer and bushing are shown in Equations (6) and (7) respectively. Where μ is the inerter–mass ratio, κ is the frequency ratio of isolation layer and bushing, ζ_{eq} is the damping ratio of the isolation layer, ζ_b is the damping ratio of bushing; ω_b is the circlar frequency of bushing, μ_b is the mass ratio of bushing and tank, ω_{eq} is the circular frequency of the isolation layer.

$$\zeta_{eq} = \frac{c_{eq}}{2m_t\omega_{eq}}, \omega_{eq} = \sqrt{\frac{k_{eq}}{m_t}}, \kappa = \frac{\omega_{eq}}{\omega_b}, \mu = \frac{b}{m_t} \tag{6}$$

$$\zeta_b = \frac{c_b}{2m_b\omega_b}, \omega_b = \sqrt{\frac{k_b}{m_b}}, \mu_b = \frac{m_b}{m_t} \tag{7}$$

Simultaneous solving of Equations (3)–(7) and Laplace transformation of Equation (2) can obtain the motion Equations (8) and (9) of the transformer–bushing system with isolation bearing and inerter isolation system in frequency domain respectively:

$$\begin{bmatrix} 1 & \\ & 1 \end{bmatrix}\begin{Bmatrix} s^2U_{t,IB} \\ s^2U_{b,IB} \end{Bmatrix} + \begin{bmatrix} 2\omega_b(\kappa\zeta_{eq} + \mu_b\zeta_b) & -2\mu_b\omega_b\zeta_b \\ -2\omega_b\zeta_b & 2\omega_b\zeta_b \end{bmatrix}\begin{Bmatrix} sU_{t,IB} \\ sU_{b,IB} \end{Bmatrix} + \begin{bmatrix} \omega_b^2(\kappa^2 + \mu_b) & -\mu_b\omega_b^2 \\ -\omega_b^2 & \omega_b^2 \end{bmatrix}\begin{Bmatrix} U_{t,IB} \\ U_{b,IB} \end{Bmatrix} = \begin{Bmatrix} -1 \\ -1 \end{Bmatrix}A_g \tag{8}$$

$$\begin{bmatrix} 1+\mu & \\ & 1 \end{bmatrix}\begin{Bmatrix} s^2U_{t,IIS} \\ s^2U_{b,IIS} \end{Bmatrix} + \begin{bmatrix} 2\omega_b(\kappa\zeta_{eq} + \mu_b\zeta_b) & -2\mu_b\omega_b\zeta_b \\ -2\omega_b\zeta_b & 2\omega_b\zeta_b \end{bmatrix}\begin{Bmatrix} sU_{t,IIS} \\ sU_{b,IIS} \end{Bmatrix} + \begin{bmatrix} \omega_b^2(\kappa^2 + \mu_b) & -\mu_b\omega_b^2 \\ -\omega_b^2 & \omega_b^2 \end{bmatrix}\begin{Bmatrix} U_{t,IIS} \\ U_{b,IIS} \end{Bmatrix} = \begin{Bmatrix} -1 \\ -1 \end{Bmatrix}A_g \tag{9}$$

where s is the Laplace operator, $s = i\Omega$, Ω is the ground motion excitation frequency, and $U_{t,IB}$, $U_{b,IB}$, $U_{t,IIS}$, $U_{b,IIS}$, and A_g are the Laplace transform of $u_{t,IB}$, $u_{b,IB}$, $u_{t,IIS}$, $u_{b,IIS}$, and a_g, respectively. From the linear matrix equations of Equations (8) and (9), $U_{t,IB}$, $U_{b,IB}$, $U_{t,IIS}$, and $U_{b,IIS}$ can be respectively solved. At the same time, the displacement response transfer functions of the bushing and tank with IB ($H_{Ub,IB}(s)$, $H_{Ut,IB}(s)$) and IIS ($H_{Ub,IIS}(s)$, $H_{Ut,IIS}(s)$) can be obtained as follows:

$$\begin{cases} H_{U_{t,IB}}(s) = \frac{U_{t,IB}(s)}{A_g(s)}, H_{U_{b,IB}}(s) = \frac{U_{b,IB}(s)}{A_g(s)} \\ H_{U_{t,IIS}}(s) = \frac{U_{t,IIS}(s)}{A_g(s)}, H_{U_{b,IIS}}(s) = \frac{U_{b,IIS}(s)}{A_g(s)} \end{cases} \tag{10}$$

3. Parameter Analysis of IIS

According to Parseval's Theorem, the root mean square (RMS) response σ of the system excited by white noise is obtained as follows:

$$\sigma = \int_0^T \frac{u^2(t)}{T_0} dt = \sqrt{\int_{-\infty}^{+\infty} |H(i\Omega)|^2 S_0 d\Omega} \tag{11}$$

where S_0 is the power spectrum of white noise. Therefore, the effect of the isolation system can be measured by comparing the reduction rate of the root mean square response of relative displacement of the bushing and the isolation layer with and without an isolation system. For the transformer–bushing system with IB and IIS installed, the displacement mitigation ratio of bushing $\gamma_{U,IB}$ and $\gamma_{U,IIS}$ is:

$$\gamma_{U,IB}(\omega_b, \zeta_{eq}, \kappa) = \frac{\left|\sigma_{U_{b,IB}} - \sigma_{U_{t,IB}}\right|}{\sigma_{U_{b,0}}}, \gamma_{U,IIS}(\omega_b, \zeta_{eq}, \kappa, \mu) = \frac{\left|\sigma_{U_{b,IIS}} - \sigma_{U_{t,IIS}}\right|}{\sigma_{U_{b,0}}} \tag{12}$$

where $\sigma_{Ub,0}$ is the root mean square response of the displacement of the original bushing. At the same time, the ratio of the root mean square of relative displacement response after base isolation can be calculated to compare the isolation effects of the two isolation systems; secondly, the ratio of the root mean square of displacement response of the isolation layer of IIS and IB can be calculated to reflect the stability of the isolation layer of the two isolation systems; finally, the ratio of root mean square of base shear force response of IIS and IB can be calculated to reflect the force output of the isolation layer. For a transformer–bushing structure with IB and IIS installed, the relative displacement response ratio of the two isolation systems γ_U, the displacement response ratio of isolation layer γ_{ID}, and base shear force response ratio γ_{SF} are in the Formula (13), where σ_{SF} is the root mean square of base shear force response of the transformer–bushing system.

$$\gamma_U(\omega_b, \kappa, \zeta_{eq}, \mu) = \frac{\gamma_{U,IIS}}{\gamma_{U,IB}}, \gamma_{ID}(\omega_b, \zeta_{eq}, \kappa, \mu) = \frac{\sigma_{U_{t,IIS}}}{\sigma_{U_{t,IB}}}, \gamma_{SF}(\omega_b, \zeta_{eq}, \kappa, \mu) = \frac{\sigma_{SF,IIS}}{\sigma_{SF,IB}} \tag{13}$$

At the same time, it is also necessary to analyze the interaction between the bushing and the tank. Assuming that the stiffness coefficient and damping ratio of transformer–bushing are constants, then μ_b can reflect the relationship between the flexibility and mass of the bushing. The greater the μ_b, the higher the height of the bushing and the greater the flexibility. This is used to describe bushings with high center of gravity and low engineering frequency (1–3 Hz). The smaller the μ_b, the smaller the height and the greater the stiffness of the bushing. This is used to describe bushings with low center of gravity and high engineering frequency (3–10 Hz). The predominant frequency of ground motion is between 1–10 Hz, so the size of value of engineering frequency of the bushing is relative to the seismic action. Table 1 shows the parameters of bushings with different engineering frequencies, where LB represents bushings with low frequency, HB represents bushings with high frequency, and f_b is the engineering frequency of the transformer–bushing.

Table 1. Parameters of bushings.

Parameters of Bushing	Low Frequency Bushing		High Frequency Bushing	
	LB–1	LB–2	HB–1	HB–2
μ_b	0.060	0.220	0.005	0.020
$h(m)$	5.0	10.0	1.2	2.5
$f_b(Hz)$	2.1	1.2	7.1	3.5

Parameter analysis of isolation layer select inerter–mass ratio μ, frequency ratio κ and damping ratio ζ_{eq}. The index of parameter analysis is the relative displacement response ratio of bushing γ_U, displacement ratio of isolation layer γ_{ID}, and base shear force response ratio γ_{SF}. The parameter range of inerter–mass ratio μ in parameter analysis is [0.01,10], frequency ratio κ, and damping ratio ζ_{eq} is [0.01,1].

3.1. Relative Displacement Response Ratio γ_U

The relative displacement between the bushing and tank can directly reflect the seismic response level of the bushing, and it is also the most important index in the vibration control of the transformer–bushing system. Therefore, the relative displacement of the bushing should be taken as the primary control index, and the random vibration analysis of the transformer–bushing isolation system under white noise input should be carried out. First, the inerter–mass ratio μ is analyzed. When the inerter–mass ratio is fixed, the influence of inerter–mass ratio on the isolation layer can be seen directly. Figure 4 shows the relative displacement response ratio γ_U when changing the damping ratio ζ_{eq} with different types of bushing systems.

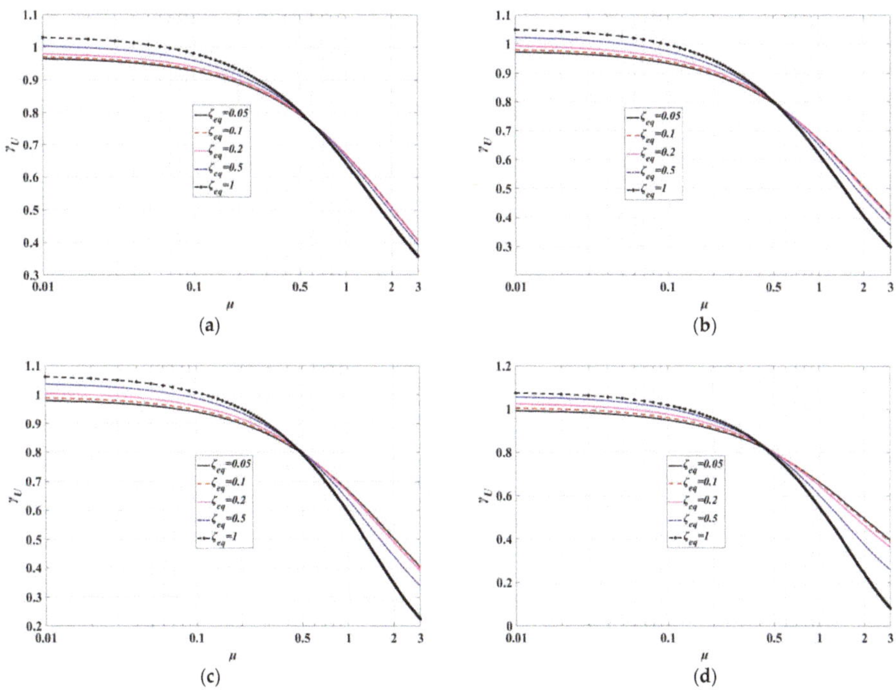

Figure 4. Curves of γ_U under different damping ratio ζ_{eq} (κ = 0.1, ζ_b = 0.02): (**a**) *HB–1*; (**b**) *HB–2*; (**c**) *LB–1*; (**d**) *LB–2*.

In the isolation design, the frequency ratio of the isolation layer should not be too large, so $\kappa = 0.1$ is a fixed value for analysis. The variety of bushings with different frequencies with inerter–mass ratio is the same, γ_U decreases with the increase of inerter–mass ratio, and the greater the damping ratio of isolation layer, the greater the reduction of γ_U in larger value of μ. The influence of the inerter–mass ratio on low–frequency bushings is more obvious. When the inerter–mass ratio is greater than 0.5, the decline of γ_U becomes faster. When the inerter–mass ratio is between [1,2], the RMS of relative displacement response of bushings with IIS is about 0.5 times of those with IB, and the displacement control effect of the inerter element is better. If the inerter–mass ratio continues to increase, we will see yet lower values of γ_U. However, cost will greatly increase, and the increase of inerter–mass ratio will also increase the base shear force of the isolation layer. Considering that the linear IB system has satisfactory isolation effects on the transformer–bushing system, the target γ_U can be set between 0.5–0.7, so inerter–mass ratio is between [0.5,2].

Second, the frequency ratio of isolation layer of IIS is considered. The inerter–mass ratio is taken as 0.5, 1, and 2, respectively, and the damping ratio of isolation layer ζ_{eq} is 0.1, 0.5, and 1. The influence of frequency ratio on γ_U is shown in Figure 5. The γ_U of bushings with different frequencies is almost a fixed value when it is within [0.001,0.05], which is less affected by the frequency ratio. When κ is [0.1,1], it decreases first and then increases significantly; the greater the inerter–mass ratio and damping ratio, the greater the influence of the frequency ratio. For bushings with high frequency, the stability range of γ_U is κ is equal to [0.001,0.1]; for bushings with low frequency, the stability range of γ_U is κ is equal to [0.001,0.0.5]. Hence, to ensure the robustness of the isolation layer of IIS on the displacement control of the upper bushing, the value of κ should not be greater than 0.1.

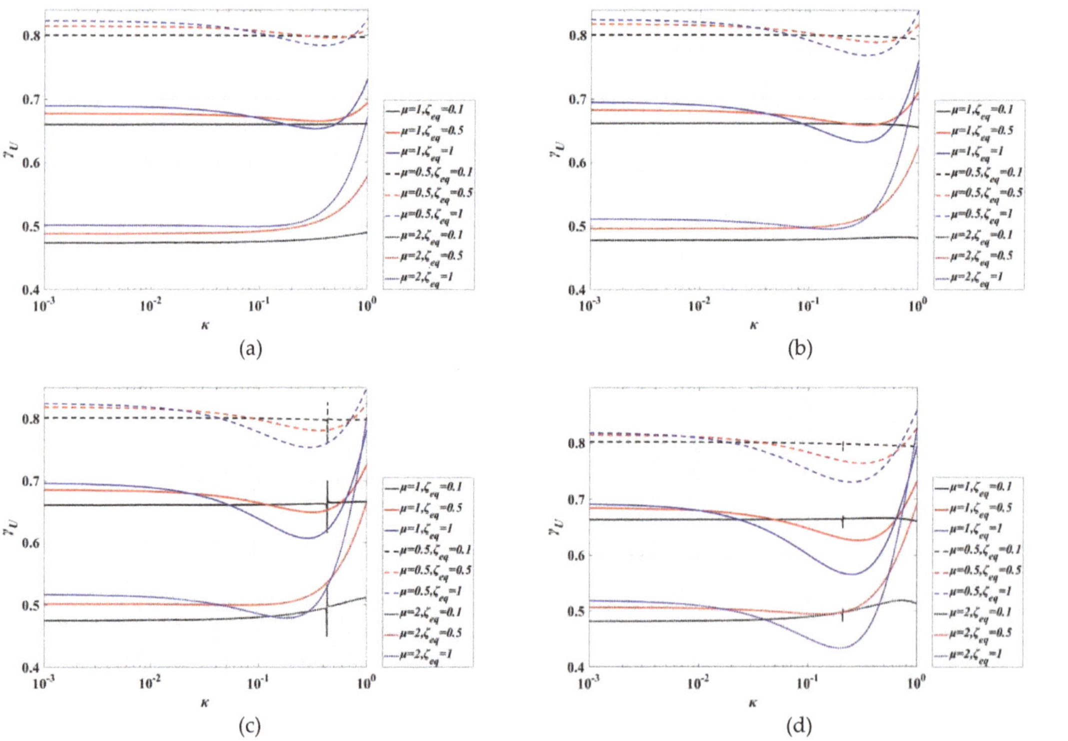

Figure 5. Curves of γ_U: (**a**) *HB–1;* (**b**) *HB–2;* (**c**) *LB–1;* (**d**) *LB–2.*

3.2. Displacement Response Ratio of Isolation Layer γ_{ID}

The smaller the relative displacement between the bushing and the tank, the more that the response of the tank and the bushing tends towards being the same, but excessive displacement of the isolation layer can easily cause overturning of the superstructure. Therefore, the displacement of the isolation layer is also an index that can not be ignored in vibration control of the transformer–bushing system. As the displacement of the isolation layer in a normal IB system is large, the displacement response ratio γ_{ID} of the isolation layer can be used in the analysis as an index to seek the trend of each parameter. Figures 6–9 show displacement response ratios of isolation layer γ_{ID} under different frequency ratios κ of four different types of transformer–bushing systems.

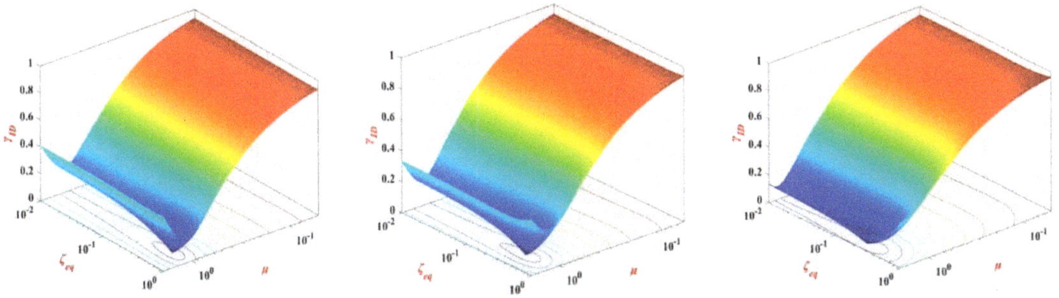

(a) (b) (c)

Figure 6. Contour plot of γ_U of *HB–1*: (**a**) κ = 0.05; (**b**) κ = 0.1; (**c**) κ = 0.3.

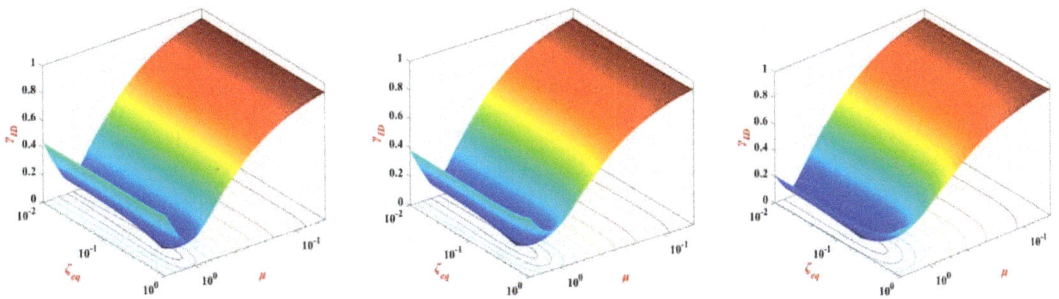

(a) (b) (c)

Figure 7. Contour plot of γ_U of *HB–2*: (**a**) κ = 0.05; (**b**) κ = 0.1; (**c**) κ = 0.3.

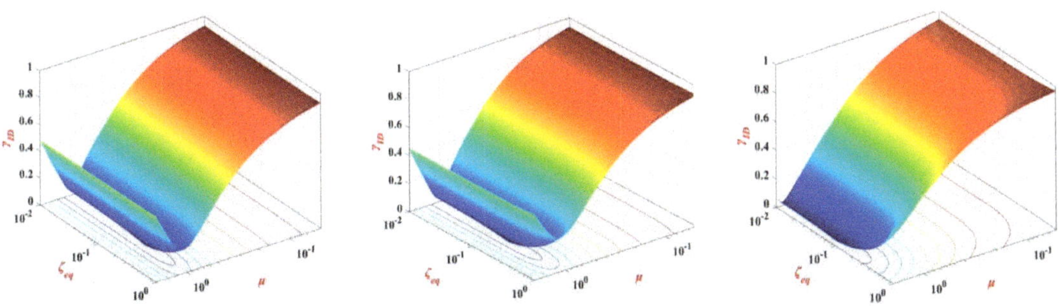

(a) (b) (c)

Figure 8. Contour plot of γ_U of *LB–1*: (**a**) κ = 0.05; (**b**) κ = 0.1; (**c**) κ = 0.3.

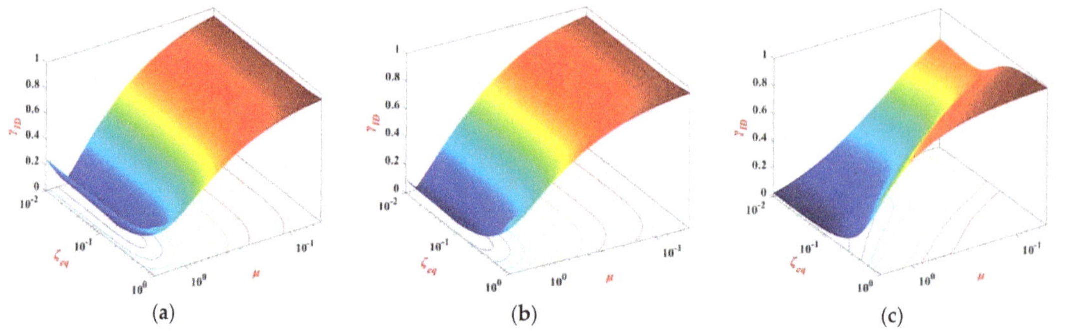

Figure 9. Contour plot of γ_U of *LB–2*: (**a**) κ = 0.05; (**b**) κ = 0.1; (**c**) κ = 0.3.

The γ_{ID} is the ratio of the displacement ratios of the isolation layer of the two isolation systems, so the damping ratio has little effect. Only low–frequency bushings (LB–2) show large fluctuation at κ = 0.3. However, γ_{ID} is greatly affected by the inerter–mass ratio, and like γ_U, will decrease with the increase of inerter–mass ratio within a certain range. However, when the inerter–mass ratio is between [1,2], the four kinds of bushing γ_{ID} will increase with the increase of the inerter–mass ratio. The smaller the frequency ratio is, the more flexible the isolation layer is, and the more 'steep' the surface of γ_{ID} is when the inerter–mass ratio is between [1,2], indicating that the restoring force of the isolation layer can not quickly mitigate the additional inertia force generated by the inerter element. Although increasing the stiffness can reduce this phenomenon, the excessive stiffness of thesolationn layer will increase the displacement of the upper bushing. So, when the value of κ is not greater than 0.1, we should try to increase it as much as possible to meet the displacement demands of the isolation layer. As can be seen from Figures 6–9, when the γ_{ID} is greater than 0.4, the demands for the values of various parameters of the isolation layer are low, and the surface of γ_{ID} is stable. At the same time, the damping ratio ζ_{eq} should not be too large. For bushings with low frequency, increasing the damping ratio will increase the displacement of isolation layer. When ζ_{eq} is less than 0.2, the surface of γ_{ID} is relatively stable and robust. The displacement index of isolation layer γ_{ID} is determined to be about 0.4. Meanwhile, when meeting the demands of the displacement mitigation of the isolation layer, the values of various parameters shall be reduced as much as possible.

3.3. Base Shear Force Response Ratio γ_{SF}

The base shear force ratio γ_{SF} is also an important index of the isolation system. We can use it to seek the trend of each parameter and compare it with the trend of displacement index to comprehensively select the final isolation layer parameters. Figures 10–13 show the base shear force ratio of four types of transformer–bushing systems under different frequency ratios κ.

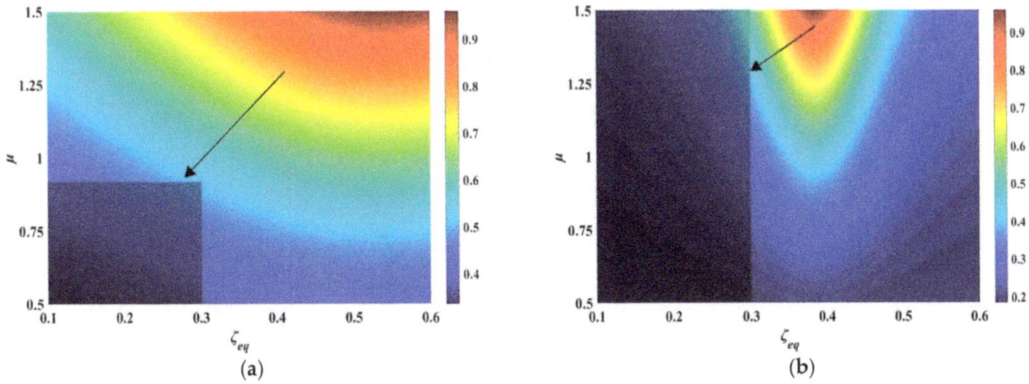

Figure 10. Contour plot of γ_{SF} of *HB–1*: (**a**) $\kappa = 0.05$; (**b**) $\kappa = 0.1$.

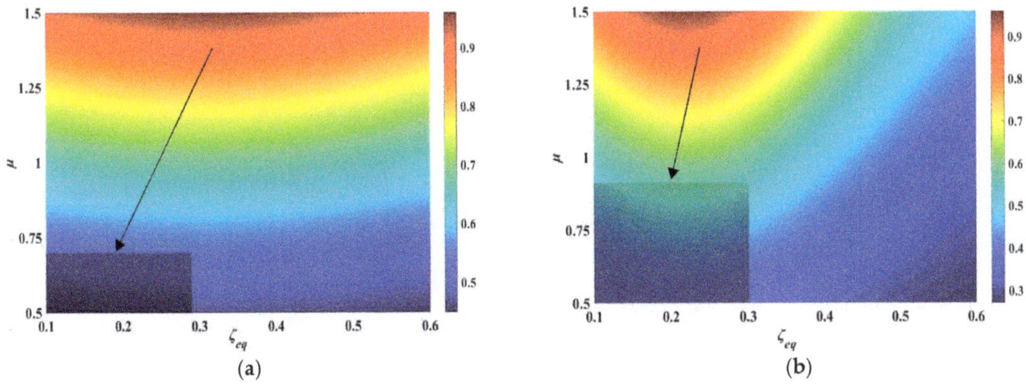

Figure 11. Contour plot of γ_{SF} of *HB–2*: (**a**) $\kappa = 0.05$; (**b**) $\kappa = 0.1$.

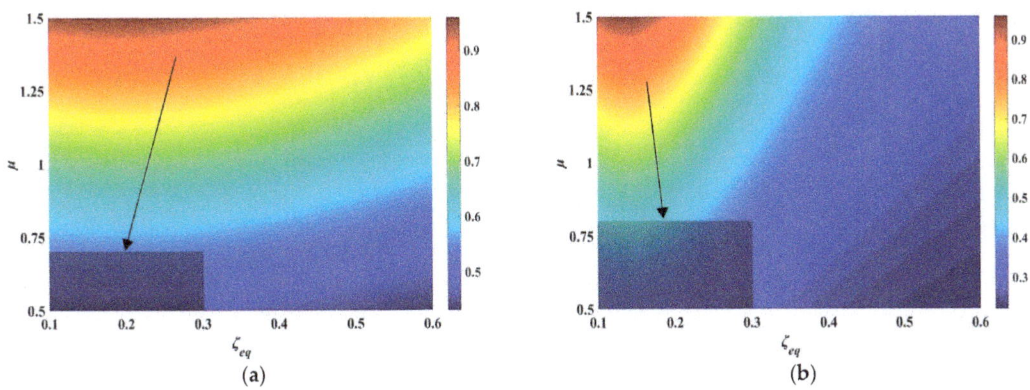

Figure 12. Contour plot of γ_{SF} of *LB–1*: (**a**) $\kappa = 0.05$; (**b**) $\kappa = 0.1$.

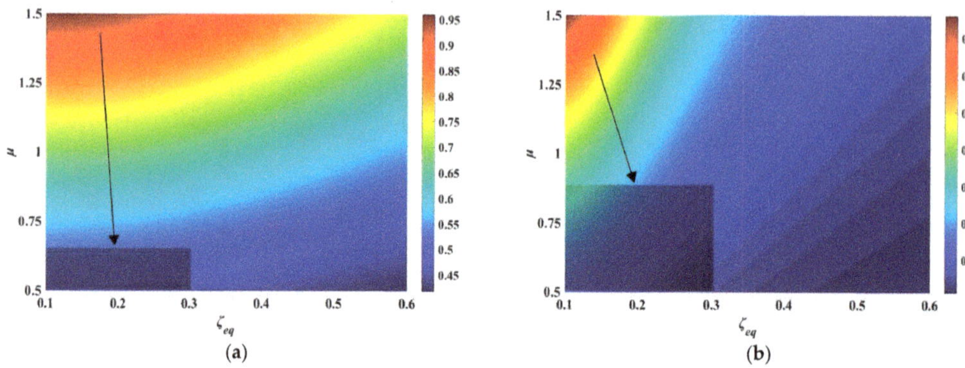

Figure 13. Contour plot of γ_{SF} of *LB–2*: (**a**) $\kappa = 0.05$; (**b**) $\kappa = 0.1$.

The base shear force of the tank of the IIS is lower than IB when the inerter–mass ratio is within [0.5,1.5], but when the inerter–mass ratio is greater than 1.5, the base shear force of IB is less than IIS, indicating that the large inertance coefficient will increase the force output of the isolation layer. This trend is opposite to γ_U and γ_{ID}. Therefore, under the condition of meeting the displacement index (γ_U and γ_{ID}), the inerter–mass ratio should be reduced as much as possible to reduce the base shear force.

Fix $\gamma_{SF} = 0.5$ and the damping ratio is less than 0.5 as the target area in Figures 10–14 (rectangular shadow). It can be seen that when the frequency ratio is 0.05, the target area is smaller and the corresponding optional parameter range becomes smaller for both bushings with low frequency and bushings with high frequency; when the frequency ratio increases to 0.1, the target area increases significantly, and the range of selectable reasonable inerter–mass ratios also expands. At the same time, it can also be seen that the target area of bushings with high frequency is larger than bushings with low frequency, so the bushings with high-frequency difficulty provide easier control over base shear force.

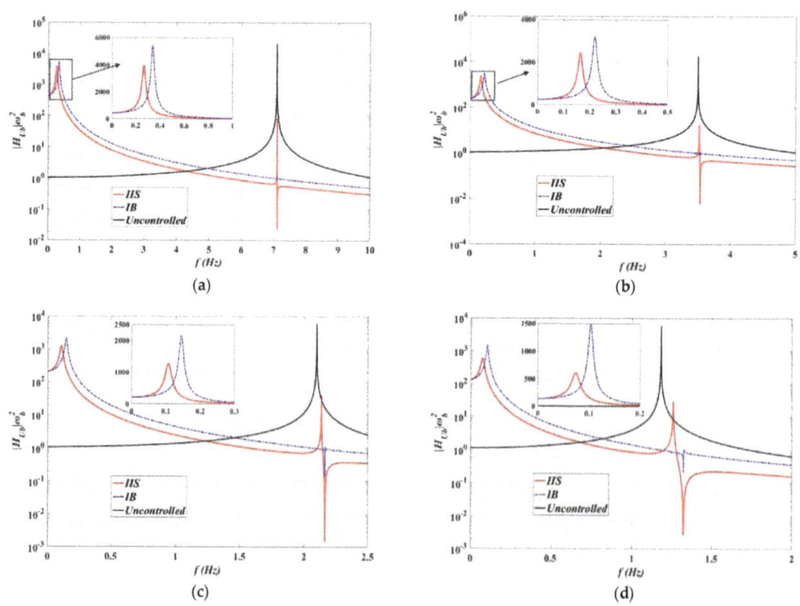

Figure 14. Displacement amplification factor of bushings: (**a**) *HB–1*; (**b**) *HB–2*; (**c**) *LB–1*; (**d**) *LB–2*.

For bushings with high frequency, the inerter–mass ratio can be reduced as much as possible because the amplitude of base shear force is low and the target area is large; due to the fact that the displacement index of bushings with low frequency are more stringent than the base shear force index, the displacement index of bushings with low frequency can be considered first when a larger base shear force ratio is fixed.

3.4. Parameters Optimization Design of IIS

The variation trend of the parameters of the displacement index (γ_U and γ_{ID}) is opposite to that of the base shear force index (γ_{SF}), which is mainly reflected in the damping ratio ζ_{eq} and inerter–mass ratio μ. The determination of IIS parameters is preferentially suggested to be based on the specified extent of vibration mitigation effect in terms of the relative displacement of transformer and bushing γ_U, which is essential to performance. Following the parametric analysis results, in the case of given κ, large μ will lead to larger shear force (shown in Figure 13), and small μ and ξ_{eq} will lead to larger displacement of bushing (shown in Figure 5). We need to obtain a balanced parameter combination in the IIS to mitigate all three indexes.

Hence, the overall design idea of IIS for transformer–bushing isolation control is to meet the demands of bushing relative displacement first, and then consider the displacement of the isolation layer and base shear force. At this point, γ_{ID} and γ_{SF} are considered as additional performance indexes; the relative displacement index is considered first. We can fix the target relative displacement response ratio γ_U according to the performance demands of the transformer–bushing system, and optimize the parameters with extremum conditions [30], that is:

$$\gamma_U(\zeta_{eq}, \kappa, \mu) = \gamma_{U,t} \tag{14}$$

$$\frac{\partial \gamma_U(\zeta_{eq}, \kappa, \mu)}{\partial \mu} = 0, \frac{\partial \gamma_U(\zeta_{eq}, \kappa, \mu)}{\partial \zeta_{eq}} = 0 \tag{15}$$

where $\gamma_{U,T}$ is the target of the relative displacement response ratio. The constraint condition of Equation (15) is to meet the relative displacement response ratio and makes the damping ratio and inerter–mass ratio as small as possible to meet the engineering needs at the same time. This optimization method is concerned with controlling both the performance of the relative displacement of the transformer and bushing γ_U and base shear force γ_{SF}, so that the excessive shear force can be reduced while simultaneously achieving a desired relative displacement mitigation ratio γ_U.

4. Seismic Response

Parameter analysis in the frequency domain was carried out under white noise excitation; to further verify the isolation effect of IIS, a dynamic response analysis of the transformer–bushing system with IIS under non-stationary ground motion was carried out as well. Different values of isolation indexes were selected for four types of bushings, and the optimization parameters were selected according to the optimization method of Equation (14). The best parameters of the isolation layer were selected for time history analysis. The values of the IIS design parameters and optimization indexes of different types of transformer–bushing systems are shown in Table 2. The primary optimization principle is that the bushing with lower frequency should keep a smaller value of γ_U. This reflects the greater control over relative displacement afforded by highly flexible bushings.

Table 2. Results of parameters of IIS obtained after optimization.

	Parameters	LB–1	LB–2	HB–1	HB–2
Tank	m_t(kg)	10.0 × 10⁴	15.0 × 10⁴	5.0 × 10⁴	7.5 × 10⁴
Bushing	μ_b	0.060	0.220	0.005	0.020
	h(m)	5.00	10.00	1.20	2.50
	f_b(Hz)	2.10	1.20	7.10	3.50
IIS	μ	0.854	1.056	0.618	0.791
	κ	0.071	0.095	0.048	0.063
	ζ_{eq}	0.113	0.171	0.055	0.070
Primary index	γ_U	0.600	0.500	0.700	0.650
Additional index	γ_{ID}	0.461	0.353	0.621	0.588
	γ_{SF}	0.452	0.598	0.275	0.364

Figure 14 shows the displacement amplification factors of bushings corresponding to Table 2. It can be seen that the natural frequency of the isolation system bushing is reduced, and the displacement responses are also decreased. Compared with the IB system, the displacement amplification factor of IIS is lower, and the response of bushings with low frequency (LB–2) decreases the most. The natural period of the bushing with IIS is further extended, far away from the predominant period of ground motion. At the same time, as the isolation effect of IIS is more obvious, the bandwidth of bushing response after isolation also increases. Figure 15 shows four types of transformer–bushing systems corresponding to Table 2. Similar to the trend in Figure 14, compared with IB system, the amplitudes of transfer function of the base shear force of IIS are lower, and the shear force response of bushing with low frequency is reduced the most.

El Centro wave, Taft wave, Chi Chi wave, and Kobe wave were selected as seismic waves; the predominant frequencies of the four seismic records were different. Figure 16 shows the acceleration response spectra of the four seismic waves. The higher the structural height and weight of the bushing, the higher the seismic vulnerability, and the difficulty of vibration control will increase. Therefore, this paper selects the LB–2 transformer–bushing system, which is the most vulnerable to earthquakes, as the analysis object for analyzing the dynamic time history under four kinds of ground motion inputs. Figure 17 shows the relative displacement time history of the bushing and the transformer, which can reflect the seismic response of the bushing itself; Figure 18 shows the **hysteresis loops** of the isolation layer, which can reflect the displacement response of the isolation layer and the energy dissipation capacity of dampers of IB and IIS.

It can be seen from Figure 17 that the control effect of IB isolation system on the bushing was considerable, and the maximum relative displacement of the bushing can be controlled between 30–50% of the original structure. IIS has a better isolation control effect on the bushing than IB, which is only about 50% of the displacement controlled by IB. The displacement response level is reduced at the same time, and the time history curve is smoother and steadier. It indicates that the overall sloshing speed of the tank and bushing is also reduced.

It can be seen from Figure 18 that IIS has smaller displacement of the isolation layer than IB, and the peak displacement is about 45–60% of IB. Based on the relative displacement response of the bushing in Figure 17, IIS can reduce the displacement of the tank and bushing at the same time and complete the overall isolation control of the transformer–bushing system. At the same time, while reducing the displacement of the isolation layer, the damping element of IIS has a larger hysteresis loop due to the amplification of the inerter element. Compared with IB, the damping force is larger and the energy dissipation effect is amplified.

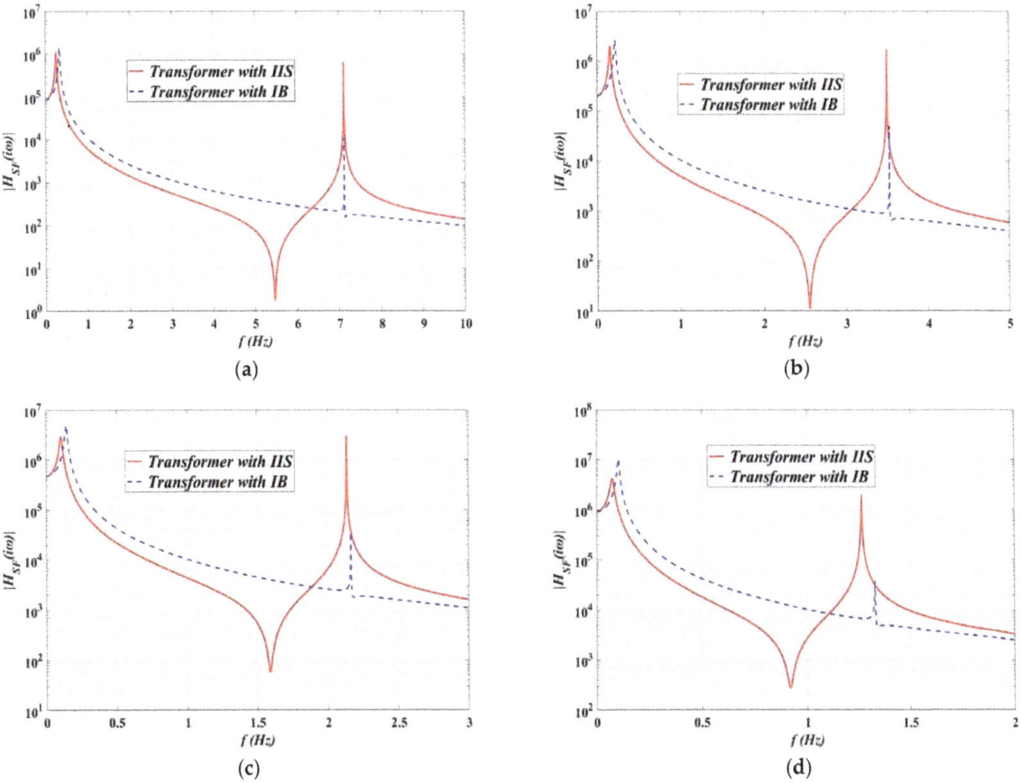

Figure 15. Transfer function curves of base shear force: (**a**) *HB–1*; (**b**) *HB–2*; (**c**) *LB–1*; (**d**) *LB–2*.

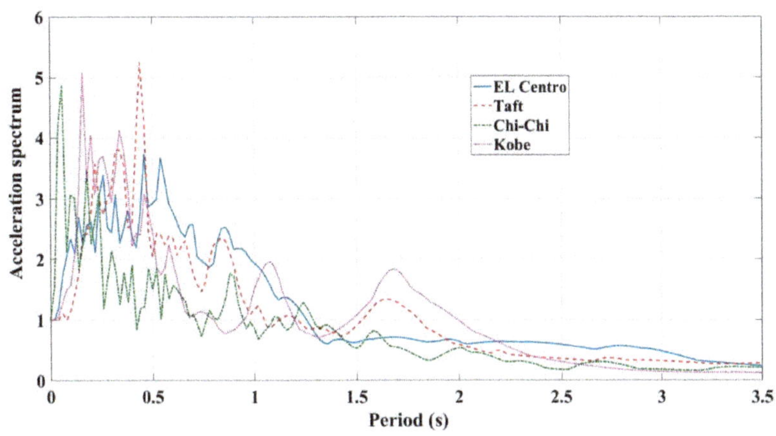

Figure 16. Normalized acceleration spectra of earthquake records.

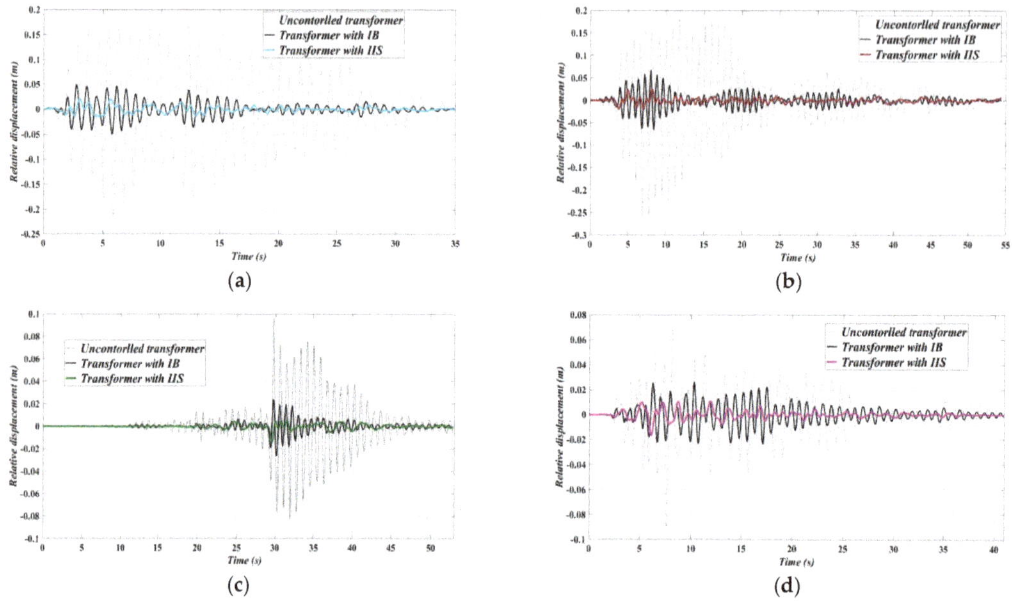

Figure 17. Relative displacement responses of bushing (LB–2): (**a**) EL Centro; (**b**) Taft; (**c**) Chi–Chi; (**d**) Kobe.

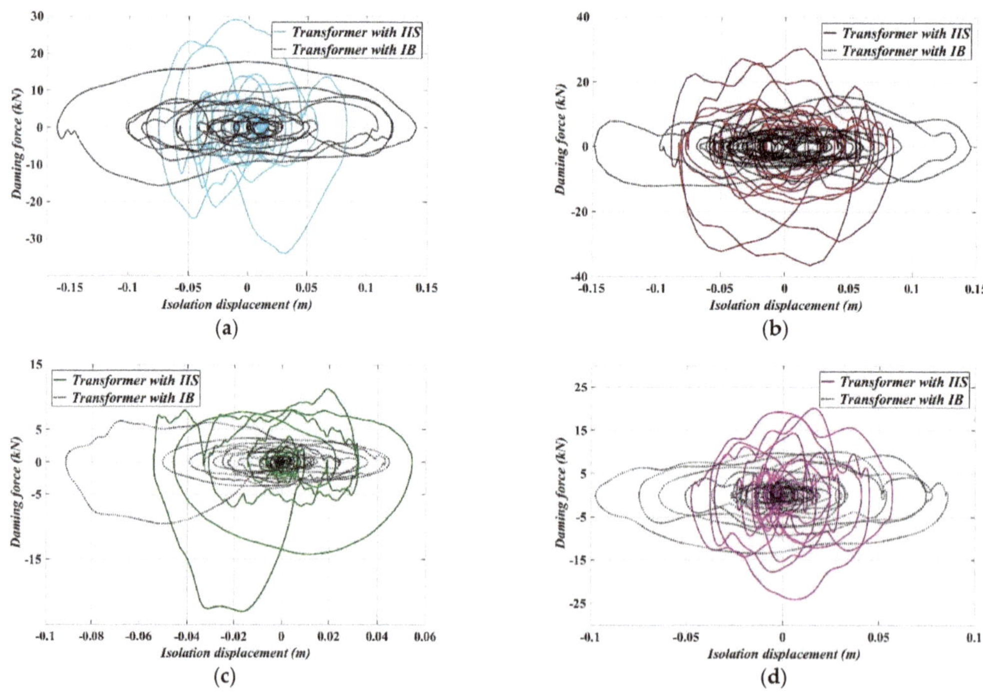

Figure 18. Hysteresis loops of isolation layer (LB–2): (**a**) EL Centro; (**b**) Taft; (**c**) Chi–Chi; (**d**) Kobe.

The isolation system with inerter element increases the mass effect and viscous damping effect of the isolation layer, but it does not mean that the system needs greater apparent mass and damping coefficient. On the contrary, the inerter system can change the original small apparent mass into a larger equivalent mass (such as moment of inertia), which can be dozens or even hundreds of times the original mass, so as to reduce the actual mass and volume of the isolation layer. However, the natural period of the superstructure can still be extended, and the displacement of the isolation layer can be reduced under the condition of ensuring the isolation rate. This has high practical value for engineering.

5. Conclusions

In this paper, the influence of the inerter isolation system (IIS) on the response of the transformer–bushing system was studied, and the parameter optimization design of IIS was carried out. Finally, the isolation performance of IIS under different input was analyzed. The main conclusions are as follows:

1. The equivalent mass coefficient and damping coefficient of IIS can be amplified by an inerter element, and the inerter–mass ratio and damping ratio are reduced simultaneously under the condition of meeting the performance demand after parameter optimization.
2. The proposed optimal design utilizes the most efficient parameter set of inerter–mass ratio and damping ratio for the relative displacement ratio of the bushing and tank in the extremum condition. The parameter optimization design method proved to be effective in meeting the target demand of relative displacement response of the bushing and tank, while base shear force and isolation displacement were reduced simultaneously.
3. Based on results from response history analysis under ground motion records, IIS can significantly suppress the resonance response of the structure and the continuous vibration response in the stable state and its peak displacement can be reduced by 50% compared with IB. IIS has a smaller displacement of the isolation layer than IB, and the peak displacement is about 45–60% of IB.

In conclusion, IIS has considerable vibration mitigation effect. This study mainly focuses on the macro-structural design parameters of typical transformer–bushing systems, but there is no detailed analysis on the design parameters of any specific IIS device. At present, the IIS device is in the development stage. Subsequent research will carry out mechanical performance tests according to the specific device with inerter, damper, and spring. More combinations of isolation system with inerter and parameters optimization methods will also be discussed. Furthermore, a shaking table test will be carried out on the specific structure with IIS installed.

Author Contributions: R.Z.: Methodology, Conceptualization, Software, Investigation, Writing—original draft, Writing—review & editing; M.C.: Conceptualization, Supervision, Funding acquisition, Writing—review & editing; J.H.: Validation, Writing—review & editing. All authors have read and agreed to the published version of the manuscript.

Funding: This research was supported by 2020 open fund project of the State Key Laboratory of Disaster Prevention & Reduction for Power Grid Transmission and Distribution Equipment (B316AF190007) and State Grid Science & Technology Project (5200-201919121A-0-0-00).

Institutional Review Board Statement: Not applicable.

Informed Consent Statement: Not applicable.

Data Availability Statement: The data used to support the findings of this study are included within the article.

Conflicts of Interest: The authors declare that there is no conflict of interest regarding the publication of this paper. The authors declare no conflict of interest.

References

1. Meigen, C.; Fulin, Z.; Ping, P.; Guangping, Z.; Zhengguo, G. Seismic response of transformer and bushing isolation system and parameter analysis of isolation layer. *Proc. CSEE* **2012**, *32*, 166–174.
2. Li, T.; Hai–Yang, P.; Rui–Sheng, M. Probabilistic seismic demand model and fragility analysis of transmission tower subjected to near–field ground motions. *J. Constr. Steel Res.* **2019**, *156*, 266–275.
3. Meirovitch, L. *Dynamics and Control of Structures*; John Wiley & Sons: New York, NY, USA, 1990.
4. Soong, T.T.; Dargush, G.F. *Passive Energy Dissipation Systems in Structural Engineering*; Wiley: Chichester, UK, 1997.
5. Cao, M.; Fan, R.; Li, S.; Cao, C.; Lu, Z.; Zhang, X. Design and application of seismic isolation system of large power transformer and bushings system. *Power Syst. Technol.* **2011**, *35*, 130–135.
6. Fujita, T.; Fujita, S.; Yoshizawa, T.; Suzuki, S. Base Isolation Support of Heavy Equipment with Laminated Rubber Bearings–2. *Jpn. Soc. Mech. Eng.* **1985**, *51*, 461.
7. Bonacina, G.; Bonetti, P.; Martelli, A.; Bettinali, F.; Serino, G. Seismic base isolation of gas insulated electrical substations: Design, experimental and numerical activities, evaluation of the applicability. In Proceedings of the 10th European Conference on Earthquake Engineering, Vienna, Austria, 28 August–2 September 1995.
8. Pham, T. *Two–Dimensional Shaking Table Test of Transformer Bushing with Seismic Isolation Device*; MCEER Report; Department of Civil and Environmental Engineering, University of California at Irvin: Irvine, CA, USA, 2005.
9. Murota, N.; Feng, M.Q.; Liu, G.Y. Earthquake Simulator Testing of Base–Isolated Power Transformers. *IEEE Trans Power Deliv* **2006**, *21*, 1291–1299. [CrossRef]
10. Murota, N.; Feng, M.Q.; Liu, G.Y. *Experimental and Analytical Studies of Base Isolation Systems for Seismic Protection of Power Transformers*; Technical Report MCEER-05-0008; Multidisciplinary Center for Earthquake Engineering Research: Buffalo, NY, USA, 2005.
11. Jiyu, L.; Junxiong, L.; Ruimin, L.; Peimin, L. Experiment and system identification of base–isolated electric power transformer. *Earthq. Eng. Eng. Vib.* **2001**, 109–116.
12. Lee, S.H.; Min, K.W.; Hwang, J.S.; Kim, J. Evaluation of equivalent damping ratio of a structure with added dampers. *Eng. Struct.* **2004**, *26*, 335–346. [CrossRef]
13. Smith, M.C.; Wang, F.C. Performance benefits in passive vehicle suspensions employing inerters. *Veh. Syst. Dyn.* **2004**, *42*, 235–257. [CrossRef]
14. Smith, M.C. Synthesis of mechanical networks: The inerter. *IEEE Trans. Autom. Control* **2002**, *47*, 1648–1662. [CrossRef]
15. Swift, S.J.; Smith, M.C.; Glover, A.R.; Papageorgiou, C.; Gartner, B.; Houghton, N.E. Design and modelling of a fluid inerter. *Int. J. Control* **2013**, *86*, 2035–2051. [CrossRef]
16. Ikago, K.; Saito, K.; Inoue, N. Seismic control of single–degree–of–freedom structure using tuned viscous mass damper. *Earthq. Eng. Struct. Dyn.* **2012**, *41*, 453–474. [CrossRef]
17. Pan, C.; Zhang, R. Design of structure with inerter system based on stochastic response mitigation ratio. *Struct. Control Health Monit.* **2018**, *25*, e2169. [CrossRef]
18. Pan, C.; Zhang, R.; Luo, H.; Li, C.; Shen, H. Demand–based optimal design of oscillator with parallel–layout viscous inerter damper. *Struct. Control Health Monit.* **2018**, *25*, e2051. [CrossRef]
19. Hwang, J.S.; Kim, J.; Kim, Y.M. Rotational inertia dampers with toggle bracing for vibration control of a building structure. *Eng. Struct.* **2007**, *29*, 1201–1208. [CrossRef]
20. Zhang, R.F.; Zhao, Z.P.; Dai, K.S. Seismic response mitigation of a wind turbine tower using a tuned parallel inerter mass system. *Eng. Struct.* **2019**, *180*, 29–39. [CrossRef]
21. Zhao, Z.P.; Zhang, R.F.; Wierschem, N.E.; Jiang, Y.Y.; Pan, C. Displacement mitigationoriented design and mechanism for inerter–based isolation system. *J. Vib. Control* **2020**, *27*, 1991–2003. [CrossRef]
22. Gao, H.; Wang, H.; Li, J.; Wang, Z.; Liang, R.; Xu, Z.; Ni, Y. Optimum design of viscous inerter damper targeting multi–mode vibration mitigation of stay cables. *Eng. Struct.* **2021**, *226*, 111375. [CrossRef]
23. Sugimura, Y.; Goto, W.; Tanizawa, H.; Saito, K.; Nimomiya, T. Response control effect of steel building structure using tuned viscous mass damper. In Proceedings of the 15th World Conference on Earthquake Engineering, Lisbon, Portugal, 25–28 September 2012.
24. Kurata, M.; Leon, R.T.; Desroches, R. Rapid seismic rehabilitation strategy: Concept and testing of cable bracing with couples resisting damper. *J. Struct. Eng.* **2012**, *138*, 354–362. [CrossRef]
25. Xie, L.; Ban, X.; Xue, S.; Ikago, K.; Kang, J.; Tang, H. Theoretical Study on a Cable–Bracing Inerter System for Seismic Mitigation. *Appl. Sci.* **2019**, *9*, 4096. [CrossRef]
26. Xue, S.; Kang, J.; Xie, L.; Zhang, R.; Ban, X. Cross–Layer Installed Cable–Bracing Inerter System for MDOF Structure Seismic Response Control. *Appl. Sci.* **2020**, *10*, 5914. [CrossRef]
27. Wang, H.; Gao, H.; Li, J.; Wang, Z.; Ni, Y.; Liang, R. Optimum design and performance evaluation of the tuned inerter–negative–stiffness damper for seismic protection of single–degree–of–freedom structures. *Int. J. Mech. Sci.* **2021**, *212*, 106805. [CrossRef]
28. Luo, H.; Zhang, R.F.; Weng, D.G. Mitigation of liquid sloshing in storage tanks by using a hybrid control method. *Soil Dyn. Earthq. Eng.* **2016**, *90*, 183–195. [CrossRef]
29. Jiang, Y.Y.; Zhao, Z.P.; Zhang, R.F.; De Domenico, D.; Pan, C. Optimal design based on analytical solution for storage tank with inerter isolation system. *Soil Dyn. Earthq. Eng.* **2020**, *129*, 105924. [CrossRef]

30. Losanno, D.; Calabrese, A.; Madera–Sierra, I.E.; Spizzuaco, M.; Marulanda, J.; Thomson, P.; Serino, G. Recycled versus Natural–Rubber Fiber–Reinforced Bearings for Base Isolation: Review of the Experimental Findings. *J. Earthq. Eng.* **2022**, *26*, 1921–1940. [CrossRef]
31. Losanno, D.; Palumbo, F.; Calabrese, A.; Barrasso, T.; Vaiana, N. Preliminary Investigation of Aging Effects on Recycled Rubber Fiber Reinforced Bearings (RR–FRBs). *J. Earthq. Eng.* **2021**, 1–18. [CrossRef]
32. Kelly, J.M. Seismic isolation for earthquake–resistant design. In *Earthquake–Resistant Design with Rubber*; Springer: London, UK, 1997; pp. 1–18.
33. Vaiana, N.; Spizzuoco, M.; Serino, G. Wire rope isolators for seismically base–isolated lightweight structures: Experimental characterization and mathematical modeling. *Eng. Struct.* **2017**, *140*, 498–514. [CrossRef]
34. Nagarajaiah, S.; Reinhorn, A.M.; Constantinou, M.C. Nonlinear Dynamic Analysis of 3–D–Base–Isolated Structures. *J. Struct. Eng.* **1991**, *117*, 2035–2054. [CrossRef]
35. Pellecchia, D.; Feudo, S.L.; Vaiana, N.; Dion, J.L.; Rosati, L. A procedure to model and design elastomeric–based isolation systems for the seismic protection of rocking art objects. *Comput.-Aided Civ. Infrastruct. Eng.* **2021**. [CrossRef]

buildings

MDPI

Article

Site Measurement Study on Mechanical Properties of SMW Piles of Building Structures in Sandy Soil Areas

Qingjun Xian [1,*], Zhe Wang [1], Xiaosong Liu [2], Shaokui Ma [3] and Zhaoran Xiao [1]

1 School of Civil Engineering, Henan University of Technology, Zhengzhou 450001, China
2 Shandong Hengye Engineering Consulting Co., Ltd., Binzhou 256600, China
3 Infrastructure Department, Henan Agricultural University, Zhengzhou 450046, China
* Correspondence: qingjunxian@haut.edu.cn

Abstract: SMW (soil mixing wall) piles have been widely used in soft soil areas such as Jiangsu, Shanghai, Tianjin and so on, and they have many advantages, such as retaining the structures of foundation pits. In order to promote the application of SMW piles in sandy soil areas such as Henan province, SMW piles were used in a deep foundation pit project of a high-rise building in Zhengzhou. Three SMW piles in the middle area of the foundation pit were selected for site measurement to determine the mechanical properties of SMW piles in sandy soil areas. Several typical test sections were determined along the height of the pile. The vibrating string type of the reinforcement dynamometers were set on the H-shaped steel of each test section, and the stress distribution of the H-shaped steel along the depth of the pit was obtained via testing. The axial force, bending moment and shearing force of the H-shaped steel were further calculated, and the affecting factors and development laws of the internal force distribution of the H-shaped steel were analyzed in detail. The research shows that, at the stage of foundation pit excavation, the overall stress of H-shaped steel increases gradually. The axial force of H-shaped steel in an SMW pile is mainly affected by such factors as the weight of the H-shaped steel, the weight of the crown beam and the first support system, the weight of the breast beam and the second support system, and the frictional resistance of the cemented soil. The bending moment and shearing force of H-shaped steel are mainly affected by such factors as the lateral soil pressure and the concentrated forces of the two support systems. When the foundation pit was excavated to the base, the development of and changes in the law of internal force with regard to the H-shaped steel was analyzed. When the overall internal force of the H-shaped steel is at its maximum, the maximum absolute values in terms of the axial force, bending moment and shearing force are −481 KN, 371 KN·m and 123 KN. In the process of foundation pit excavation and backfilling, the point of contraflexure of the H-shaped steel moves down gradually, and the fixed end of corresponding SMW pile also moves down and stabilizes below the base. These results may provide a reference for the design and construction of SMW piles of building structures in sandy soil areas.

Keywords: high-rise building; sandy soil area; Soil Mixing Wall pile; site measurement; mechanical property; H-shaped steel; internal force

Citation: Xian, Q.; Wang, Z.; Liu, X.; Ma, S.; Xiao, Z. Site Measurement Study on Mechanical Properties of SMW Piles of Building Structures in Sandy Soil Areas. *Buildings* **2022**, *12*, 1733. https://doi.org/10.3390/buildings12101733

Academic Editors: Liqiang Jiang, Jihong Ye and Wei Guo

Received: 10 September 2022
Accepted: 8 October 2022
Published: 19 October 2022

1. Introduction

The SMW (soil mixing wall) pile is a kind of composite support structure which is integrated with soil retaining and seepage prevention after H-shaped steel is inserted into a continuous lapping three-axis cemented soil mixing pile [1,2]. The SWM pile not only has a high strength, high stiffness and good anti-seepage effect, but also has the advantages of a small impact on the surrounding strata and environment, small construction space and speedy construction. After the supporting effect of SMW piles is achieved, the H-shaped steel can be pulled out and recycled according to requirements. This not only saves the steel, but it also achieves sustainable development. SMW piles are very suitable for deep foundation pit engineering in large and medium-sized cities with dense buildings. Therefore, SWM piles have been widely used in soft soil areas such as Jiangsu, Zhejiang, Shanghai and Tianjin. At the same time, there are also many studies on the mechanical characteristics of cemented soil based on soft soil and corresponding SMW piles [3–6].

As SWM piles have many advantages as foundation pit supporting structures, and in order to further promote their application to sandy soil areas such as Henan, SMW piles were adopted in the deep foundation pit supporting project of a high-rise building in Zhengzhou. According to the design requirements of the "Technical Specification for Soil Mixed Wall" (JGJ/T 199-2010) [7], the internal forces, such as the bending moment and shearing force, generated by the lateral soil and water pressure of the foundation pit are borne by the H-shaped steel in the SMW pile, while the cemented soil only plays the role of forming a sealing curtain [8–10].

In order to determine mechanical properties of H-shaped steel in SMW piles in sandy soil areas, the following works were carried out in the process of constructing a foundation pit supporting structure for the high-rise building. Firstly, three SMW piles with specific mechanical behaviors were selected, and the stress distribution for the H-shaped steel was monitored under different construction conditions by setting vibrating chord reinforcement dynamometers on several typical test sections of the H-shaped steel. Then, the measured normal stress of H-shaped steel was decomposed into uniform compressive stress and bending normal stress, and the axial pressure and bending moment of the H-shaped steel were calculated, respectively. The development of and variation in the distribution curves of axial force and bending moment of the H-shaped steel were discussed under different construction conditions, and the influencing factors were analyzed. Finally, the bending moment fitting curve was obtained using the polynomial fitting method, and the distribution curves for the shearing force were obtained under different construction conditions by derivation. The development trends for and influencing factors behind the shearing force curves were further analyzed.

The stress state of the H-shaped steel in the SMW piles was monitored on site, and the axial force, bending moment and shearing force of the H-shaped steel were analyzed and studied. Thus, the mechanical properties of SMW piles could be determined and used as a reliable basis for the design and construction of SMW piles in sandy soil areas.

2. Retaining Structures of the Foundation Pit

Concerning the foundation pit of the high-rise building, its north–south length is 48.04 m, its east–west width is 25.76 m, and the excavation depth is 11.46 m. Three SMW piles were selected for site measurement in the western edge of the pit [11,12]. There are six soil layers within a depth of 39.6 m below ground level at the proposed site. The name and burial depth of each soil layer are as follows. The first layer is miscellaneous fill, with a buried depth of 1.0–3.9 m; the second layer is clayey silt, with a buried depth of 2.1–6.7 m; the third layer is silt, with a buried depth of 4.0–5.7 m; the fourth layer is clayey silt, with a buried depth of 5.8–10.3 m; the fifth layer is fine sand, with a buried depth of 9.3–20.7 m; the sixth layer is medium sand, with a buried depth of 24.2–39.6 m. It can be seen that the six soil layers are mainly sandy soils. The upper five soil layers within the range of the pile length were directly involved in forming the cemented soil of the SMW piles.

The diameter of the SMW piles is 850 mm, their center distance is 600 mm, and there are two kinds of pile lengths of 15.55 m and 18.25 m. The H-shaped steel, H700 × 300 × 13 × 24, with the steel grade of Q235B, was inserted at intervals or partly but densely inserted, and the two H-shaped steel lengths corresponding to the pile lengths were 16.65 m and 19.35 m. Two layers of steel supports were set from top to bottom along SMW piles as inner supports to form the retaining structures of the foundation pit. The first layer of steel supports was set on the concrete crown beam at the top of the piles 1.61 m below the ground. The second layer of steel supports was set on the steel waist beam at 6.11 m below the ground. The steel supports were made of circular steel tubes and specific axial forces were applied in advance to reduce the lateral displacement of the SMW piles towards the foundation pit.

3. Site Measurement Design

3.1. *Selection of Measured SMW Piles*

The following two factors should be considered when selecting SMW piles measured on site. First, the piles should be located at the edge of the foundation pit where the mechanical behavior is relatively clear. Second, the piles corresponding to the support positions should be selected, if at all possible, to investigate the influence of the supports on the mechanical behavior of the piles. Therefore, in the middle of the west edge of the foundation pit, the three piles corresponding to the middle three steel supports were selected and measured. The piles were numbered Z1, Z2 and Z3 consecutively from north to south. The length of the three piles is 18.25 m, and the length of H-shaped steel inside is 19.35 m. There are two piles between Z1 and Z2 and between Z2 and Z3, so as to avoid the influence of mechanical behavior between the measured piles.

3.2. *Layout of Stress Monitoring Points*

To determine the stress of H-shaped steel in SMW piles, it is necessary to use vibrating chord type reinforcement dynamometers and a frequency reading instrument for monitoring. The vibration frequency of the vibrating string is obtained by the frequency reading instrument, and the stress of the measured structure can be obtained via calculations.

The SMW pile can be regarded as a compression-bending member, and the lower end of the pile is inserted into a certain depth below the base, which can be used as the fixed end. The upper end of the pile is connected by the crown beam, and the inner steel supports act on the crown beam, which can be used as the elastic support of the upper end. The middle of the pile is connected by the steel waist beam, and the internal steel supports act on the waist beam, which can be used as the intermediate elastic support. The soil and water pressure from the outside of the foundation pit can be considered as the distributed load acting on the pile, and the dead weight of the pile and the vertical force transmitted to the pile by the supporting system through the crown beam and the waist beam can be considered as the axial compression load. When considering the pile as a compression-bending member, the test sections need to be set in the following parts: the base elevation, which the intermediate steel support is acting on, and the junction of two adjacent soil layers. Additionally, the uniformity in terms of test section distribution should be considered to obtain more monitoring data.

As shown in Figure 1, a total of eight test sections were set along the measured pile from top to bottom, and the distance from each section to the ground is listed in Table 1. Two reinforcement dynamometers were arranged at the connections between one side of the web and the upper and lower flanges of the H-shaped steel in each section, and the measured stresses can be regarded as the stresses at the upper and lower ends of the web. A full length of steel strip was welded at the connection between the web and the flange of the H-shaped steel to protect the dynamometers and corresponding wires. The bottom of the steel strip was sealed to prevent the entry of cemented soil during the insertion of the H-shaped steel.

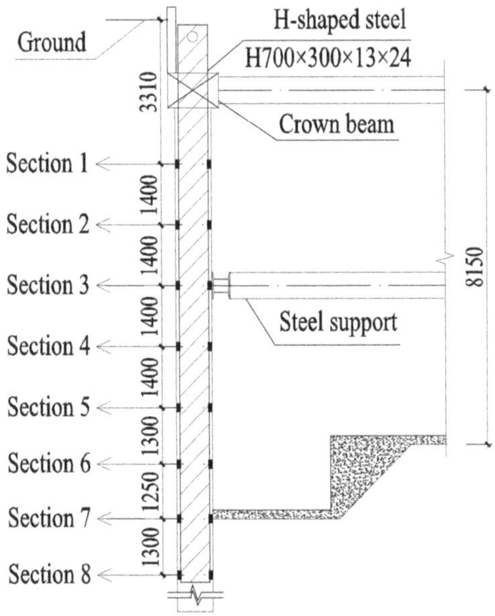

Figure 1. Test section layout.

Table 1. Position of test section.

Test Section	Distance to the Ground (m)
Section 1	3.31
Section 2	4.71
Section 3 (second layer of supports)	6.11
Section 4	7.51
Section 5	8.91
Section 6	10.21
Section 7 (the base)	11.46
Section 8	12.76

3.3. *Main Construction Conditions*

The stress frequency of the H-shaped steel in the measured pile was monitored as follows. The monitoring frequency was once a day during the excavation process of the foundation pit and once every two days during substructure construction and foundation pit backfilling, resulting in the obtainment of a large amount of measured data. In order to facilitate the analysis of the main mechanical characteristics of the piles, the following representative construction conditions were selected for research.

Construction condition 0: The foundation pit is excavated to 2.01 m below the ground and reaches the elevation of the bottom surface of the crown beam. This is the initial loading state when the initial value data for reinforcement dynamometers are collected.

Construction condition 1: On the basis of construction condition 0, pour the reinforced concrete crown beam, and set the first layer of steel supports.

Construction condition 2: Excavate the foundation pit to 6.41 m below the ground, install the steel waist beam, and set the second layer of steel supports.

Construction condition 3: Excavate the foundation pit to 11.46 m below the ground and pour the concrete cushion of the base.

Construction condition 4: Build the foundation pit floor and the internal structure, backfill the foundation pit, and remove the second layer of steel supports.

Construction condition 5: Continue to construct the internal structure and the roof, backfill the foundation pit, and remove the first layer of steel supports.

4. Stress Analysis of H-Shaped Steel in Measured SMW Piles

It can be seen from the analysis in Section 3.2 that the SMW pile is used as a compression-bending member and the internal forces of the H-shaped steel have both axial pressure and bending moment. The normal stress obtained by dynamometers is the superposition of uniform compressive stress corresponding to axial pressure and bending normal stress corresponding to bending moment. Figure 2 shows the stress distribution law for the H-shaped steel in the SMW pile Z1 along the depth of the foundation pit under five construction conditions. In the figure, the abscissa represents the section stress of the H-shaped steel. The tensile stress is positive, while the compressive stress is negative. The ordinate represents the depth of the foundation pit below the ground. The "inner side" in the legend indicates the stress measured by the dynamometer at the connection of the flange and web of the H-shaped steel near the foundation pit, while the "outer side" indicates the stress measured by the dynamometer on the section of the H-shaped steel at a distance from the foundation pit.

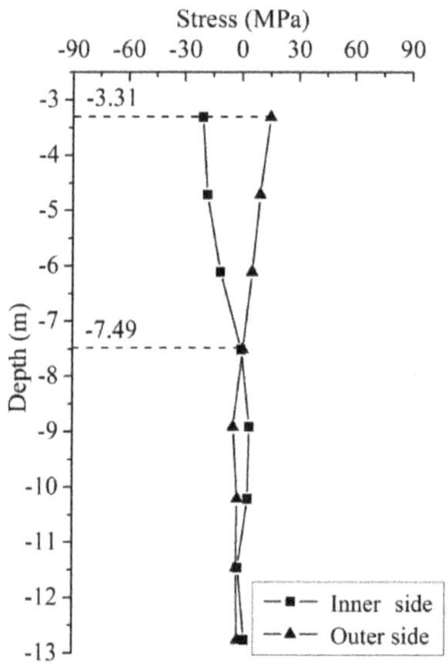

(a)

Figure 2. *Cont.*

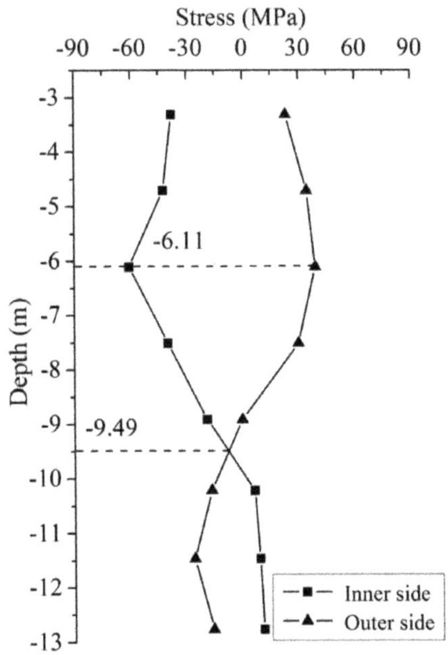

(b)

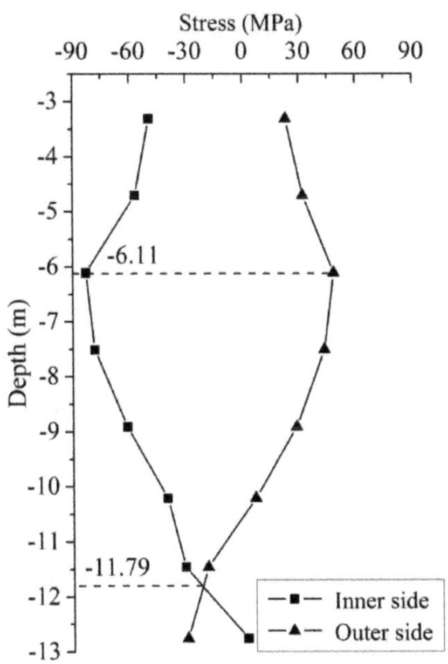

(c)

Figure 2. *Cont.*

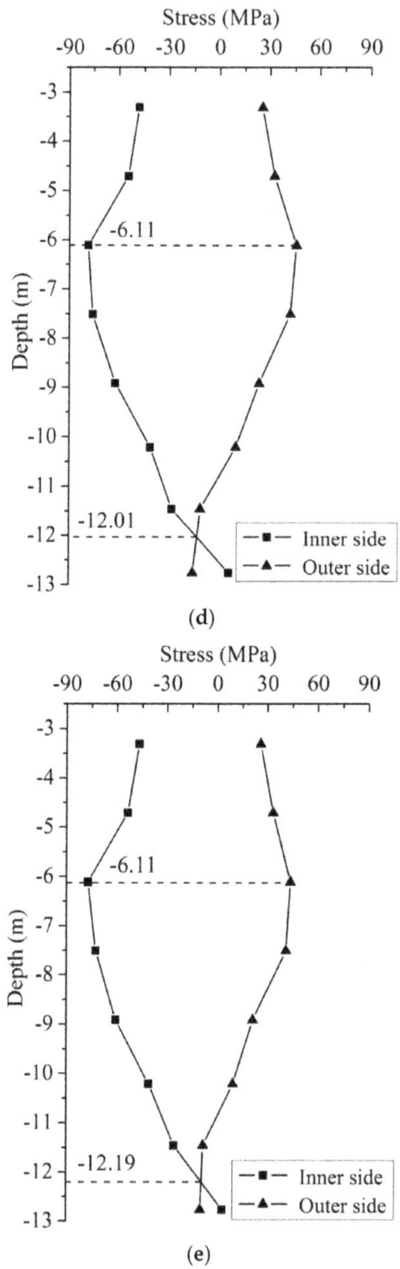

Figure 2. The stress distribution characteristics of the H-shaped steel in the SMW pile Z1 under five construction conditions. (**a**) The stress distribution of the H-shaped steel under construction condition 1; (**b**) the stress distribution of the H-shaped steel under construction condition 2; (**c**) the stress distribution of the H-shaped steel under construction condition 3; (**d**) the stress distribution of the H-shaped steel under construction condition 4; and (**e**) the stress distribution of the H-shaped steel under construction condition 5.

The following stress analysis can be recognized from Figure 2:

(1) Under construction condition 1, the excavation depth of the foundation pit was only −2.01 m, and the first layer of supports was set. The lateral soil pressure was small, so the overall stress of the H-shaped steel was small. The maximum tensile and compressive stresses occurred in Section 1 at −3.31 m, and the stress values were 14.9 MPa and −20.1 MPa, respectively. The reason why the absolute value of compressive stress was larger is that in addition to the normal stress generated by the bending moment, the compressive stress generated by axial pressure was also superimposed on this section. The axial pressure can be considered to be mainly caused by the vertical force transmitted by the first supporting system to the SMW pile through the crown beam. At the section of −7.49 m, which is very close to Section 4 (−7.51 m), the inner and outer stress curves intersect at a point, and the compressive stress value was −0.11 MPa. It indicates that this section only had axial pressure and the bending moment was zero; thus, the section was the contraflexure point of the bending moment of H-shaped steel. According to the bending moment of the H-shaped steel, the outside of the H-shaped steel above this section was tensile, while the inside of the H-shaped steel below this section was tensile. The tensile and compressive stresses of the H-shaped steel below the contraflexure point were close to zero, indicating that this construction condition had little influence on mechanical performance of this part of the SMW pile.

(2) Under construction condition 2, the excavation depth of the foundation pit further increased to −6.41 m, and the second layer of supports was set. The lateral soil pressure was larger, so the overall stress of the H-shaped steel noticeably increased. The maximum tensile and compressive stresses occurred in Section 3 at −6.11 m, and the stress values were 39.3 MPa and −60.8 MPa, respectively. This section corresponds to the position of the second layer of supports, and there is a large difference between the algebraic values for tensile and compressive stresses, indicating that this section was subjected to a large bending moment. This is because the bending moment diagram of the H-shaped steel produced a turning point and the bending moment reached an extreme value under the action of the pre-added axial pressure of the second layer of supports. At the section −9.49 m, the inner and outer stress curves intersected, and the compressive stress value was −7.67 MPa, indicating that this section was the contraflexure point of the bending moment of the H-shaped steel. The increase in compressive stress indicates that the axial pressure of the H-shaped steel increased. This was caused by the vertical force further transmitted to the SMW pile through the waist beam by the second layer of the support system. Below the contraflexure point, there was still a difference between the tensile and compressive stresses, indicating that the H-shaped steel still bore a bending moment.

(3) Under construction condition 3, the excavation depth of the foundation pit increased to −11.46 m, the base was reached, and the concrete cushion of the base was poured. The lateral soil pressure was at its greatest at this stage, so the overall stress on the H-shaped steel was at its greatest. The greatest values in terms of tensile and compressive stresses also appeared in Section 3 at −6.11 m, with these being 48.6 MPa and −82.3 MPa, respectively, corresponding to the position of the second layer of supports. At this time, the difference between the tensile and compressive stresses was at its most significant, indicating that the bending moment of this section was also the largest. At the base section of −11.46 m, the outer and inner compressive stresses were −18.38 MPa and −29.97 MPa, respectively. The two stress values were different, but the difference was not large, indicating that the H-shaped steel also bore a bending moment, with the bending moment value being small. The contraflexure point appeared at the section −11.79 m, slightly downward, and the H-shaped steel below the contraflexure point still bore a bending moment. This part of the SMW pile can be used as the embedded end.

(4) Under construction conditions 4 and 5, the internal structure and the backfilling of the foundation pit were in the process of being constructed. Due to the offsetting

part of the soil and water pressure outside of the foundation pit, the shape of the inner and outer stress curves of the H-shaped steel was similar to that for construction condition 3, but the overall stress on the H-shaped steel was reduced. The maximum tensile and compressive stresses under construction condition 4 were 45.86 MPa and −78.01 MPa, respectively, which are lower than those under construction condition 3. The contraflexure point appeared at the section −12.01 m below the base, which is further down when compared with construction condition 3. The maximum tensile and compressive stresses under construction condition 5 were 43.76 MPa and −76.99 MPa, respectively, which show a continued decrease from the basis of construction condition 4. The contraflexure point appeared at the section −12.19 m below the base, which is further down when compared with construction condition 4.

A comparison of the stress distribution curves for the H-shaped steel in the SMW pile Z1 under five construction conditions in Figure 2 is shown in Figure 3. The comparative analysis shows that:

(1) Construction conditions 1 to 3 correspond to foundation pit excavation, and the overall stress of the H-shaped steel increased gradually under these conditions. With the exception of when the first layer of supports was set, the maximum tensile and compressive stresses of the H-shaped steel under construction condition 1 appeared at the uppermost Section 1. After setting the second layer of supports, the maximum tensile and compressive stresses of the H-shaped steel under construction conditions 2 and 3 appeared at Section 3, corresponding to the position of the second layer of supports.

(2) The excavation depths of the foundation pit from construction conditions 1 to 3 were 2.01 m, 6.41 m and 11.46 m (the base), respectively, and the corresponding contraflexure point depths of the H-shaped steel were −7.49 m, −9.49 m and −11.79 m, respectively. The contraflexure points were all below the excavation face, and it can be considered that the embedded end of the SMW pile gradually moved below the base.

(3) The SMW pile can be regarded as a compression-bending member. Taking construction condition 3 as an example, the bending moment at the base section was very small, and the SMW pile below the contraflexure point can be regarded as the embedded end. The first layer of supports can be used as the elastic support of the upper end. Because this section is still below the ground, there is a bending moment, but the value of that bending moment is not large. The second layer of supports can be used as an intermediate elastic support, and the pre-added axial pressure of the supports makes the bending moment produce an extreme value, which is also the maximum bending moment value. Even if the effect of axial pressure is further considered, the H-shaped steel at this section still produces maximum tensile and compressive stresses. The maximum tensile and compressive stresses were 48.6 MPa and −82.3 MPa, respectively. The steel grade used for the H-shaped steel was Q235 B. The maximum stress value was far less than the designed strength value for the steel, and the safety of the SMW pile met the requirements.

(4) Construction conditions 4 to 5 correspond to the construction of the internal structure and the backfilling of the foundation pit. Compared with construction condition 3, under these conditions, the overall stress of the H-shaped steel was reduced, the maximum tensile and compressive stresses were gradually reduce and the position of the contraflexure point was still within a small range below the base and gradually moved down, indicating that this stage had little influence on the mechanical performance of the SMW pile.

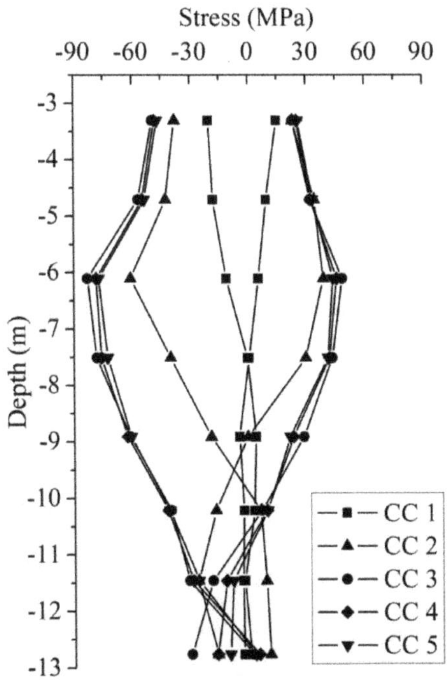

Figure 3. A comparison of the stress distribution of the H-shaped steel in the SMW pile Z1 under five construction conditions (construction condition is abbreviated as "CC").

The stress distribution curves of the H-shaped steel of the SMW piles Z2 and Z3 are approximately the same as that of the SMW pile Z1, and the mechanical properties are approximately the same. The corresponding curves are no longer given here.

5. Axial Force Analysis of H-Shaped Steel in SMW Piles

Since the SMW pile can be regarded as a compression-bending member, the measured normal stress of the H-shaped steel in Figure 3 can be decomposed into uniform compressive stress and bending normal stress. The axial pressure of the H-shaped steel can be calculated from the uniform compressive stress, and the bending moment can be calculated from the bending normal stress.

Figure 4 shows the calculation results for the axial forces of the H-shaped steel in the SMW pile Z1 under five construction conditions. The abscissa represents the axial force of the H-shaped steel, which was positive under tension and negative under compression. The ordinate represents the depth of the foundation pit. Each construction condition corresponds to an axial force distribution curve.

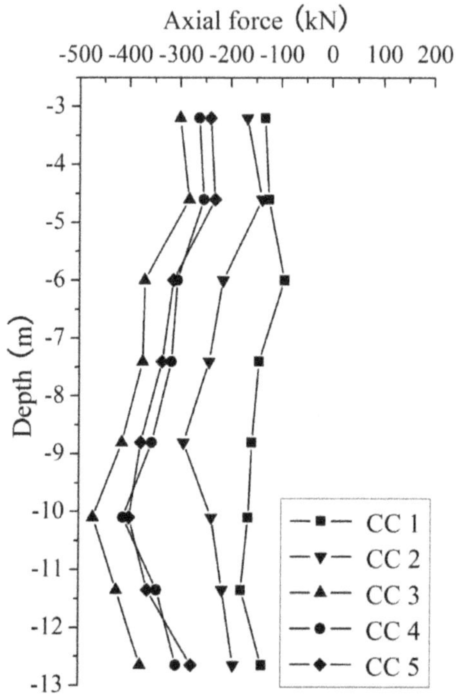

Figure 4. The axial force distribution curves of the H-shaped steel in the SMW pile Z1 under five construction conditions (construction condition is abbreviated "CC").

According to the analysis in Figure 4, it can be seen that:

(1) Under construction condition 1, the foundation pit was excavated to −2.01 m, the reinforced concrete crown beam was poured and the first layer of steel supports was set. All test sections, including Section 1 at −3.31 m, were below the first layer of supports. The axial pressure of the H-shaped steel was mainly generated by the dead weight of the SMW pile, crown beam and the first layer of supports. Since the dead weight of crown beam and supports was constant, and the dead weight of the SMW pile increased gradually with the increase in depth, the axial pressure had a trend of increasing gradually. The maximum axial pressure occurred in Section 7 at −11.46 m, and the axial force value was −176 KN.

(2) Under construction condition 2, the foundation pit was excavated to −6.41 m, a steel waist beam was installed and a second layer of steel supports was set. Section 3 and all the following test sections were below the second layer of supports. As the cemented soil at one side of the foundation pit was removed, the corresponding frictional resistance decreased. The axial pressure of Section 1 (−3.31 m) and Section 2 (−4.71 m) increased compared with that of construction condition 1. Due to the dead weight of the waist beam and corresponding supports, the axial pressure of Section 3 (−6.11 m) suddenly increased compared with that of Section 2. The axial pressure in the following sections gradually increased, and the maximum axial pressure occurred in Section 5 at −8.91 m, with the axial force value being −293 KN. The soil buried at a depth of 9.3 m to 13.6 m is fine sand. The frictional resistance of cemented soil formed by fine sand is relatively large, resulting in a decrease in the axial pressure of the H-shaped steel in the range of Section 6 (−10.21 m) to Section 8 (−12.76 m).

(3) Under construction condition 3, the foundation pit was excavated to −11.46 m, the base was reached and the concrete cushion was poured. The test sections, including

Section 7 and Section 8, were below the base. The frictional resistance provided by the cemented soil greatly reduced because all the cemented soil of the H-shaped steel at the side of the foundation pit was removed. Compared with construction condition 2, the axial pressure of all sections of H-shaped stee noticeably increased. The axial force of the H-shaped steel gradually increased with an increase of depth. The maximum axial pressure appeared in Section 6 at −10.21 m, and the axial force value was −481 kN. Due to the large frictional resistance provided by the cemented soil formed based on fine sand in the range of Section 7 (−11.46 m) to Section 8 (−12.76 m), the corresponding axial pressure of the H-shaped steel decreased.

(4) Construction conditions 4 and 5 correspond to the construction of the internal structure and the backfilling of the foundation pit. The second layer of supports and the first layer of supports were removed. The original weight of the two layers of supports passed through the waist beam and crown beam was eliminated. Compared with construction condition 3, the distribution law of axial force with depth remained unchanged, while the axial pressure value decreased to a certain extent. The maximum axial pressure occurred in Section 6 at −10.21 m under construction condition 4, and the axial force value was −413 kN. The change in the axial force of the H-shaped steel under construction conditions 4 and 5 was not obvious.

6. Bending Moment Analysis of H-Shaped Steel in SMW Piles

Based on the bending normal stress obtained from the measured normal stress decomposition in Figure 3, the calculation results for the bending moment of the H-shaped steel in the SMW pile Z1 under five construction conditions are given in Figure 5. The abscissa represents the bending moment of the H-shaped steel, which is positive when the flange of H-shaped steel away from the foundation pit is under tension. The ordinate represents the depth of the foundation pit. Each construction condition corresponds to a bending moment distribution curve.

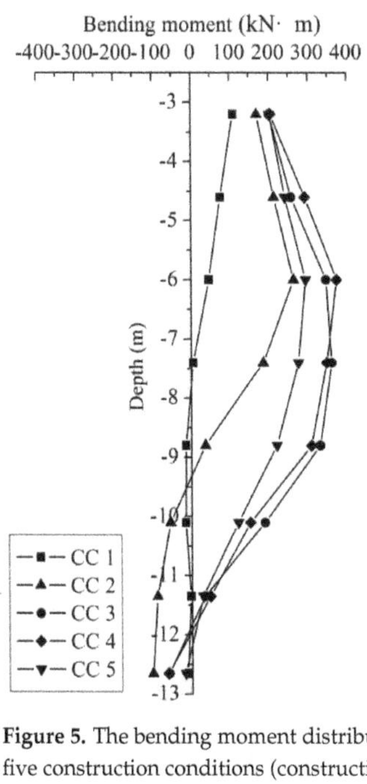

Figure 5. The bending moment distribution curves of the H-shaped steel in the SMW pile Z1 under five construction conditions (construction condition is abbreviated "CC").

According to the analysis in Figure 5, it can be seen that:

(1) Under construction condition 1, the excavation depth of the foundation pit was not large and only the first layer of supports was set. The corresponding lateral soil pressure was small, resulting in a small bending moment in the H-shaped steel along the height direction. The maximum positive bending moment occurred in Section 1 (−3.31 m), and the bending moment value was 111 KN·m. It can be considered that the H-shaped steel bent to the side of the foundation pit, and the bending moment of H-shaped steel below this section gradually decreased. The bending moment distribution curve intersected the vertical axis at a point adjacent to Section 4 (−7.51 m). The bending moment here was zero, which can be regarded as the contraflexure point of the H-shaped steel. In the portion below this point, the bending moment made the H-shaped steel bend to the outside of the foundation pit, and the bending moment value was close to zero, indicating that this construction condition had little influence on mechanical performance of this portion.
(2) Under construction condition 2, the excavation depth of the foundation pit was increased and the second layer of supports was further set. The corresponding lateral soil pressure increased, and the bending moment of the H-shaped steel noticeably increased. According to the distribution state of the bending moment curve, most of the upper range of the H-shaped steel bore a positive bending moment, and the H-shaped steel continued to bend to the inner side of the foundation pit. The maximum positive bending moment occurred in Section 3 (−6.11 m), with bending moment value of 261 KN·m. This is where the second layer of supports was set. The large axial pressure pre-added in the supports was the equivalent to a large, concentrated force applied to the H-shaped steel. The bending moment curve bent at this section, and the bending moment value reached its extreme value, which was the maximum value. At the depth of −9.49 m, the bending moment curve also intersected the vertical axis at a point. Here, the bending moment was zero, which was the contraflexure point of the H-shaped steel. Below this section, the bending moment caused the H-shaped steel to be strained at the inner side of the foundation pit. The H-shaped steel still bore a bending moment, and the bending moment gradually increased. At the bottom Section 8 (−12.76 m), the bending moment value reached −101 KN·m.
(3) Under construction condition 3, the foundation pit was excavated to the base. The lateral soil pressure reached its maximum, and the bending moment of the H-shaped steel also reached its maximum. In most of the above ranges, the H-shaped steel bears positive bending moments, and the bending moment values were large. The maximum positive bending moment still occurred in Section 3 (−6.11 m), where the second layer of supports is set, with a bending moment value of 371 KN·m. At the base, Section 7 (−11.46 m), the H-shaped steel still bore a positive bending moment, which was 49 KN·m. At its lower depth of −11.79 m, the bending moment curve intersected the longitudinal axis and the bending moment value was zero. This was the contraflexure point of H-shaped steel, and there was a certain of negative bending moment in the lower portion of the H-shaped steel.
(4) Under construction conditions 4 and 5, two layers of supports were removed successively from bottom to top in the process of foundation pit backfilling. The shape of the bending moment distribution curve is practically the same as that of construction condition 3, but the overall bending moment of the H-shaped steel decreased gradually. The maximum positive bending moments under construction conditions 4 and 5 were 340 KN·m and 287 KN·m, respectively, which were further reduced compared with construction condition 3. The positions of the contraflexure points under construction conditions 4 and 5 appeared at −12.01 m and −12.19 m below the base, which further gradually moved down compared with construction condition 3.
(5) From construction conditions 1 to 5, the positions of the contraflexure points of the H-shaped steel are −7.49 m, −9.49 m, −11.79 m, −12.01 m and −12.19 m, respectively. It can be seen that during the whole construction process, from foundation pit excavation to backfilling, the position of the contraflexure point gradually moves down. It can

be considered that the embedded end of the SMW pile gradually moves down and stabilizes below the base.

7. Shearing Force Analysis of H-Shaped Steel in SMW Piles

The axial force and bending moment of the H-shaped steel in the SMW pile Z1 are analyzed above. In order to further analyze the shearing force of the H-shaped steel, the distribution curve for the shearing force along the depth of foundation pit should be obtained first. According to the relationship between the bending moment and shearing force of the bending member in material mechanics, the derivative of the bending moment to the position coordinate of the member section is equal to the shearing force. In other words, the shearing force distribution curve can be obtained from the slope of the bending moment distribution curve. Figure 5 just shows the line between the bending moments of each test section under each construction condition. In order to facilitate the derivation, the fitting curve of each bending moment curve should be obtained.

Here, the method of polynomial fitting was adopted to determine the polynomial whose highest degree term is of degree 5 and whose form is $M(x) = ax^5 + bx^4 + cx^3 + dx^2 + ex + f$. Each bending moment curve shown in Figure 5 was fitted to obtain the bending moment fitting curve under each construction condition. By taking the derivative of each bending moment fitting curve, the shearing force distribution curve under each construction condition could be obtained. Figure 6 shows the shearing force distribution curves of the H-shaped steel in the SMW pile Z1 under five construction conditions. The abscissa represents the shearing force of the H-shaped steel, which was positive in the clockwise direction determined from the angle of view in Figure 1. The ordinate represents the depth of the foundation pit.

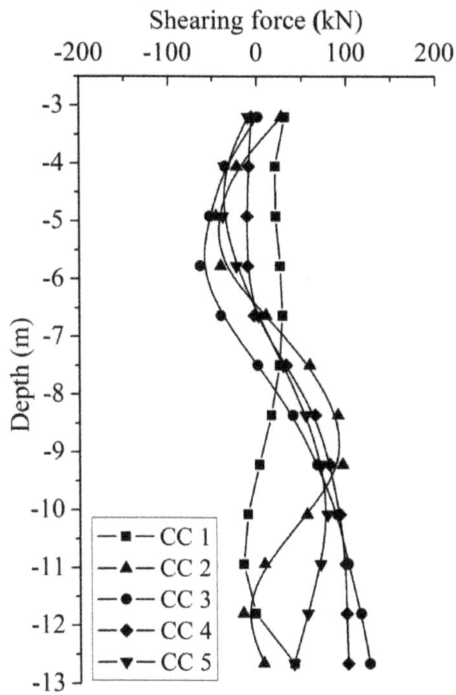

Figure 6. The shearing force distribution curves of the H-shaped steel in the SMW pile Z1 under five construction conditions (construction condition is abbreviated "CC").

According to the analysis in Figure 6, it can be seen that:

(1) Under construction condition 1, the first layer of supports was set and the overall shearing force of the H-shaped steel was small. The maximum positive shearing force occurred in Section 1 (−3.31 m), with a value of 32 KN. The shearing force below this section underwent changes, with the overall trend gradually decreasing. At a depth of −9.25 m, the shearing force decreased to zero. From this depth down, the shearing force was very small and changed little.

(2) Under construction condition 2, the second layer of supports was set and the shearing force of the H-shaped steel increased significantly. The upper portion of the H-shaped steel mainly bore the negative shearing force. The negative shearing force with the largest absolute value occurred at the depth of −5.25 m, with the value of −48 KN. The shearing force value was zero at the depth of −6.45 m, which is just near the position of the second layer of support (−6.11 m). Due to the concentrated force provided by the supports, the shape of the shearing force curve changed, with indications that the shearing force value changed both positively and negatively. If an ideal concentrated force is applied, the shearing force curve should be abrupt here. Considering that the shearing force curve is obtained by taking the derivative of the bending moment fitting curve, the shearing force curve slowly changed from the above negative shearing force to the below positive shearing force. The lower portion of the H-shaped steel mainly bore the positive shearing force. The maximum positive shearing force occurred at the depth of −8.82 m, with a value of 81 KN.

(3) Under construction condition 3, the foundation pit was excavated to the base and the shearing force of H-shaped steel reached its maximum. The upper portion of the H-shaped steel bore the negative shearing force. Compared with construction condition 2, the shearing force further increased. The negative shearing force with the largest absolute value occurred at the depth of −5.68 m, with a value of −63 KN. This is also the negative shearing force with the largest absolute value among the five construction conditions. The shearing force value was zero at the depth of −7.23 m, which is also near the position of the second layer of supports. The shearing force curve slowly changed, and the shearing force value changed from the above negative value to the below positive value. The lower portion of the H-shaped steel bore the positive shearing force, and the shearing force increased as the depth increased. In the bottom Section 8 (−12.76 m), the shearing force reached its maximum, with a value of 123 KN. This is the maximum positive shearing force among five construction conditions and also the shearing force with the largest absolute value.

(4) Under construction conditions 4 and 5, the second supporting system and the first supporting system were successively removed along the foundation pit depth from bottom to top. Compared with construction condition 3, the overall shearing force of the H-shaped steel was significantly reduced. The upper portion of the H-shaped steel still bore negative shearing force. The negative shearing force with the largest absolute value was −41 KN. The lower portion of the H-shaped steel still bore positive shearing force. The maximum positive shearing force was 99 KN.

8. Conclusions

Three SMW piles in the foundation pit supporting the structure of a high-rise building in Zhengzhou were monitored on site. The stress distribution law of the H-shaped steel under significant construction conditions was obtained. On this basis, the internal forces of the H-shaped steel, such as its axial force, bending moment and shearing force, were calculated and analyzed in detail. The mechanical properties of the SMW piles in sandy soil areas were discussed. The conclusions are as follows:

(1) Three SMW piles with clear mechanical properties were selected on site. Reinforcement dynamometers were arranged in several typical test sections along the H-shaped steel. The stress distribution state of the H-shaped steel with foundation pit depth and the development of and changes in the related laws according to the construction

process were monitored. These works laid the foundation for the subsequent internal force analysis of H-shaped steel in an SMW pile.

(2) The axial force of the H-shaped steel is mainly caused by the dead weight of the SMW pile, the dead weight of the crown beam and the first supporting system, the dead weight of waist beam and the second supporting system and the frictional resistance of the cemented soil, etc. Under several construction conditions, the H-shaped steel only bears axial pressure, which varies with the depth of foundation pit, the setting and removal of the two supporting systems and the frictional resistance of the cemented soil. The axial pressure generally increases with the increase in foundation pit depth under the influence of various factors. The maximum axial pressure appeared in Section 6 (-10.21 m) under construction condition 3, with an axial force value of -481 kN. The frictional resistance of cemented soil based on fine sand near the base is large and the axial pressure is small.

(3) The bending moment of the H-shaped steel is mainly caused by the lateral soil pressure and the concentrated forces provided by the first supporting system and the second supporting system. Under construction condition 3, the overall bending moment reached its maximum, and the upper most portion of the H-shaped steel bore positive bending moment. At the same time, the portion of the SMW pile below the contraflexure point slightly below the base can be regarded as the embedded end, which this bearing a small bending moment at the base section. The first supporting system can be used as the elastic support of the upper end, and there is still a bending moment at the corresponding section. The second supporting system can be used as the intermediate elastic support, with a maximum positive bending moment occurring in the corresponding section, with the value of 371 KN·m.

(4) With an increase in the excavation depth of the foundation pit, the contraflexure point of the bending moment of the H-shaped steel in the SMW pile gradually moves down. The contraflexure points are all below the excavation face. In the process of foundation pit backfilling, the contraflexure point continues to move down slightly. It can be considered that in the whole construction process, the embedded end of the SMW pile gradually moves down and stabilizes below the base.

(5) The shearing force of the H-shaped steel is mainly affected by the lateral soil pressure and the concentrated forces of the first and second supporting systems. The overall shearing force reaches its maximum under construction condition 3. At the same time, the upper portion of H-shaped steel bears a negative shearing force, which its largest absolute value being -63 KN. The lower portion of the H-shaped steel bears a positive shearing force, which produces a maximum positive shearing force value of 123 KN. Near the middle second supporting system, the shearing force is zero, the value changes from positive to negative and the shearing force curve is in a transition stage.

Author Contributions: Conceptualization, Q.X.; methodology, Q.X.; investigation, Z.W. and X.L.; resources, S.M.; data curation, Z.W. and X.L.; writing—original draft preparation, Q.X.; writing—review and editing, Z.X.; supervision, S.M.; project administration, S.M.; funding acquisition, Z.X. All authors have read and agreed to the published version of the manuscript.

Funding: This research was funded by the Natural Science Foundation of China, the Cultivation Programme for Young Backbone Teachers in Henan University of Technology and the Fundamental Research Funds for the Henan Provincial Colleges and Universities in Henan University of Technology through the grant Nos. 51708181, 21420045 and 2018RCJH11, respectively.

Conflicts of Interest: The authors declare no conflict of interest.

References

1. Russell, A.R.; Chapman, M.; Teh, S.H.; Wiedmann, T. Cost and Embodied Carbon Reductions in Cutter Soil Mix Walls through Fibre Reinforcement. *Geosynth. Int.* **2017**, *24*, 280–292. [CrossRef]
2. Zhang, J.W.; Chen, Y.L. Experimental Study on Mitigations of Seismic Settlement and Tilting of Structures by Adopting Improved Soil Slab and Soil Mixing Walls. *Sustainability* **2018**, *10*, 4069. [CrossRef]
3. Fatahi, B.; Khabbaz, H.; Fatahi, B. Mechanical Characteristics of Soft Clay Treated with Fibre and Cement. *Geosynth. Int.* **2012**, *19*, 252–262. [CrossRef]
4. Yi, Y.; Liu, S.; Puppala, A.J. Laboratory Modelling of T-shaped Soil-cement Column for Soft Ground Treatment under Embankment. *Geotechnique* **2016**, *66*, 85–89. [CrossRef]
5. Rashid, A.S.A.; Kueh, A.B.H.; Mohamad, H. Behaviour of Soft Soil Improved by Floating Soil-cement Columns. *Int. J. Phys. Model. Geotech.* **2018**, *18*, 95–116. [CrossRef]
6. Ter-Martirosyan, A.; Sidorov, V.; Sobolev, E. Dynamic Properties of Soil Cements for Numerical Modelling of the Foundation's Basis Transformed under the Technology of Deep Soil Mixing: A Determination Method. *Buildings* **2022**, *12*, 1028. [CrossRef]
7. Ministry of Housing and Urban-Rural Development of the People's Republic of China. *Technical Specification for Soil Mixed Wall: JGJ/T 199-2010*; China Architecture & Building Press: Beijing, China, 2010. (In Chinese)
8. Ministry of Housing and Urban-Rural Development of the People's Republic of China. *Technical Code for Testing of Building Foundation Piles: JGJ 106-2014*; China Architecture & Building Press: Beijing, China, 2014. (In Chinese)
9. Liu, Y.; Hu, J.; Li, Y.P.; Li, L.H. Statistical Evaluation of the Overall Strength of a Soil-cement Column under Axial Compression. *Constr. Build. Mater.* **2017**, *132*, 51–60. [CrossRef]
10. Shu, B.; Gong, H.; Chen, S.; Ren, Y.; Li, Y.; Yang, T.; Zeng, G.; Zhou, M.; Barbieri, D.M.; Li, Y. Case Study of Solid Waste Based Soft Soil Solidifying Materials Applied in Deep Mixing Pile. *Buildings* **2022**, *12*, 1193. [CrossRef]
11. Ministry of Housing and Urban-Rural Development of the People's Republic of China. *Technical Specification for Retaining and Protection of Building Foundation Excavations: JGJ 120-2012*; China Architecture & Building Press: Beijing, China, 2012. (In Chinese)
12. Hasheminezhad, A.; Bahadori, H. Seismic Response of Shallow Foundations over Liquefiable Soils Improved by Deep Soil Mixing Columns. *Comput. Geotech.* **2019**, *110*, 251–273. [CrossRef]

buildings

MDPI

Article

Seismic Performance of Panel Connectors with Steel Frame Based on Autoclaved Lightweight Concrete (ALC)

Chikun Zhang [1,2], Kewei Ding [1,*] and Shulin He [1]

1 College of Civil Engineering, Anhui Jianzhu University, Hefei 230601, China; chikzhang4-c@my.cityu.edu.hk (C.Z.); heshulin4@gmail.com (S.H.)
2 Architecture and Civil Engineering, City University of Hong Kong, Hong Kong 999077, China
* Correspondence: dingkw@ahjzu.edu.cn

Abstract: This paper proposes new prefabricated autoclaved lightweight concrete (ALC) connectors which allow limited sliding. The seismic performance of the proposed connectors and a traditional connector (L-hooked bolt) were investigated in depth. Three specimens of full-scale steel frames with ALC panel and connectors were subjected to horizontal low-cyclic loading. The failure modes, hysteretic behavior, strength and deterioration of stiffness and energy-dissipating performance of all specimens were analyzed and discussed. The experimental results indicated that the frames of the new connectors were more reliable than the traditional connector. The energy-dissipating capacity of the specimen frames with new connectors increased by 23.0% and 24.4%, and deterioration of stiffness increased by 10.6% and 13.0%. The new ALC panel connectors not only increased elastic stiffness in the early stage, but also provided some stiffness in the elastoplastic and plastic stages. Using relevant construction measures, the frames with new connectors displayed good cooperative behavior and safety reliability. To summarize, the new ALC connectors tested showed excellent performance in resisting seismic loading and had good assembly efficiency and could provide a basis for the development and application of a new type of ALC steel frame connector.

Keywords: prefabricated buildings; steel structures; autoclaved lightweight concrete (ALC) panel; seismic behavior; connector type

Citation: Zhang, C.; Ding, K.; He, S. Seismic Performance of Panel Connectors with Steel Frame Based on Autoclaved Lightweight Concrete (ALC). *Buildings* **2022**, *12*, 372. https://doi.org/10.3390/buildings12030372

Academic Editors: Liqiang Jiang, Jihong Ye and Wei Guo

Received: 28 January 2022
Accepted: 15 March 2022
Published: 17 March 2022

1. Introduction

Prefabricated structure is a form of modern industrial building structure. Compared to cast-in-place structure, it has advantages of convenient construction, less wet work on site, lower cost, energy saving and environmental protection [1–3]. With the rapid development of building industrialization, prefabricated reinforcement structures have been widely used in many countries [4]. Autoclaved lightweight concrete (ALC) panel has advantages of light weight, heat insulation, sound insulation, fire protection, energy saving and environmental protection, and easier curing than traditional concrete, and has been widely used in prefabricated reinforcement structures [5–8]. Therefore, the use of ALC panel in the enclosure walls of steel structure buildings is increasing by the day. At present, research on ALC panels mainly focuses on their performance, including the stress of the panel with a multi-layer H-shaped light steel structure. Qu carried out structural testing and numerical simulation of four ALC panels, using ABAQUS to simulate the test panels [9]. Zhang Guowei carried out four-point loading tests on ALC panels, finding that ALC panel damage occurred mainly in the 45° zone, near the joints, in panel-panel and frame-panel areas [10]. De Matteis and Llandolfo carried out numerical simulation of a steel frame structure with cladding light wall panels. The results showed that steel consumption in low earthquake zones could be reduced by 20%, while ensuring stability, by replacing the support system with wall panels [11]. The connector between the ALC panel and the main structure is a critical part in the steel structure; its performance directly affects the stiffness, stability and bearing capacity of the structural system [12,13]. Wang Bo conducted low-cycle repeated loading tests of ALC panel or block-filled CFST frame structures. It was demonstrated that u-shaped steel clamps, swing joints and angles were necessary to ensure that frames could work well under seismic action (in accordance with the requirement of the plastic limit displacement angle: 1/50) [14]. In the process of using the wall panel and the steel frame cladding connector, not only should the structure be firmly connected, but the deformation of the two parts should also be coordinated under several kinds of impact. If the connector is broken during an earthquake, the wall panel will fall off and cause serious secondary disasters [15].

In China, there is little research on the connection of cladding panels—the main connection method used is the hook bolt. Although the hook bolt is easy to assemble, and is seismically resistant when the panel is not deformed greatly, the bolts are connected to a steel angle which is welded to the steel beams through an opening that penetrates the wall. During long periods of suspension or seismic loading, the hole will be damaged, which will reduce the connection between the panels and even cause the panels to fall off [16,17]. Although the Nanjing Xujian Company innovated a kind of wall panel connection connector for NALC panels, it requires special processing in the factory, and mortar needs to be poured on site—in contradiction to the prefabricated building approach currently being promoted [18]. Cao Shi et al. evaluated a new type of ALC connector, which consists of steel angle, stiffening floor, and embedded parts of the wall panel. The experimental results showed that the new type of node had good mechanical properties and broad potential practical application [19]. The embedded node compensates for the defect of the bolt connection of the hook head to some extent, but its structure is intricate, field installation speed is slow, and cost is high. In addition, errors in production, construction and transportation may lead to size changes, resulting in insufficient precision, preventing utilization [20,21].

In order to solve the above problems, based on our previous work, two new types of ALC connectors were proposed. In these, the ALC panel is supported from the bottom by a new connector instead of being hooked by a traditional L-hooked bolt; at the same time, the sliding holes are reserved to offset the seismic load. This kind of connector overcomes some of the defects of the existing connector and has advantages of no wet work, simple structure, convenient construction, high construction speed and good seismic performance. To guarantee the reliability of the connector, experimental investigations of full-scale steel frame models incorporating two panel connection types subjected to low cyclic reversed load test were conducted.

2. Details of Research

2.1. Connector Design

When a building structure has a large lateral shift, damage to panels can be reduced depending on their sway which is driven by the node. Therefore, two new types of connectors were designed to enable the ALC panel and steel frame to connect stably to form an integrated structural system, such that deformation due to external forces can be managed. The design of the connectors followed the design concept of "keep stable in small earthquakes, can be repaired after medium earthquakes, and no collapse in a large earthquake" [22]. The traditional L-hooked bolt is connected by angles which are as long as a steel beam, so high temperature in one place quickly affects the functioning of the whole ALC wall. This creates high risk of fire. However, the new connector is an independent component applied to the whole ALC wall. This means that, in a fire environment, even if one part suffers from high temperature, it will not immediately affect the working condition of other connections.

The new ALC crossing panel connector combines load-bearing holes and limitation holes to form a cross-shaped connector. It includes an I-beam, an ALC panel, a cross-shaped connector, a bolt and a nut. The new ALC pendulous Z-panel connector can be divided into an upper node and a lower node. An I-beam, ALC panel, full-length steel angle, common bolts and nuts, and a Z-node panel are the main components.

In order to facilitate installation, a slotted hole is opened in the upper part of the connector as the lower load-bearing node of the upper exterior wall panel, which bears the vertical, outward horizontal and inward horizontal loads of the upper exterior wall panel; a long circular hole is opened in the lower part of the connector in the horizontal direction, with a hole diameter slightly larger than the diameter of the bolt; the type of upper bolt hole is a long circle hole, which allows a sliding displacement of 75 mm. As the upper limit node of the lower exterior wall panel no longer bears the load in the inward horizontal direction, it can enable the panel and the main structure to work together. The ALC panel connectors are shown in Figure 1.

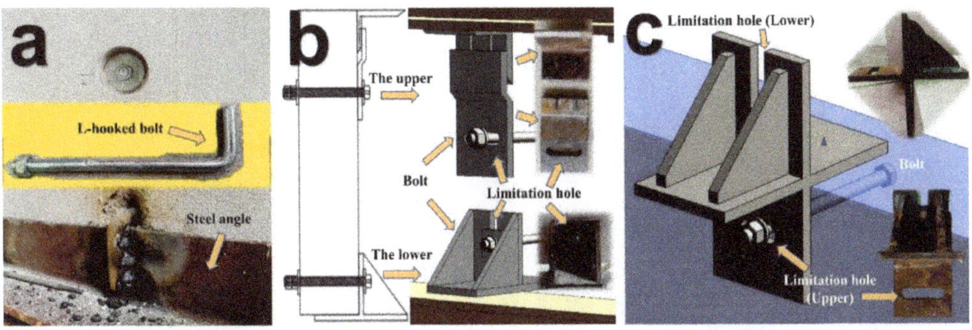

Figure 1. ALC panel connectors: (**a**) L-hooked bolt; (**b**) Pendulous Z-panel connector; (**c**) Crossing panel connector.

2.2. Experimental Design

Full-sized steel frames were selected for the low cyclic reversed load tests. According to the standards for residential building height, column span, beam and column section size, the final design for a story height of 3000 mm, requires a 3400 mm single span plane steel frame. In this test, there were three sets of specimens connected to the panel and the main frame, respectively, as shown in Table 1.

Table 1. Information about the test specimens.

Specimen	Connector Type
SFW1	L-hooked bolt
SFW2	Crossing panel connector
SFW3	Pendulous Z-panel connector

All specimens were prefabricated at the factory according to dimensional requirements. The panels were ALC panels, of dimensions, 600 mm wide, 200 mm thick and 3000 mm high. The wall was made of five external ALC panels, and the seams of the panels were made of ALC panel special patchwork binder. The length of the L-hooked bolt was 200 mm and the specification was M12. The beams were designed as HM244 mm × 175 mm × 7 mm × 11 mm, beam length 3800 mm; the column was designed as HW200 mm × 200 mm × 8 mm × 12 mm, column height: 3888 mm. The connectors were Q345, and the other steel members were Q235 hot-rolled H-beams; the ALC panels were A3.5 B05 standard [23]. After all the specimens were produced and maintained, they were then delivered to and assembled on the test site. The setup process and overview of the test specimens are shown in Figures 2 and 3 below.

Figure 2. Assembling process: (**a**) Construction of steel frame; (**b**) Connectors are welded on beam; (**c**) ALC panels are assembled in order; (**d**) Bolts are connected to the connectors; (**e**) Special patchwork binder filled in the seam between ALC panels; (**f**) MTS device is connected to column.

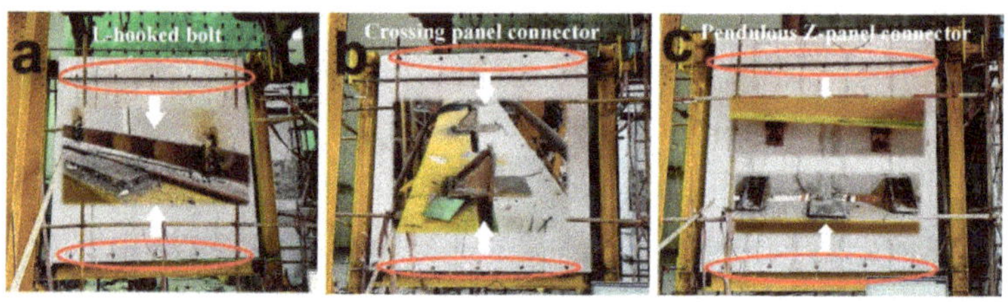

Figure 3. Overview of the test specimens: (**a**) SFW1; (**b**) SFW2; (**c**) SFW3.

2.3. Experimental Facility and Setup

In this test, an American MTS servo-loading system with actuator displacement stroke of 500 mm was used. In this experiment, a quasi-static loading scheme was adopted in which a horizontal, low-cycle reciprocating load was applied to the top of the column, and the top panel of the column was connected with the hydraulic servo actuator. The test loading device is shown in Figure 4.

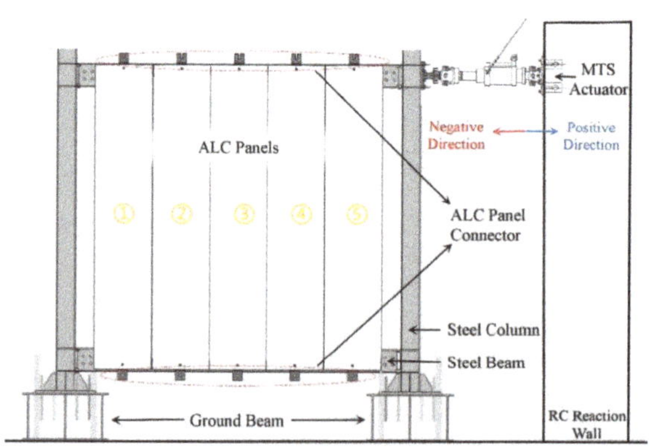

Figure 4. Test loading devices.

As illustrated in Figure 5, the experimental loading device consisted of an electro-hydraulic servo-loading system, hydraulic jack, hydraulic actuator, connecting rod, anchor bolt, and reaction frame. For the loading test, the electro-hydraulic servo loading system was attached to the actuator.

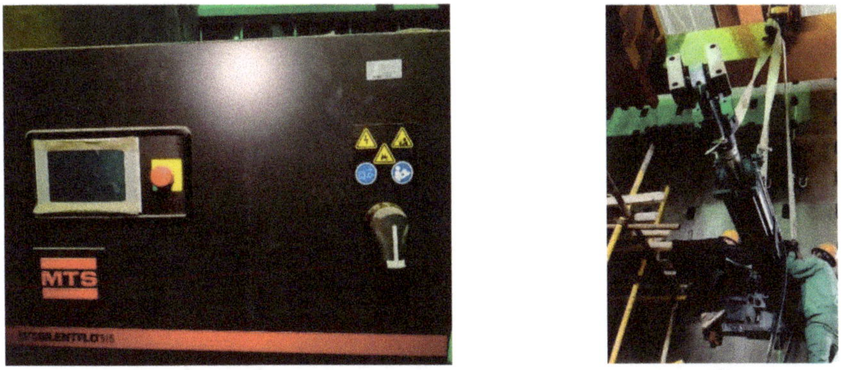

(a) (b)

Figure 5. Photograph of the test devices: (**a**) MTS servo-loading system; (**b**) the actuator of MTS with column.

2.4. Material Properties

Six 100 mm × 100 mm × 100 mm and three 100 mm × 100 mm × 300 mm ALC cube test blocks were poured under the same conditions, as shown in Figure 6. All samples were manufactured and maintained under the same conditions, according to "The test methods for autoclaved aerated concrete" (GB/T 11969-2020) [24]. For the steel properties, steel coupons, according to GB/T 228.1-2010 [25], were cut from steel tubes and sheets and assessed to determine the tensile strength, modulus of elasticity, as well as breaking elongation. Table 2 shows the results of the steel material tests. Table 3 shows the ALC panel mechanical performance index.

(a) (b) (c)

Figure 6. Material properties test: (**a**) Universal testing machine, (**b**) ALC test blocks, (**c**) Test steel coupons.

Table 2. Material properties of steel.

Specimen	Thickness (mm)	Yield Stress (N/mm^2)	Ultimate Stress (N/mm^2)	Elongation Stress (%)
Steel beam flange	11	263.4	401.6	25.2
Steel beam web	7	275.3	411.3	22.3
Steel column flange	12	289.5	435.4	24.7
Steel column web	8	278.2	409.8	20.8
Connector	10	376.6	510.1	19.6

Table 3. Material properties of ALC panel.

Specimen	Specimen Dimension (mm)	Measured Compressive Strength (MPa)	Elastic Modulus (GPa)
Sac1	100 × 100 × 100	3.89	
Sac2	100 × 100 × 100	2.97	
Sac3	100 × 100 × 100	3.26	
Sac4	100 × 100 × 100	3.78	
Sac5	100 × 100 × 100	3.96	
Sac6	100 × 100 × 100	3.49	
Average		3.56	
Sae1	100 × 100 × 300		1640
Sae2	100 × 100 × 300		1880
Sae3	100 × 100 × 300		1790
Average			1770

2.5. The Experiment Loading System

For loading, the displacement control method was used. Figure 7 illustrates the loading system. After yielding, displacement was applied, the process was repeated three times for each displacement. The push-down was positive, while the pull-up was negative. According to the inter-story displacement angle index of the GB50011-2010 "Code for Seismic Design of Buildings" [26], the limit value of the interlayer displacement angle for light and medium earthquakes is 1/250, while the limit value of the interlayer displacement angle for rare earthquakes is 1/50.

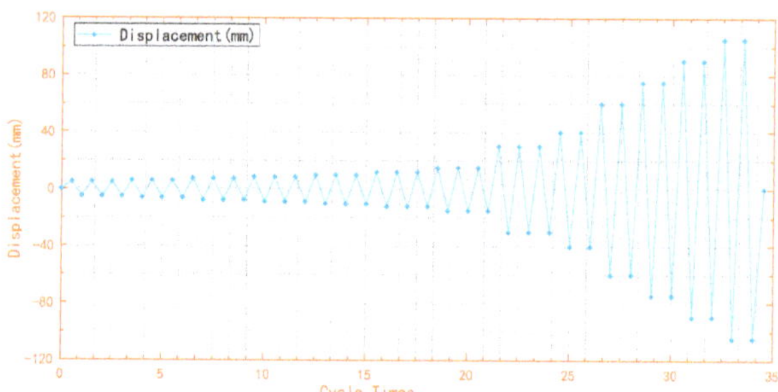

Figure 7. Loading system.

The test was preloaded with 2 mm, loaded for three turns, to ensure that the steel frame loading version and the splice gap of the booster were fully adjusted. Then the load was controlled by the displacement angle, where the maximum displacement angle was 3/10 (90 mm). Three turns were loaded at each stage before 1/250, 1/600, 1/500, 1/400, 1/350 and 1/300 cycles, and two turns were loaded at each stage afterwards (1/250, 1/200, 1/100, 1/75, 1/50, 1/40, 3/10 cycles).

3. Experimental Results

3.1. Specimen SFW1

Specimen SFW1 was a traditional L-hooked bolt which connects to the external ALC panel with a steel frame. When the displacement angle reached 1/500 (±6 mm cycle), cracks appeared in the joint mortar at the bottom of the No.1 and No.2 panel joints, as shown in Figure 8a. When the displacement angle was between 1/300 and 1/250 (±10–12 mm cycle), small fragments of ALC panels were constantly falling off, caused by the pressure between the panels, as shown in Figure 8b. When the displacement angle reached 1/200 (±15 mm cycle), there was obvious dislocation between the panels which had fallen off the patchwork joints, which made a sliding sound between the retaining steel bars and the edge of the panel, as shown in Figure 8c. When the displacement angle reached 1/75 (±40 mm cycle), angle steel welds on the upper part of No.2 panel broke, and in the hole of the L-hooked bolt under No.3 panel an extended crack appeared, as shown in Figure 8d,e. When the displacement angle reached 1/40 (±75 mm cycle), the L-hooked bolt became flexible and started swinging, the corners of each panel broke, and at the holes of the L-hooked head bolts under No.4 and No.5 panel extended cracks appeared. Meanwhile, the holes of the bolts began to expand, the bolts swinging with the steel angle as the displacement was applied, as shown in Figure 8f,g. When the displacement angle reached ±90 mm, the left upper beam-column joints made a loud sound because of fracture of the weld, as shown in Figure 8h.

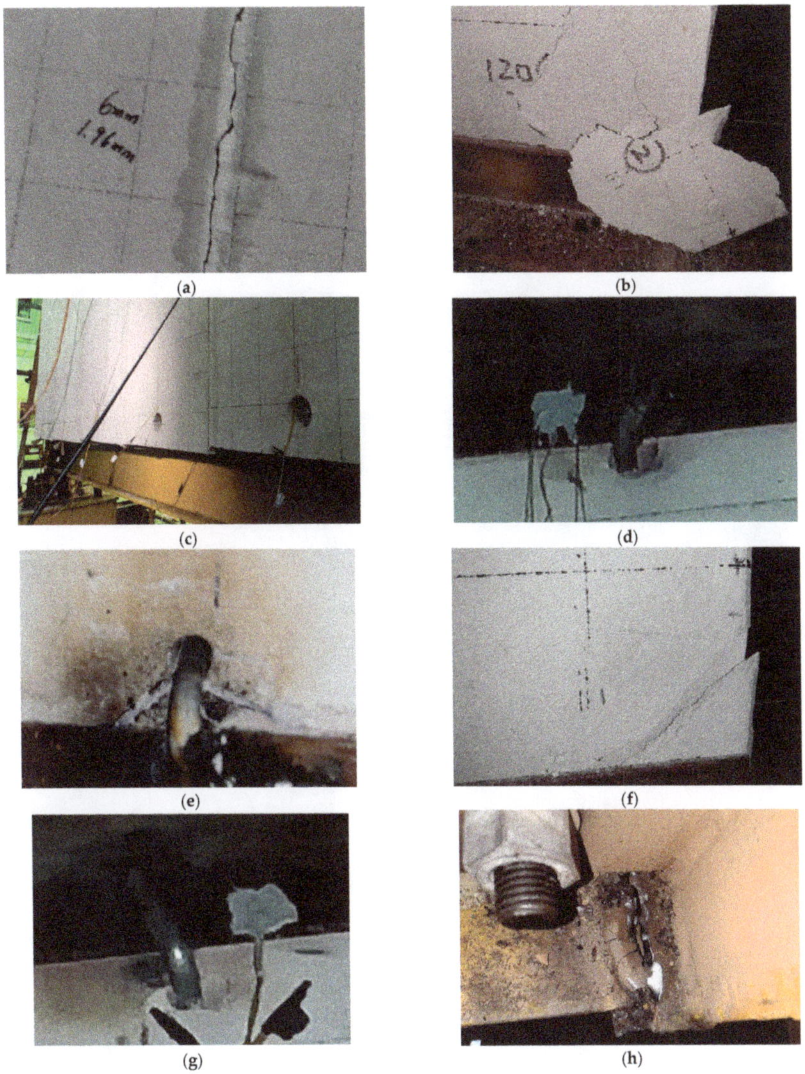

Figure 8. Experimental results of SFW1: (**a**) cracks appeared in the joint mortar panel joint; (**b**) ALC debris falling off; (**c**) obvious dislocation between panel; (**d**) angle steel weld fracture; (**e**) extended cracks at bolt holes; (**f**) the broken phenomenon of the panel angle; (**g**) the upper bolt holes expand and bolt wobbled; (**h**) weld fracture.

3.2. Specimen SFW2

Specimen SFW2 was the new crossing panel connector. In the process of loading to the displacement angle 1/100 (±30 mm cycle), a friction sound between the enclosure reinforcement and the end panel occurred. There was still no significant change in the frame and panel, although the bolt at the upper panel connector slipped within its bolt hole, implying that the follower performance of the junction was quite good, as shown in Figure 9a. When loading to 40 mm, small pieces of debris came off, and the corner of the panel behind panel No. 5 was broken, as shown in Figure 9b. When loading to 60 mm, the maximum limitation of the limitation hole was reached, the stiffness significantly increased. the splicing mortar between No.1 and No.2 panels broke, cracks appeared in the lower left corner panel corner of No.1 panel, cracks appeared in the back panel corner of No.2 panel with steel frame, and the relative movement between No.1 and No.2 panels could be seen, as shown in Figure 9c. When loading to 75 mm, cracks appeared at the upper bolt holes of No.2 and No.3 panels, and the lower node of the second panel was loose. Obvious interpanel dislocation could be seen, as shown in Figure 9d,e. At 90 mm loading, the mortar of the joints collapsed between No.2 and No.3 and No.3 and No.4 panels, and the corner of the back of the No.1 panel was broken off, as shown in Figure 9f.

Figure 9. Experimental results of SFW2: (**a**) upper connector slipped within bolt hole; (**b**) corner panel was broken; (**c**) splicing mortar broke and relative movement; (**d**) cracks appeared at the upper bolt holes; (**e**) obvious interpanel dislocation; (**f**) corner of the panel back was broken off.

3.3. Specimen SFW3

Specimen SFW3 was the new pendulous Z-panel connector. Before test loading to the displacement angle 1/250 (±12 mm cycle), there was no obvious change in the frame and no cracks appeared in the panel. However, the bolt at the upper panel connector slipped within its bolt hole, which made a friction sound, indicating that the slipping mechanism of the panel connector did work, as shown in Figure 10a. When the displacement angle reached 1/200 (±15 mm cycle), a vertical crack appeared in the bottom of the No.2 panel, as shown in Figure 10b. When the displacement angle reached 1/100 (±30 mm cycle), cracks in the bottom of No.2 panel extended, and the back corner of No.4 panel broke, as shown in Figure 10c, because of the further enhancement of the panel-panel displacement. When the displacement angle reached 1/75 (±40 mm cycle), the bonding mortar between No.1, No.2 and No.4 and No.5 cracked. A small amount of shedding of fragments, as shown in Figure 10d, occurred. When the displacement angle reached 1/40 (±75 mm cycle), extended cracks at bolt holes in No.1. Obvious dislocation between panels could be seen, shown in Figure 10e. At 60 mm displacement angle, the maximum limitation of the limitation hole was reached, and stiffness was significantly increased. When the displacement angle reached 90 mm, new vertical cracks appeared in the bottom of the No.2 panel, and the fragments of back corners of No.3 and No.4 panels fell off, as shown in Figure 10f,g.

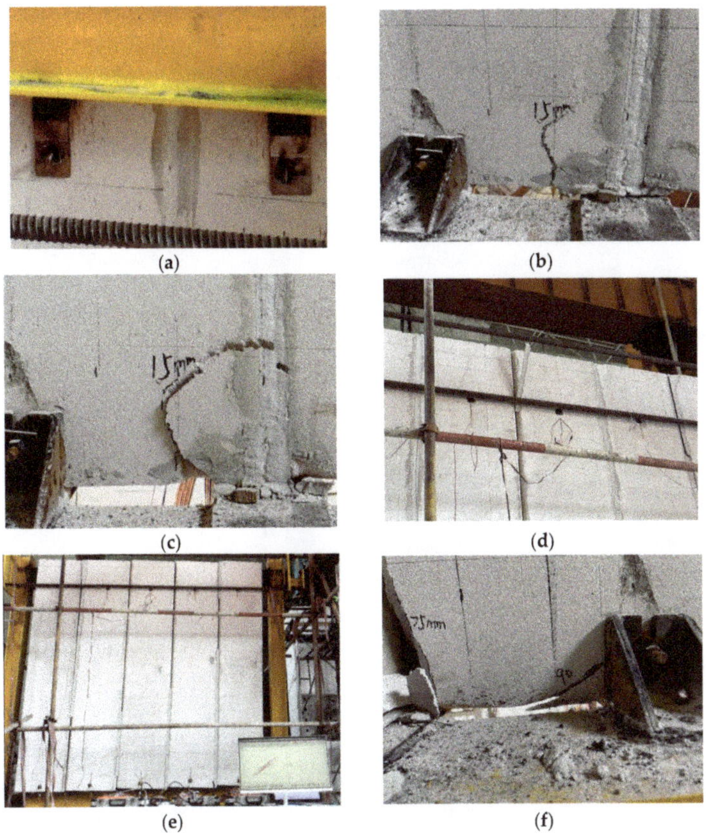

Figure 10. *Cont.*

(g)

Figure 10. Experimental results of SFW3: (**a**) slip in bolt holes and bolts; (**b**) crack in the bottom of panel No. 2; (**c**) cracks extended and corner broken; (**d**) mortar crack; (**e**) obvious dislocation in panels; (**f**) extended cracks at bolt holes; (**g**) fragments of back corners.

4. Discussion

4.1. Load-Displacement Hysteretic Behavior

Measured cyclic response curves (hysteretic curves) obtained from all specimens are given in Figure 11. With increasing column end displacement, the overall stiffness degradation and the strength degradation of SFW1 at the same level loading were observed; the same phenomenon occurred with SFW2 and SFW3, though this was relatively less obvious when compared to SFW1. This was mainly attributed to: the frame gradually moving from the elastic stage to the elastic-plastic and plastic stage, mortal splitting between ALC panels or blocks, cracking and crushing on the ALC panels plastic deformation, and welding fracture at the beam-to-column connections and wall-to-frame connections, which resulted in composite frame destruction and partial transmission of mutual forces. The reason why this was less obvious for SFW3 was that the new connectors gave more stiffness to the frame and did not allow it to go through the elastic-plastic and plastic stages for a loaded displacement of 90 mm, as occurred with SFW1.

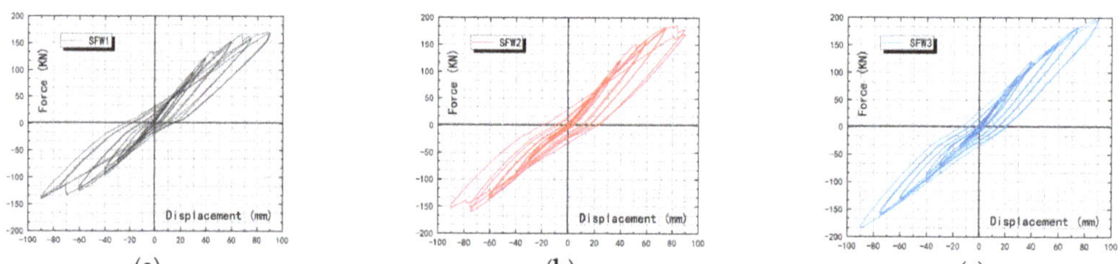

Figure 11. Hysteric curves of SFW1 (**a**), SFW2 (**b**) and SFW3 (**c**).

The hysteresis curves of all three types of ALC panel connected frames were inverse S-shaped. As shown by B. Wang [14], the contribution of the facade to the antiliteral load is greater than that of the pure steel frame. At the same displacement level in the late stage, the load carrying capacity of the cross-panel connector group and the pendulous Z panel connector group (specimens SFW2 and SFW3) was greater than that of the L-hooked bolt group (specimen SFW1). More importantly, specimen SFW1 was more brittle than specimens SFW2 and SFW3 in terms of the experimental evaluation of stiffness degradation and strength degradation, under the same horizontal load. This was due to the fracture of the weld between the L-hooked bolt group (SFW1) and the angle at the 75 mm stage, which resulted in degradation in stiffness and loss of load-carrying capacity after 75 mm, although its load-carrying capacity was previously higher than the other groups. In contrast, specimens SFW2 and SFW3 were in swaying energy dissipation in the early stage (before

30 mm) and provided higher stiffness and load capacity for the frame in the later stage due to the restricted displacement of the slip on the bolt hole.

4.2. Skeleton Curve

The skeleton curve connecting the peak points of each cyclic loading curve of the hysteresis curve of the structure and members reflects the link between peak loads and corresponding displacements from the specimen's hysteresis loops, such as the characteristics of the different stages of force and deformation of the specimens. As can be seen from Figure 12, the trends in the skeleton curve of the three groups of specimens were basically the same. Each specimen experienced the three stages of elasticity, elastomer-plasticity and damage. The skeleton curves were approximately S-shaped, and the trend in stiffness degradation was more obvious.

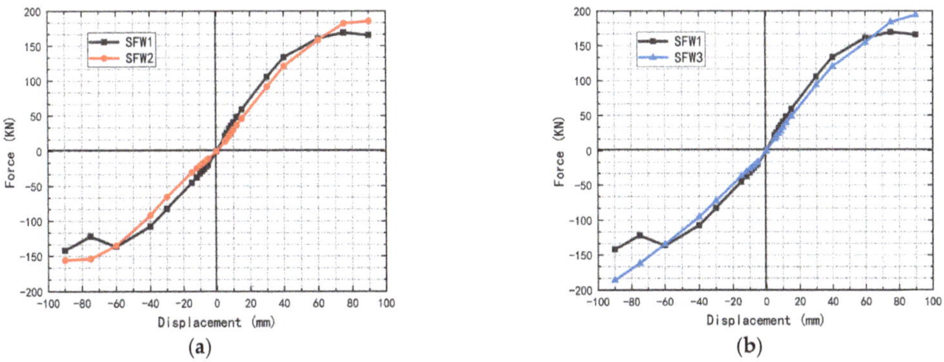

Figure 12. Skeleton curves of SFW1 and SFW2 (**a**) and SFW1 and SFW3 (**b**).

In the early stage (5–40 mm cycle), the trends in the skeleton curves of the three groups basically overlapped. SFW1 was larger than SFW2 and SFW3, indicating that both specimens were in the elastic stage. However, because the bolts of the new connectors of SFW2 and SFW3 slid in the limitation holes with loading, their load capacity and stiffness were less than that of the L-hooked bolt connector (SFW1), which means the new connector effectively released the energy. In the intermediate stage of the test (40–75 mm cycle), the partial rigidity of the new connector (SFW2 and SFW3) was only demonstrated after the displacement of the new connector bolt at the limitation hole had reached its limit. Due to the rigid connection of SFW1 entered the yielding stage first, the rise in the load-carrying capacity of SFW1 became flat as the displacement loading was exceeded by SFW2 and SFW3. At this point, the new connectors (SFW2 and SFW3) started to contribute stiffness to the frame, making the frames' load capacity higher. During the later 75–90 mm cycle stage, each of the specimens was in the yielding stage. The curve for SFW1 fluctuated suddenly, then rapidly started to decrease to a value much smaller than those for SFW2 and SFW3. This was because as the weld of the L-hooked bolt (SFW1) with the angle steel broke at 75 mm loading, the connection type of SFW1 became too flexible overall, and the stiffness contribution of the frame decreased rapidly. The bearing capacity of the L-hooked bolt started to reach its peak, then rapidly entered the damage phase, which was obviously smaller than for the situation with SFW2 and SFW3. The connection types of specimens by displacement loading is show in Table 4.

Table 4. Connection types of specimens by displacement loading.

Loading Cycle	Signs of Test Phenomena	Connection Type		
		SFW1	SFW2	SFW3
Early stage 0–40 mm cycle	The displacement of the new connector bolt at the limitation hole had reached its limit	Rigid	Flexible	Flexible
Intermediate stage 40–75 mm cycle		Rigid	Partial rigid	Partial rigid
Late stage After 75 mm cycle	Weld of L-hooked bolt (SFW1) with angle steel broken	Flexible	Major rigid	Major rigid

Table 5 summarizes characteristic yielding and peak points of specimens based on envelope curves. Three typical characteristic points were introduced in accordance with Figure 12 shown in the JGJ/T 101-2015 [27]. Point 1 represents the yield point and point 2 shows the ultimate load and corresponding displacement of the composite structure. With the benefit of the sliding holes of the new connections, the yield loads of SFW2 and SFW3 increased by 13.8% and 10.6% compared to SFW1, respectively. The maximum bearing capacity of specimen SFW1 decreased by 8.9% compared to specimen SFW2 and decreased by 13.3% compared to specimen SFW3. In terms of yield displacement, SFW2 and SFW3 were delayed by 13.6% and 12.7%, respectively. With respect to ultimate displacement, SFW2 and SFW3 were delayed by 22.7% and 30.0%, respectively. This showed that the new ALC panel connectors (SFW2 and SFW3) not only increased elastic stiffness in the early stage, but also provided some stiffness after the early stage, with enhanced ultimate bearing capacity compared to SFW1. The results indicated that the new ALC panel connectors (SFW2 and SFW3) greatly improved the bearing capacity and stiffness of the composite frame when compared to the L-hooked bolt (SFW1).

Table 5. Skeleton curves of specimens.

Specimens	Yielding Point		Peak Point	
	P_y (kN)	Δ_y (mm)	P_m (kN)	Δ_m (mm)
SFW1	137.65	51.75	169.19	68.58
SFW2	156.59	58.77	185.68	84.17
SFW3	152.3	58.32	195.15	89.13

4.3. Deterioration in Stiffness

Stiffness degradation is another metric for describing the seismic behavior of structures [28]. Stiffness steadily deteriorates as the lateral drift in the quasi-static test load increases—this degradation is typically regarded as a critical index for determining the extent of structure collapse. In this investigation, the equivalent stiffness method was used to characterize the stiffness of the connectors. Equivalent stiffness is given by the slope of the line connecting the maximum value of the reaction force (positive and negative) of each step of loading. The secant stiffness of each frame during the mean value of cycles of each amplitude loading is shown in Figure 13. The formula for secant stiffness is as follows:

$$K_j = \sum_{i=1}^{n} P_j^i / \sum_{i=1}^{n} u_j^i \tag{1}$$

As for the 40 mm cycle, the deterioration in stiffness of SFW2 and SFW3 appeared to increase briefly because the bolts of the new connectors reached the limitation of the limit holes. This was also the reason that the initial deterioration in stiffness of SFW2 and SFW3 was slower than SFW1 during the beginning of loading. As for the 40 mm cycle, the deterioration in stiffness of SFW2 and SFW3 appeared to increase briefly because the bolts of the new connectors reached the limitation of the limit holes. Then, in the 60–90 mm stage, the deterioration of stiffness of SFW1 decreased sharply, while the SFW2 and SFW3 were relatively flat and exceeded SFW1 at 60 mm. The averages of the three groups of deterioration of stiffness after the elastic stage were 2.15, 2.38 and 2.43, respectively. Compared to the traditional L-hooked bolt (SFW1), SFW2 increased by 10.6% and SFW3 increased by 13.0%. This indicated that the bearing capacity of the new ALC connectors was larger than for the L-hooked bolt. The new ALC connectors showed good stiffness characteristics.

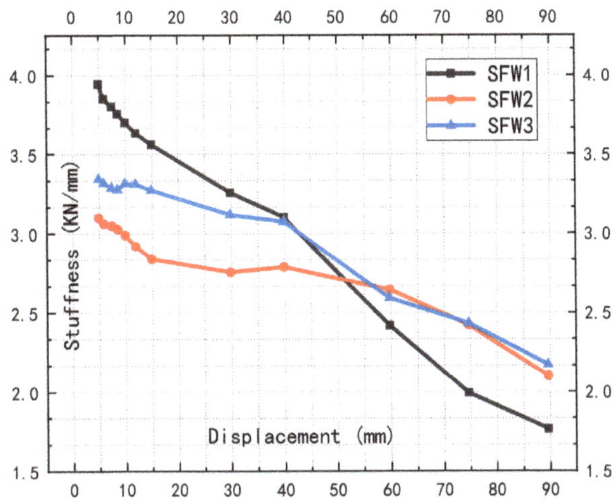

Figure 13. Deterioration of stiffness of specimens.

4.4. Energy Dissipation

The dissipation energy capacity of the structure at the drift angle was measured by the area of stress to strain enclosed. The energy dissipation capacity of the structure is closely related to the area of the hysteresis loop [29]. The more the structure dissipates energy, the safer the structure and the less likely it is to be damaged.

The equivalent damping factor (ξ_e) was defined by Equation (2) [30]. The area diagram is shown in Figure 14. The area surrounded by the ABCF curve is the energy dissipation capacity of the structures.

$$\xi_e = \frac{1}{2\pi} \cdot \frac{S_{ABC} + S_{CDA}}{S_{OBE} + S_{ODF}} \tag{2}$$

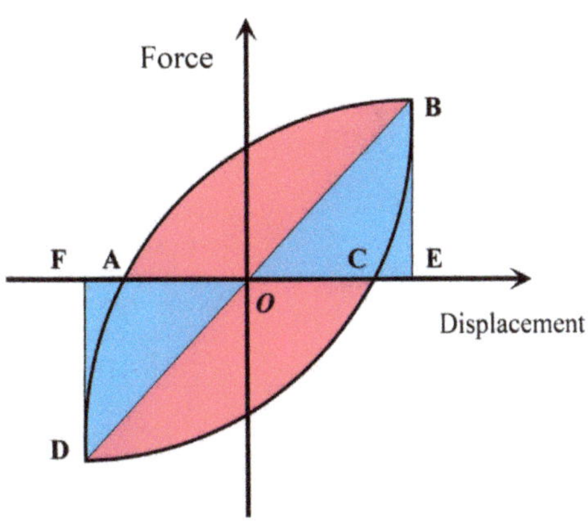

Figure 14. The dissipation capacity area.

The energy dissipation parameters of specimens at ultimate state are listed in Table 6. The total dissipation energy W_{total} is defined as the areas of the hysteresis curves when specimens are under ultimate state. The energy dissipation capacity of the specimen is expressed by the energy dissipation factor DE. The energy dissipation coefficient D_E can be defined as:

$$D_E = \frac{S_{ABC} + S_{CDA}}{S_{OBE} + S_{ODF}} \quad (3)$$

The energy-dissipating capacity (D_E) of specimens is shown in Figure 15. This shows the equivalent damping factor (ξ_e) versus the relative horizontal displacement relationship of the specimens. The overall trend of the curves was that the equivalent damping factor (ξ_e) of SFW1 reduced rapidly with increasing relative displacement (0–10 mm), then increased rapidly (after 10 mm). The new connectors (SFW2 and SFW3) also showed this phenomenon, except with a delay starting in a slow decline until 40 mm, and a rapid raise after 60 mm. The difference was that the equivalent damping factor (ξ_e) of SFW2 and SFW3 rose quickly in the 0–40 mm stage because the bolts of the new connectors were dissipating energy by sliding in the limitation holes. At a displacement less than 52.7 mm, the equivalent damping factors (ξ_e) of SFW2 and SFW3 were larger than for SFW1. Subsequently, at each equivalent level load, SFW1 was slightly larger than SFW2 and SFW3, and was overtaken by them after 80 mm.

Table 6. Energy dissipation parameters of specimens at ultimate state.

Specimen	Displacement (mm)	W_{toal} (kN·mm)	ξ_e	D_E
SFW1	68.58	5522.7	0.06916	0.4346
SFW2	84.17	11,008.2	0.08505	0.5344
SFW3	89.13	12,776.5	0.08594	0.5405

The total dissipation energy W_{total}, the equivalent damping factor (ξ_e) and the energy-dissipating capacity (D_e) of the frame with new connectors (SFW2 and SFW3) with ALC panels were obviously larger than that of the L-hooked bolt (SFW1) at the ultimate limit stage. At the ultimate state, the total dissipation energy W_{total} of SFW2 and SFW3 improved by 99.3% and 131.3%, respectively, compared to SFW1. For the equivalent damping factor (ξ_e) and the energy-dissipating capacity (D_e), SFW2 increased by 23.0% and SFW3 increased by 24.4%. Generally, the frame with new connectors with ALC panel had good dissipated-energy capacity. Compared with Wang J.'s study [7], the new connectors had better seismic performance and energy dissipation performance, as shown in Table 7.

Table 7. Comparison of the energy dissipation parameters of specimens at ultimate state.

Specimen	J. Wang Specimen [7]	Type	ξ_e	D_E
SFW1		L-hooked bolt	0.06916	0.4346
SFW2		Pendulous Z-connector	0.08505	0.5344
SFW3		Crossing connector	0.08594	0.5405
	SFW2	L-hooked bolt	0.077	0.484
	SFW3	Rocking connector	0.079	0.496
	SFW4	U-typed connector	0.082	0.515
	SFW5	Angle steel	0.084	0.528

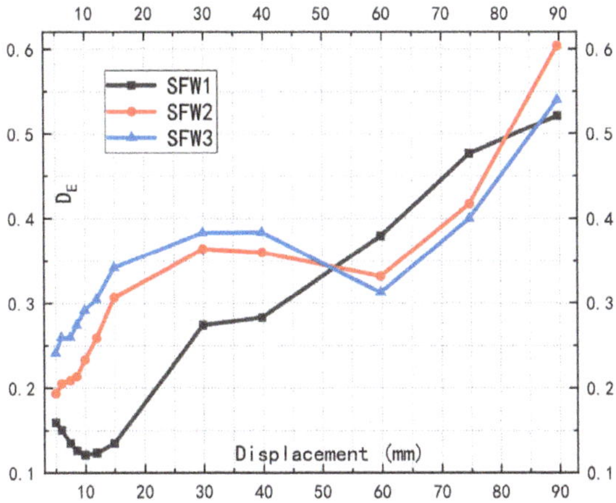

Figure 15. Energy-dissipating capacity (D_E) of specimens.

5. Conclusions

This paper proposed two new types of ALC panel connectors, based on the shortcomings of L-hooked bolts at construction sites. Cyclic loading tests of the steel frames with ALC panel were carried out. The new connectors meet the requirements of the design and are easy to assemble. The test results obtained can be summarized as follows:

1. The new connectors and traditional L-hooked bolt conformed to the seismic design, and all of them remained undamaged and in good working status at an inter-story displacement angle of 1/50 (60 mm). However, the traditional L-hooked bolt broke the weld at 1/40 (75 mm), while the new connectors showed good hysteretic performance and ductility. Even at larger displacement (90 mm), the new connectors were still reliable.
2. In terms of hysteresis loops, each specimen approached an inverse S-shape and "rheostriction" phenomenon was obvious. The hysteresis loops of the new connectors were fuller than for the L-hooked bolts. The maximum bearing capacity of the new connectors increased by 8.9% and 13.3%, respectively, compared to the L-hooked bolt. As for the yield loads, the pendulous Z-panel connector increased by 13.8% and the crossing panel connector increased by 10.6%. In terms of yield displacement, the new connectors delayed by 13.6% and 12.7%, and for the ultimate displacement delayed by 22.7% and 30.0%, individually.
3. The new connectors had better energy consumption capacity (D_e). Compared to the traditional L-hooked bolt, deterioration of stiffness after the elastic stage for the pendulous Z-panel connector increased by 9.7% and for the crossing panel connector increased by 11.5%. In the early stages of loading, the De of the new connectors were lower than the L-hooked bolt. When the bolt at the limitation hole reached its limit, it provided stiffness to the frame. The new connectors are suitable for areas where small earthquakes are frequent and can extend the durability of steel prefabricated buildings under high frequency earthquakes.

Author Contributions: Conceptualization, K.D.; methodology, K.D. and C.Z.; software, C.Z.; validation, C.Z. and S.H.; formal analysis, C.Z. and S.H.; investigation, K.D. and C.Z.; resources, K.D.; data curation, C.Z. and S.H.; writing—original draft preparation, K.D. and C.Z.; writing—review and editing, K.D. and C.Z.; visualization, C.Z.; supervision, K.D.; project administration, K.D.; funding acquisition, K.D. All authors have read and agreed to the published version of the manuscript.

Funding: This research was financially supported by the National Natural Science Foundation of China (11472005), the University Synergy Innovation Program of Anhui Province (GXXT-2019-005),

Anhui Provincial Universities Natural Science Research Project (KJ2020ZD43) and Anhui Provincial Natural Science Foundation Project (1908085ME144).

Institutional Review Board Statement: Not applicable.

Informed Consent Statement: Not applicable.

Data Availability Statement: Not applicable.

Conflicts of Interest: The authors declare no conflict of interest.

References

1. Khaloo, A.; Bakhtiari Doost, R. Seismic Performance of Precast RC Column to Steel Beam Connections with Variable Joint Configurations. *Eng. Struct.* **2018**, *160*, 408–418. [CrossRef]
2. Cheng, B.; Lu, K.; Li, J.; Chen, H.; Luo, X. Muhammad Shafique, Comprehensive assessment of embodied environmental impacts of buildings using normalized environmental impact factors. *J. Clean. Prod.* **2022**, *334*, 130083. [CrossRef]
3. Ghayeb, H.H.; Razak, H.A.; Sulong, N.H.R. Development and Testing of Hybrid Precast Concrete Beam-to-Column Connections under Cyclic Loading. *Constr. Build. Mater.* **2017**, *151*, 258–278. [CrossRef]
4. Ding, K.; Liu, J.; Ma, W.; Liu, Y. Experimental study on seismic performances of a new type of fabricated semi-rigid beam-to-column connection. *China Civ. Eng. J.* **2021**, *54*, 1–7. [CrossRef]
5. Zhang, D.; Kuai, G.; Xu, H.; Luo, L.; Leng, H. Application research of autoclaved sand aerated concrete wallboard in fabricated steel Structure System. *Steel Struct.* **2016**, *31*, 89–93. [CrossRef]
6. Wang, J.; Li, B. Cyclic Testing of Square CFST Frames with ALC Panel or Block Walls. *J. Constr. Steel Res.* **2017**, *130*, 264–279. [CrossRef]
7. Kalpana, M.; Mohith, S. Study on autoclaved aerated concrete: Review. *Mater. Today Proc.* **2020**, *22*, 894–896. [CrossRef]
8. Kang, J.; Shen, D.; Li, C.; Li, M.; Wang, X.; Hu, H. Effect of water-to-cement ratio on internal relative humidity and autogenous shrinkage of early-age concrete internally cured by superabsorbent polymers. *Struct. Concr.* **2022**, *23*, 1–15. [CrossRef]
9. Qu, X.; Chen, Z.; Sun, G. Experimental and Finite element Study on structural Performance of autoclaved concrete wallboard. *J. Build. Mater.* **2012**, *15*, 268–273. [CrossRef]
10. Zhang, G.; Xiao, W.; Chen, B.; Miao, Q.; Wu, H. Research on hysteretic performance of autoclaved aerated concrete exterior Wall. *Ind. Build.* **2016**, *46*, 86–92. [CrossRef]
11. De Matteis, G.; Landolfo, R. Diaphragm Action of Sandwich Panels in Pin-Jointed Steel Structures: A Seismic Study. *J. Earthq. Eng.* **2000**, *4*, 251–275. [CrossRef]
12. Ding, K.; Chen, W. Experimental Study and Restoring Force Modeling on Seismic Performance of Prefabricated Concrete Beam-column Joints. *J. Shenyang Jianzhu Univ. (Nat. Sci.)* **2021**, *37*, 51–60. [CrossRef]
13. Kim, T.S.; Kuwamura, H.; Cho, T.J. A Parametric Study on Ultimate Strength of Single Shear Bolted Connections with Curling. *Thin-Walled Struct.* **2008**, *46*, 38–53. [CrossRef]
14. Wang, B.; Wang, J.; Wan, H.; Hou, H.; Wang, J. Research on the Performance and Connection structure of filled wall in high-rise steel structure buildings under cyclic loading. *Build. Steel Struct.* **2015**, *17*, 44–50. [CrossRef]
15. Wu, J.; Lu, J.; Xu, Z.; Zheng, Z. Design method of prefabricated concrete external wall panel bearing joint. *Archit. Struct.* **2014**, *44*, 47–51. [CrossRef]
16. Jin, F.; Guo, A.; Wang, Z. Application of autoclaved lightweight aerated concrete panel inner partition and enclosure wall. *Constr. Technol.* **2001**, *30*, 24–25. [CrossRef]
17. Dai, S. *Structural Performance and Supporting Technology of Steel Frame-Concrete Tube Housing*; Wuhan University of Technology: Wuhan, China, 2004.
18. Sun, X. Exploration of NALC stacked floor slab application technology. *Jiangsu Build. Mater.* **2019**, *6*, 16–21.
19. Cao, S.; Shu, G.; Lin, K.; Fan, S.; Gao, H. Analysis and design method of a new type of prefabricated steel structure external wall panel joint. *Build. Struct.* **2017**, *47*, 46–52. [CrossRef]
20. Hu, J. *Analysis of Flexural Performance of Aerated Concrete Panel and Experimental Study of Joints*; Tongji University: Shanghai, China, 2006.
21. Wang, X.; Liu, X.; Ma, L.; Zhou, X. Experimental Study on seismic Performance of prefabricated ALC External Wall Panel. *Steel Struct.* **2017**, *32*, 22–26. [CrossRef]
22. Meng, Y. *Research on Lateral Force Resistance of Steel Frame with Multi-Layer Inner Cladding Wall Panel*; Harbin Institute of Technology: Harbin, China, 2012.
23. *JGJ/T 17-2020*; Technical Standard for Application of Autoclaved Aerated Concrete Product. China Building Industry Press: Beijing, China, 2020.
24. *GB/T 11969-2020*; Test Methods of Autoclaved Aerated Concrete. Standardization Administration of China Press: Beijing, China, 2020.
25. *GB/T 228.1-2010*; Metallic Materials-Tensile Testing. Standardization Administration of China Press: Beijing, China, 2010.
26. *GB 50011-2010*; Code for Seismic Design of Buildings. China Building Industry Press: Beijing, China, 2010.
27. *JGJ/T 101-2015*; Specification for Seismic Test of Buildings. Architecture Industrial Press of China: Beijing, China, 2015.

28. Ding, K.; Zhang, Y. Experimental Study on Seismic Performance of Fabricated Bolted Joint under Low-Cycle Reciprocating Loads. *Results Eng.* **2021**, *9*, 100208. [CrossRef]
29. Ding, K.; Ye, Y.; Ma, W. Seismic Performance of Precast Concrete Beam-Column Joint Based on the Bolt Connection. *Eng. Struct.* **2021**, *232*, 111884. [CrossRef]
30. Ding, K.; Zhang, C.; He, S.; Liu, Y. Hysteresis Behavior and Design of the New Autoclaved Lightweight Concrete (ALC) External Panel Connector with the Steel Frame. *Adv. Mater. Sci. Eng.* **2022**, *2022*, 8319044. [CrossRef]

buildings

Article

Full-Scale Testing on Seismic Performance of Surge Arrester with Retrofitted Composite Insulators

Sheng Li [1,2,*], Yongfeng Cheng [2], Zhicheng Lu [2], Nelson Lam [1], Yaodong Xue [2] and Haibo Wang [2]

1 Department of Infrastructure Engineering, The University of Melbourne, Melbourne 3010, Australia
2 China Electric Power Research Institute, Beijing 100055, China
* Correspondence: shengl1@student.unimelb.edu.au

Abstract: In electrical substations, glass-fibre reinforced polymer (GFRP) insulators provide an alternative to porcelain insulators for better seismic performance. The bonded connection between the composite tube and the metal end fitting is a weak link under earthquake actions, and the failure risk of ultra-high voltage (UHV) substation equipment with GFRP composite insulators can be considerable in places with high seismicity. This study solved the problem by retrofitting and experimentally qualifying the seismic performance of surge arresters made of composite insulators. To ensure safety under a bending moment corresponding to the seismic demand, the bonded sleeve joint between the tube and end fitting was retrofitted by shear stiffeners and higher strength end fitting. A full-scale shaking table test was carried out to verify the performance of the retrofitted surge arrester, and the seismic responses to increasing earthquake actions were analyzed. The specimens in this study were full-scale, had the largest dimensions among substation equipment of the same type, and were subjected to an extremely high earthquake action of 0.5 g; therefore, the seismic testing results produced in this study can serve as an important reference for the industry.

Keywords: GFRP insulators; seismic performance; bending; joints/joining; shaking table test

Citation: Li, S.; Cheng, Y.; Lu, Z.; Lam, N.; Xue, Y.; Wang, H. Full-Scale Testing on Seismic Performance of Surge Arrester with Retrofitted Composite Insulators. *Buildings* **2022**, *12*, 1720. https://doi.org/10.3390/buildings12101720

Academic Editors: Liqiang Jiang, Jihong Ye and Wei Guo

Received: 7 September 2022
Accepted: 14 October 2022
Published: 18 October 2022

1. Introduction

The seismic safety of electrical equipment in substations is essential to the reliability of a power system [1–3]. Substations install a range of cylindrical electrical equipment, such as surge arresters, capacitor voltage transformers, and post insulators. These items are traditionally manufactured using porcelain insulators composed of ceramic material that is brittle and fragile, resulting in a significant risk of failure in earthquakes. Figure 1 shows porcelain insulators that were damaged in the 2008 Wenchuan earthquake in China [4]; similar failures occurred in the 1977 Vrancea earthquake in Romania [5], the 1989 Loma Prieta earthquake in the United States [6], the 2011 Tohoku earthquake in Japan [7], the 2011 Christchurch earthquake in New Zealand [8], and the 2013 Lushan earthquake in China [9]. Due to insulation clearance requirements, electrical equipment in ultra-high voltage (UHV) substations, which are the substations with the highest voltage, becomes substantially taller [10,11]. The normal height of cylindrical electrical equipment in the primary circuit of a substation, such as the 1000 kV surge arrester examined in this study, is approximately 10–15 m, and the diameter of an insulator is approximately 0.3–0.7 m. In seismic hazardous areas, the usage of porcelain insulators for UHV electrical equipment may pose a significant seismic risk [12].

Figure 1. Seismic failure of porcelain insulators in the M7.9 Wenchuan earthquake in 2008.

Insulators made of glass-fibre reinforced polymer (GFRP), referred to herein as composite insulators [13], are potential alternatives for replacing porcelain insulators due to the high strength and excellent electrical insulation properties of GFRP [14]. Composite insulators in substations have a service history dating back to the 1980s [15], but there has been no report on their performance after earthquakes. In the past ten years, composite insulators have emerged in UHV substations, with the majority being used in lightweight post insulators. Several earlier studies found that UHV equipment with composite insulators still lacked seismic strength, necessitating the use of special seismic isolation devices [16,17]. As a result, the adaptability of composite insulators in substations located in regions of high seismicity is hindered.

Composite insulators are made of GFRP tubes and metal flanges at both ends. The joints between the GFRP tube and the metal flange could be vulnerable to lateral loads by earthquakes [18]. The joints are categorized into three types: crimped joint [19,20], bolted sleeve joint [21,22], and bonded sleeve joint [23,24]. For crimped joints, the procedure of crimping produces significant residual stress, which works well for solid core insulators [25,26] but is not good for hollow core insulators, as the crimping process could crack the tube. Through bolts link the sleeve of the flange to the composite tube in a bolted sleeve joint. The penetration of bolts to the hollow core beam is often used in the beam-column joints [27,28] of composite structure buildings. For electrical equipment, the cavity of the insulator often houses valve plates and should not be penetrated. A bonded sleeve joint joins the GFRP tube and the sleeve of flange with an adhesive layer, which is more appropriate for hollow core insulators. Studies on bonded connections in engineering structures mostly focus on axial loads or torque loads [29–31], but bending is a less-researched subject [32]. However, composite insulators in substation equipment are primarily subjected to bending resulting from wind or earthquake actions. Cracking of the metal flange, tearing of the composite tube at the bonding zone, and slippage at the bonding interface were identified as potential failure modes based on bending tests of composite insulators [33–37]. However, investigations on retrofitting the bonded sleeve joint of composite insulators to undertake larger seismic actions are scarce.

Clause 8.6 of the standard Q/GDW 11391 [38] recommends performing an electrical test following seismic testing to validate its electrical functionality. Thus, it is recommended to use full-scale or "real" equipment as a test specimen so that electrical testing can be performed. Currently, full-scale shaking table tests of cylindrical electrical equipment with composite insulators are still limited. For example, 230 kV switches [39] and transformer bushings in various voltage levels [40] were tested in the United States, and a 380 kV circuit breaker [41] was tested in Italy. Some kinds of UHV equipment, such as a 1000 kV composite gas-insulated switch bushing [42] and ±800 kV composite post insulator [43], have been tested. Previous studies have not dealt with a full-scale test of a 1000 kV surge arrester with GFRP composite insulators.

The objective of this study was to improve and validate the seismic performance of a 1000 kV surge arrester with GFRP composite insulators to meet safety requirements under an extreme seismic action, thereby making the newly designed equipment suitable for installation in any seismic hazard zone in China. The 1000 kV surge arrester is an essential item in UHV substations, with the heaviest weight of all types of cylindrical equipment. The decision to retrofit was made since similar equipment had only passed 0.2 g PGA seismic testing [16]. In this study, the specimen and input motion in seismic qualification are introduced in Section 2. The retrofitting design is given in Section 3, along with the bending test results of the retrofitted specimen. The seismic performance of the retrofitted equipment is validated in Section 4 using a full-scale shaking table test.

2. Input Motion, Specimen Description, and Seismic Demand Analysis

2.1. Input Motion

The 1000 kV surge arrester was designed to be installed in substations in high seismic hazard areas, including one target substation located in the northern China earthquake zone. It is a major earthquake zone, where the historic M 7.8 Tangshan earthquake happened in 1976. The design earthquake action was 0.5 g peak ground acceleration (PGA) for 2% exceedance probability in 50 years. The vital importance of the UHV substation in the electrical grid system requires the installed equipment to have a safety margin larger than 1.67 under the design earthquake action, according to Chinese standard GB 50260 [44].

The acceleration spectrum defined in the Chinese standards [44,45] was plotted on the left side of Figure 2, with a notation of RS 0.5 g. The acceleration spectrum with a resonant period range of [0.1 s, 0.9 s] was chosen, making it independent of foundation soil types. The tested equipment can be appropriate for installation in substations with various types of foundation soil. This differs from the site-specific design of buildings and bridges. The electrical industry prefers this type of equipment standardization due to the advantages of employing standard production lines and having efficient stock management.

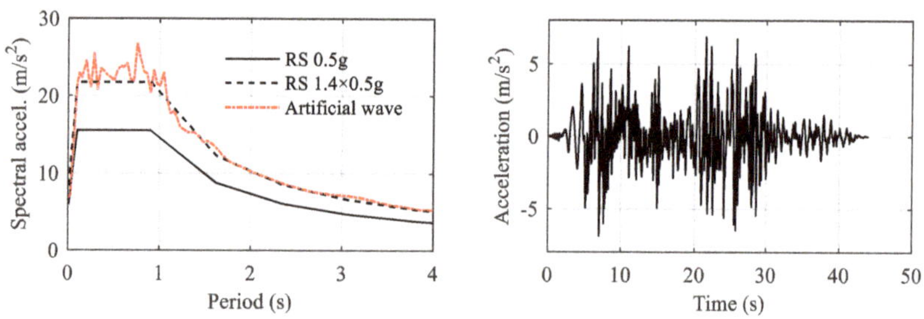

Figure 2. Acceleration spectrum of 0.5 g and 1.4 × 0.5 g PGA earthquake action in GB 50260 standard (**left**) and artificial seismic wave compatible to the spectrum (**right**).

In substations, the 1000 kV surge arrester would be mounted on the top of a 5~8 m steel supporting frame, as shown in Figure 3. Note that the height of the supporting structure is determined by the requirement of ground insulation distance and the position of nearby equipment interconnected with the surge arrester. The supporting structure can be different even in the same substation. The seismic design and qualification of the surge arrester usually consider a conservative dynamic amplification factor of 1.4, in order to take the influence of the supporting structure into consideration [46]. Conversely, the supporting structures are designed to have a dynamic amplification factor less than 1.4 [38]. During the testing, a linear amplification factor of 1.4 was directly applied to the input excitation, resulting in an input PGA of 1.4 × 0.5 g. The acceleration spectrum is plotted on the left side of Figure 2, with a notation of RS 1.4 × 0.5 g.

Figure 3. The 1000 kV surge arrester with supporting structure installed in an UHV electrical substation.

According to various standards on the seismic design of electrical equipment, including IEEE 693 [47], GB 50260 [44] and Q/GDW 11391 [38], spectral matching is the primary consideration [48] for input motion in the seismic qualification of a shaking table test. It is typically necessary that the test spectrum (TS) envelops the required spectrum (RS) within a tolerance of −10%/+50%. As a result, the artificial input motion depicted in Figure 2 was used, as suggested by Q/GDW 11132 [45]. This artificial earthquake wave was created by modifying the ground motion data from the Landers earthquake to match the RS in GB 50260 [44].

2.2. Specimen Description and Seismic Demand Analysis

An example of a hollow core composite insulator for cylindrical electrical equipment is shown in Figure 4 [33]. The main component of the insulator is a hollow core GFRP tube made by a fibre winding process. Epoxy adhesive is used to join the tube with metal flanges at both ends. The metal flange comprises a sleeve bonded to the composite tube and a flange plate bolted to a nearby insulator in the column. Silicone rubber sheds attached to the outer surface of the tube and the electrical valve plates inside the tube are non-structural components.

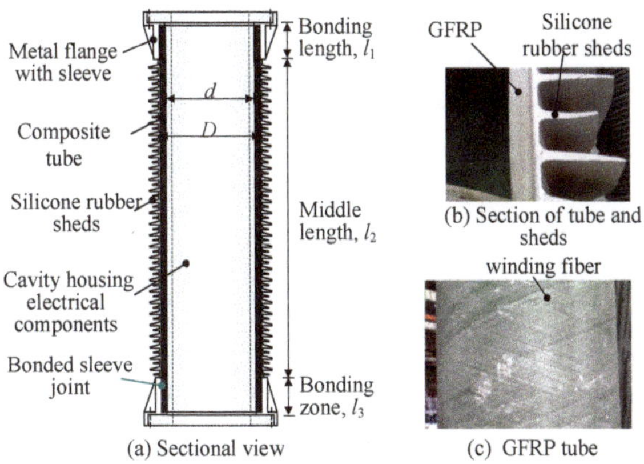

Figure 4. A section view of the hollow core composite insulator.

As depicted in Figure 5, the 1000 kV surge arrester, which was the specimen in this study, consisted of four 2.78 m long composite insulators. The inner diameters, d_i, of the GFRP tube were determined by the varistor installed inside the tube; the diameters were 535 mm, 600 mm, 600 mm, and 690 mm for insulators A1, A2, A3, and A4, respectively.

The outer diameter, D_i, was equal to $d + 2t$, in which t was the thickness of the hollow core composite tube. The density of the GFRP composite was around 1800 kg/m^3, and the elastic modulus was around 22 GPa. Components such as the metal end fitting and non-structural components were not modelled, but their masses were added to the model. It weighed 600 kg at the top, representing the corona ring and bus bar at top. The masses for insulators A1 to A4, excluding the mass of the composite tube, were 530 kg, 540 kg, 640 kg, and 640 kg.

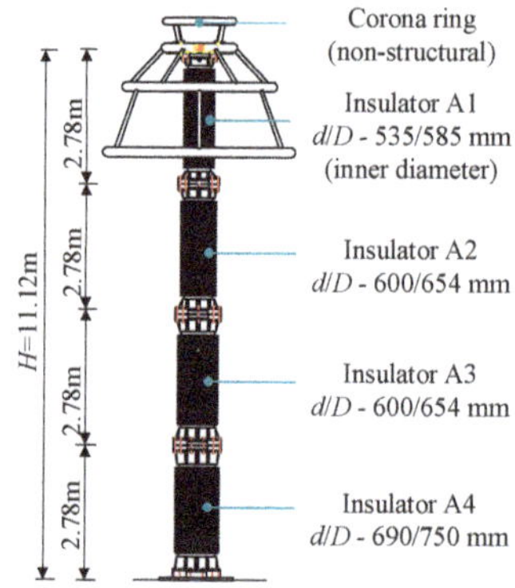

Figure 5. Dimensions of the 1000 kV surge arrester with four composite insulators.

The seismic demand on the displacement of the cantilever structure is generally related to the rigidity of composite tubes. The criteria limit for the deflection of the cylindrical electrical equipment was 1/18 of its length, or 600 mm [49]. The safety margin requirement of 1.67 limited it to 360 mm. Considering uncertainties in modelling, our target of deflection was set to 300 mm in the design of the composite tube. A finite element (FE) model of the specimen with the beam element was built, as shown on the left side of Figure 5, and the thickness of the composite tube was taken as a variable, changing from 10 mm to 50 mm. The response spectrum method was adopted in this parametric analysis. Figure 6 shows that the targeted 300 mm deflection corresponded to a 24 mm thickness of the tube and 1.58 Hz of fundamental frequency. Thus, the thickness of the tube for the insulator at the top was chosen as 25 mm, and a slightly larger thickness, 30 mm, was selected for the insulator at the bottom. The inner/outer diameters of the four insulators became 535/585, 600/654, 600/654, and 690/750 (in millimetres). These values were adopted in the production and experiments in this study.

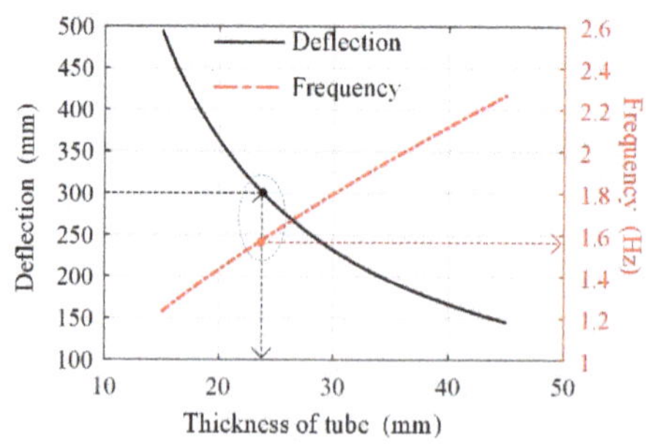

Figure 6. Relationship among tube thickness, deflection at the top, and primary frequency.

The vibration modes of the equipment are shown on the right of Figure 7, and the frequencies of the first three modes were 1.73 Hz, 8.8 Hz, and 22.5 Hz. By inputting the 1.4×0.5 g PGA seismic motion, a time history analysis was carried out to estimate the seismic demand of the specimen. As depicted in Figure 8, the bending moment at the base was 665 kN·m, and the displacement at the top was 265 mm. A safety margin of 1.67 demanded that the composite insulator should withstand a bending moment of 1112 kN·m.

Figure 7. FE model of the 1000 kV surge arrester (**left**) and the first three mode shapes (**right**).

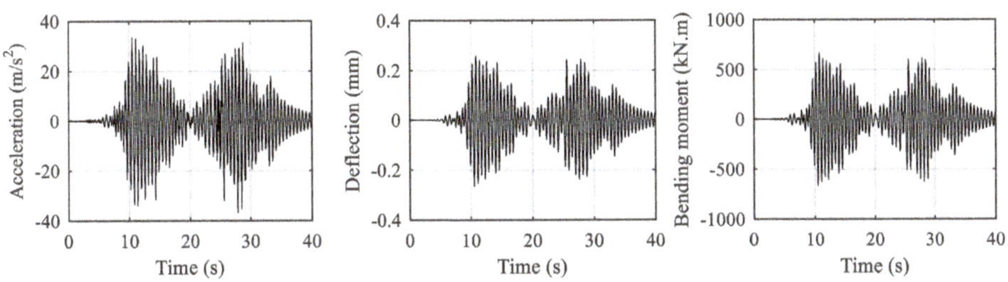

Figure 8. Seismic demand estimation of the 1000 kV surge arrester based on the linear FE model, according to the acceleration spectrum in the GB 50260 and Q/GDW 11391 standard.

3. Bending Test of Composite Insulator with Retrofitted Bonded Sleeve Joints

3.1. Bonding Failure at the Joint of Composite Insulators under Bending

As depicted in Figure 4, the bonded sleeve joint of the composite insulator utilized epoxy as the adhesive, and a layer of epoxy was created between the inner face of the metal sleeve and the outer face of the composite tube. Figure 9 illustrates that the bending moment is resisted by tangential shear and lateral pressure stress at the cohesive joint.

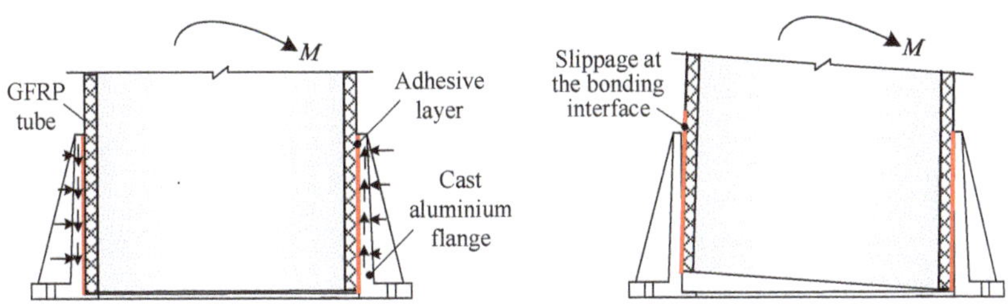

Figure 9. The bonded connection at the end of the composite insulator.

When the combined stress reaches a threshold, bonding failure may happen. Previous testing on an 800 kV composite post insulator by the authors [33], as seen on Figure 10a, revealed that the bonded joint at connection was a significant weak point. When the load reached 33 kN, slippage or pull-out occurred at the bonding interface, resulting in a significant loss of stiffness and loading capacity, as indicated by a sharp turn in the loading–deflection curve shown in Figure 10b. The maximum stress on the composite tube was approximately 80 MPa, considerably lower than the bending strength of the tube.

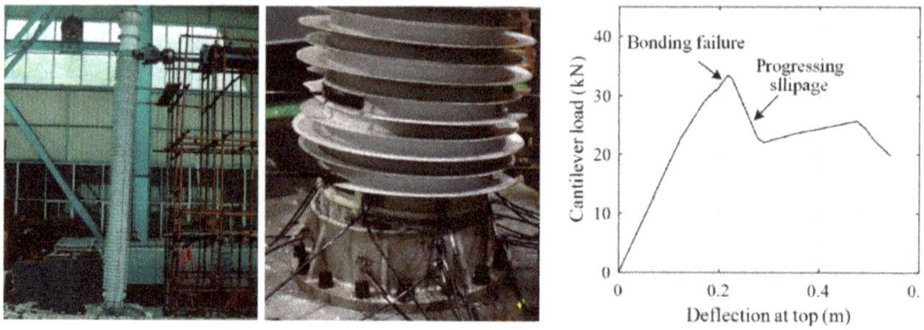

(**a**) (**b**)

Figure 10. Bonding failure of an 800 kV composite insulator under a cantilever bending test (**a**) and loading–deflection curve (**b**) [33].

3.2. *Bending Test of Composite Insulator Retrofitted with Shear Stiffeners*

The bending failure of the composite insulator was found to be localized at the joint, indicating that retrofitting may be focused on the bonded sleeve joints and the remaining components could remain unchanged. Examining the joint revealed debonding at the interface of the aluminium and epoxy layer, with no adhesion remaining on the inner face of the aluminium flange, indicating the bonding strength between the epoxy and aluminium was insufficient.

In order to enhance bonding, the manufacturer of the composite insulators designed shear stiffeners at the bonding interface. As illustrated in Figure 11, fillisters were cut into the inner surface of the sleeve and the outer surface of the GFRP tube, thus creating a sequence of ring-shaped voids that were to be filled with epoxy. When the epoxy had fully solidified, it turned into ring-shaped stiffeners inserted between the flange sleeve and GFRP tube. Figure 11 illustrates the distribution of shear stresses at the flange sleeve and the composite tube under bending moment. The stiffeners improved the shear resistance at the bonding interface, thereby retrofitting the composite insulator for the bending load.

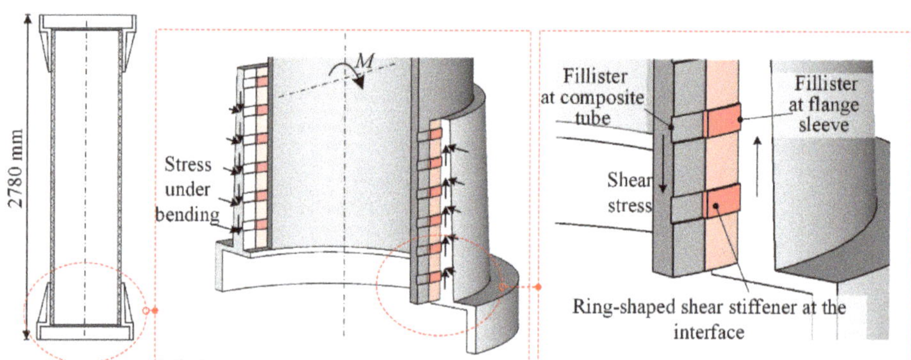

Figure 11. Section of the bonded sleeve joint for the GFRP composite insulator retrofitted by ring-shaped shear stiffeners.

A full-scale bending test was conducted to evaluate the performance of the aforementioned retrofitting on the composite insulators of the 1000 kV surge arrester. Figure 12 shows the specimen, which was made up of 4 insulators and was 11.2 m long. The joints at the insulators had ring-shaped shear stiffeners applied for retrofitting, and the flange was made of cast aluminium. Figure 12a depicts the setup for the bending test. Figure 12b illustrates the application of a cantilever force at the top, where the displacement was measured. The pull force at the top gradually increased until the bending failure of the specimen. The testing setup and loading protocol followed the procedure in [49].

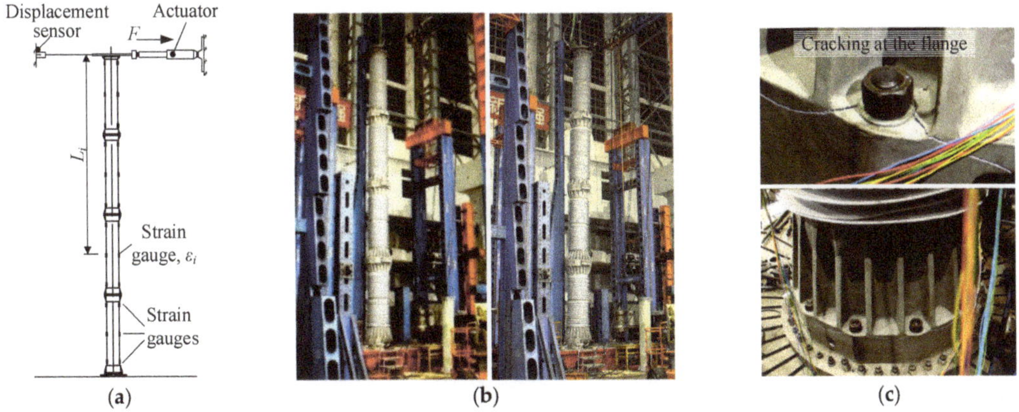

Figure 12. Cantilever bending test of composite insulators retrofitted by shear stiffeners. (**a**) Test setup. (**b**) Specimen and loading facility. (**c**) Cracking of the flange.

Figure 13 shows the loading–deflection curve of the bending test. When the loading reached 78 kN (corresponding to an 867 kN·m bending moment), audible sound emission and a minor drop in force occurred, indicating the debonding at the interface. However, as shown in the zoomed window of Figure 13, the specimen continued to undertake lateral loads, proving that the shear stiffeners were successful in preventing the development of bonding failure at the interface. Following that, an abrupt failure occurred at the base when loading reached 85.8 kN or 954 kN·m, as shown in Figure 12c. Interestingly, the break appeared at the cast aluminium sleeve, contrasted sharply with the debonding at the interface in the previous test (shown in Figure 10). By referring to the crack's location and orientation, it can be concluded the crack was brought on by tension stress, implying the cast aluminium flange's lack of tensile strength. The ultimate bending moment at the bottom was 963 kN·m, lower than the targeted 1112 kN·m predicted in Section 2.

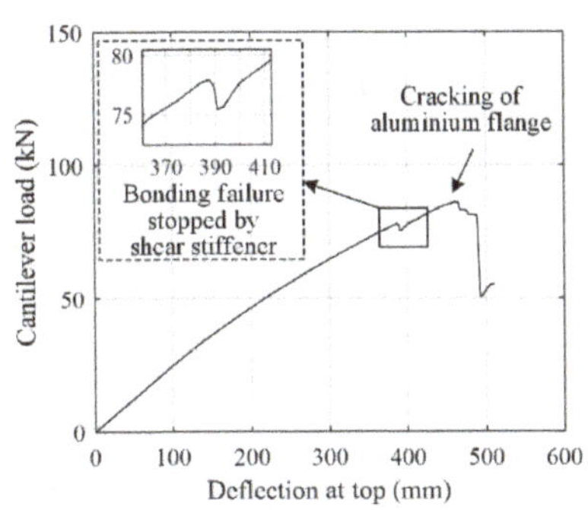

Figure 13. Cantilever loading–deflection curve recorded in bending test I.

3.3. *Bending Test of Composite Insulator Retrofitted with Cast Iron Flanges*

A second attempt to retrofit the specimen was carried out, in order to complete the partially successful design in Section 3.2. The strength of the flange could be greatly increased by exchanging the cast aluminium flange at the base for a cast iron flange. The specimen's weight only slightly increased as a result, and the cast iron flange could be made using the same mould. Therefore, the specimen was kept the same as the one in Section 3.2, with the exception that the flange at the base was made of cast iron.

Figure 14 shows a repeat of the bending test, and Figure 15 depicts the loading–deflection curve. An acoustic event was initially heard when the load was 74.7 kN. The force was close to the 78 kN in Section 3.2, which corresponded to the initial debonding at the joint that was stopped by the shear stiffeners. After that, a major acoustic emission event occurred when the load was 96 kN with 780 mm deflection. The resistance force slightly decreased, and the stiffness deterioration became more pronounced. This incident could be caused by the partial breakdown of some shear stiffeners. Notably, the specimen continued to withstand bending loads after 1067 kN·m, which indicated the structure's earthquake resilience.

Figure 14. Cantilever bending test of composite insulators retrofitted by shear stiffeners and cast iron flange.

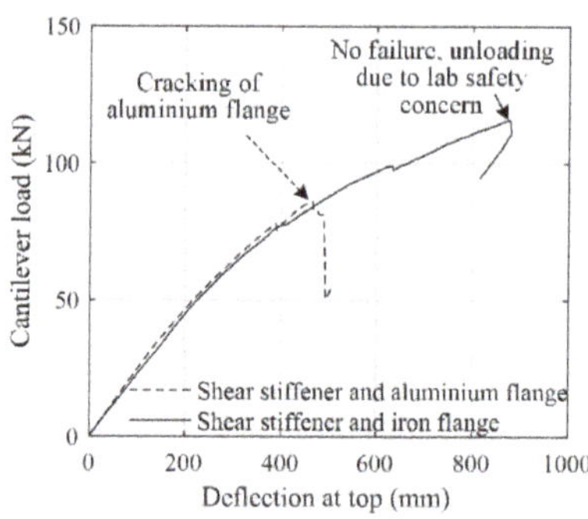

Figure 15. Cantilever loading–deflection curve recorded in bending test II.

When the load reached 117 kN or 1301 kN·m with 1080 mm deflection at the top and 115 MPa stress at the tube, the testing was ended. The testing was terminated because of safety concerns in the laboratory. This load was considerably larger than that of the estimated seismic demand in Section 2. As shown on the right side of Figure 12, the specimen was inspected after the test. The effectiveness of the retrofitting at the bonded sleeve joint was demonstrated by the absence of slippage in the bonded interface or crack on the metal flange.

4. Seismic Performance Validation by Full-Scale Shaking Table Test

4.1. Testing Design

In this section, a full-scale shaking table test was conducted as a direct method of seismic performance validation of the surge arrester with the retrofitted composite insulators. All of the electrical parts had been assembled inside the surge arrester, and the insulators utilized were the same as those in the flexural test. The specimen was fixed on a 4 m by

6 m shaking table, as shown in Figure 16. Sensors were arranged as shown in Figure 17. It included strain gauges at the bottom of each insulator and accelerometers at the top. Displacements of the specimen can be acquired by integrating the acceleration curve, and deflections of the specimen can be obtained by subtracting the displacement at the top of the specimen from the displacement at the shaking table.

Figure 16. Shaking table test of the 1000 kV surge arrester with GFRP insulators.

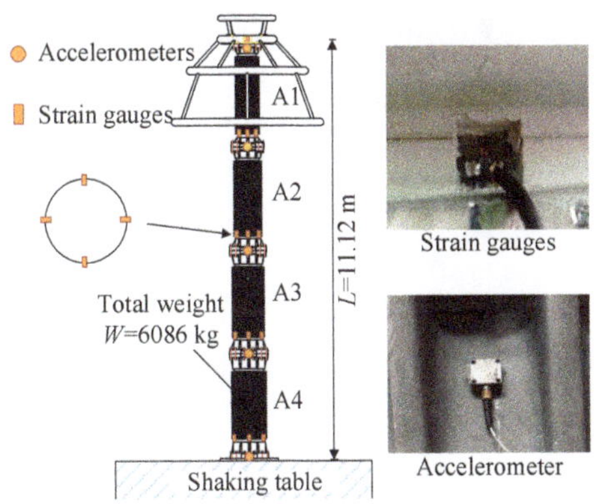

Figure 17. Sensor arrangement in the testing.

The testing schedule is listed in Table 1. The influence of vertical stimulation was minimal due to the axial symmetric shape of the equipment and upright installation position. In the test, the base excitation was input in a single horizontal direction. The GB 50260 compatible artificial ground motion forms the major testing cases, as explained in the selection of input motion in Section 2.1.

Typically, in shaking table testing, both the artificial ground motion and the recorded ground motion excitation should be used. However, over-testing should be avoided for consistency in the succeeding test since the same full-scale specimen was used in electrical performance testing after seismic qualification. The recorded seismic motion excitation test was therefore not planned. The target PGA in the testing schedule increased from 1.4×0.1 g to 1.4×0.5 g in five tests so that the trend in the earthquake response against the magnitude of excitation could be analyzed.

Table 1. Testing schedule.

No	Target PGA	Direction	Input Wave
1	0.05 g	X	White noise
2	1.4 × 0.1 g	X	GB-compatible wave
3	0.05 g	X	White noise
4	1.4 × 0.2 g	X	GB-compatible wave
5	0.05 g	X	White noise
6	1.4 × 0.3 g	X	GB-compatible wave
7	0.05 g	X	White noise
8	1.4 × 0.4 g	X	GB-compatible wave
9	0.05 g	X	White noise
10	1.4 × 0.5 g	X	GB-compatible wave
11	0.05 g	X	White noise

4.2. Testing Result

(1) Result in 1.4 × 0.4 g input test case

In this section, the result for 1.4 × 0.4 g test case is presented. Figure 18 shows the input acceleration at the table surface as well as the response acceleration spectrum. In the testing, the PGA was close to 0.60 g, and the TS could match the RS in the standard.

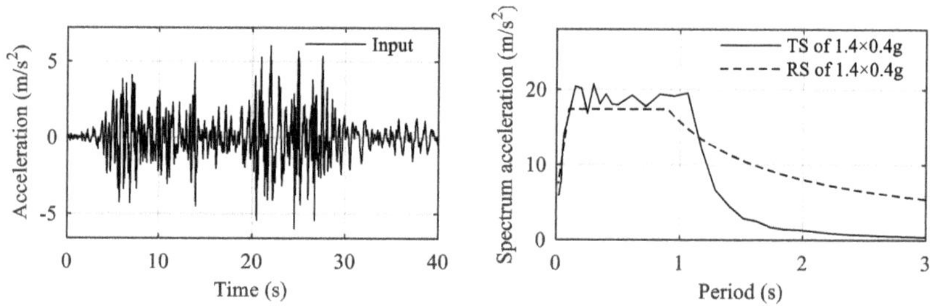

Figure 18. Input excitation and test response spectrum in the testing of the 1.4 × 0.4 g test case.

The curves of major response items are shown in Figure 19. For acceleration at the top, the testing result was 29.43 m/s^2, and the corresponding numerical result was 30.08 m/s^2, which had a different rate of −2.2%. For displacement at the top, the testing result was 0.267 m, and the corresponding numerical result was 0.224 m, which had a different rate of 16%. For bending moment, the testing results were 211 kN·m, 381 kN·m, and 525 kN·m at the bottom of insulators A2, A3, and A4, respectively, and the corresponding numerical results were 214 kN·m, 375 kN·m, and 555 kN·m, which had a different rate of 1.60%, −1.51%, and 5.80%. The comparisons demonstrated the validity of the numerical model.

The relationship between the input motion and the seismic response of the specimen was used to identify the vibration modes of the specimen. The transfer function, $T(f)$, between acceleration at the top of each insulator and acceleration at the table surface was calculated using $T(f) = P_{yx}(f)/P_{xx}(f)$ [50], in which $P_{yx}(f)$ was the cross power spectral density between the output and the input, and $P_{xx}(f)$ was the power spectral density of input. The imaginary part curves of the transfer function are shown on the left side of Figure 20. The peaks of the curves represent the coordinates of a mode shape, as shown on the right side of Figure 20. The fundamental frequency was 1.69 Hz in the test and 1.73 Hz in the analysis, which had a different rate of 2.2%. The second mode frequency was 8.7 Hz in the test and 9.3 Hz in the analysis, which had a different rate of 7%. The mode shapes in the test (as shown on the right side of Figure 20) could also match the mode shapes in numerical analysis (Figure 7).

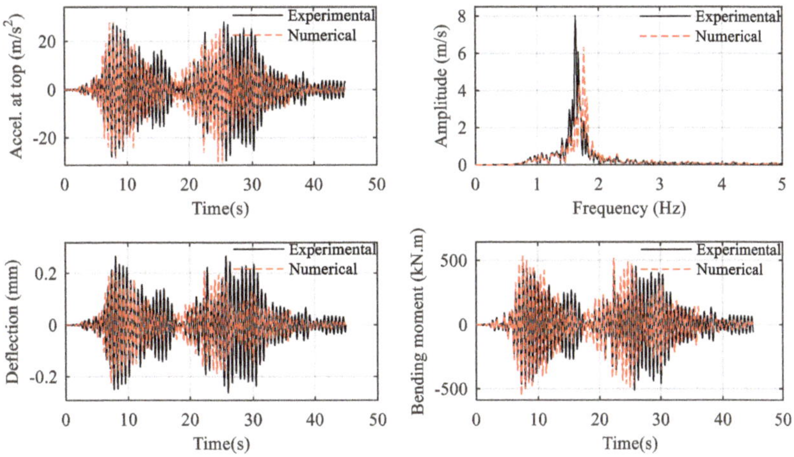

Figure 19. Comparison of testing and numerical results in the 1.4 × 0.4 g test case.

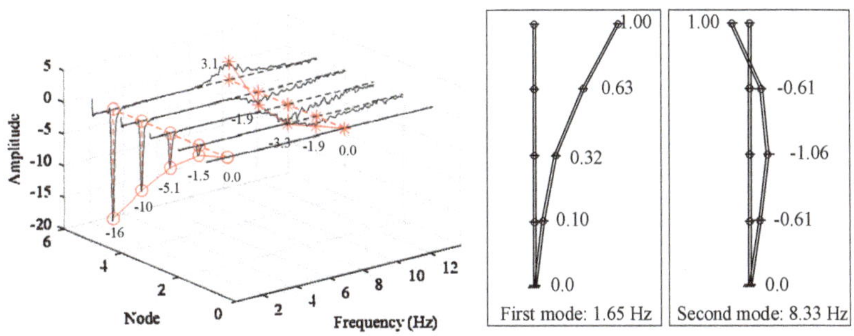

Figure 20. Transfer functions and vibration modes were obtained from the shaking table test.

(2) Result in 1.4 × 0.5 g input test case

The 0.5 g level input represented the most severe seismic action for the 1000 kV surge arrester. This was the most crucial test case for confirming the seismic performance of the GFRP insulator retrofitted in this study. Figure 21 shows the input acceleration and spectrum. The PGA was 0.74 g, and a lowpass filter eliminated the composition of base excitation with periods longer than 1.2 s to avoid driving the shaking table facility beyond the permissible displacement. In the main frequency range of 1–10 Hz, the TS could well match the RS.

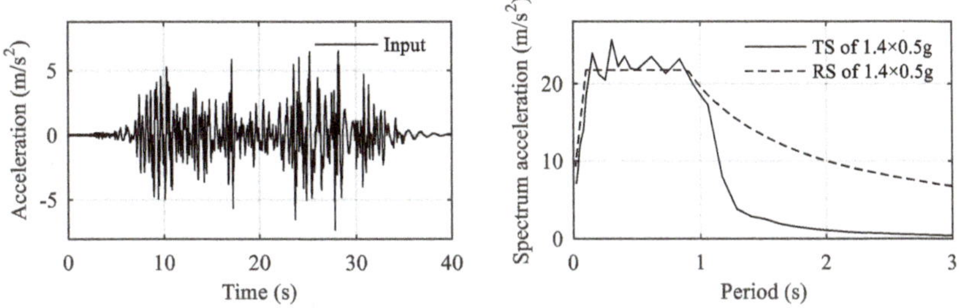

Figure 21. Input excitation and test response spectrum in the testing of 1.4 × 0.5 g test case.

The response time histories in the test are shown in Figure 22. The maximum acceleration at the top was 32.93 m/s^2, which was 4.5 times larger than the excitation. The maximum deflection at the top was 334 mm, smaller than the criterion value of 360 mm. The bending moment was obtained by strain measurement at the base of the composite tube. The maximum stress was 52.1 MPa on the left side and 48.1 MPa on the right side, corresponding to 593 kN·m of bending moment. For the flexural test result, the local bonding damage at the joint of the insulator happened when the bending moment was 1013 kN·m at the position of strain measurement (or 1067 kN·m at the bottom). The shaking table test results of 593 kN·m corresponded to a safety margin of 1.7, meeting the safety margin requirement of 1.67. Additionally, the continued load resisting capacity in the flexural test indicated its resilience in earthquakes. In contrast, the bonded joint broke at 867 kN·m, which was below the safety margin requirement, if no retrofitting design of shear stiffener was made. The joint failed in 954 kN·m, which was close to the safety margin requirement but less resilient due to the failure mode of sudden fracture, if there was no retrofitting with a cast iron flange.

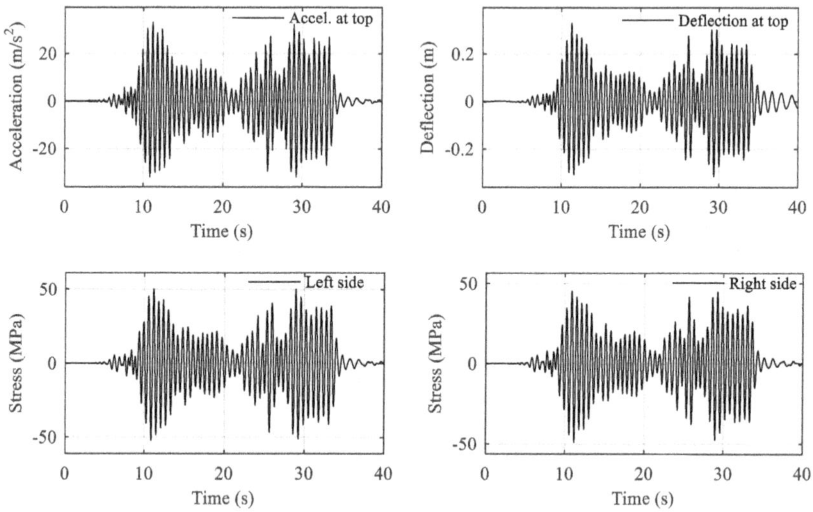

Figure 22. Seismic responses of the specimen in the 1.4 × 0.5 g input test.

In this test case, the deflection of the specimen was around 330 mm, which was 25% larger than the 265 mm estimated by seismic demand analysis but was still below the acceptable limit of 360 mm. It indicated that the specimen had seen a certain degree of rigidity deterioration. In the substation design of electrical equipment with composite insulators, a 25% larger deflection can be empirically considered in cases of high seismic action when the deflection demand is predicted using a linear elastic model. As the slackness of the interconnection conductor is essential for decoupling the earthquake response from nearby equipment, this recommendation derived from the full-scale tests is noteworthy.

4.3. Trend Analysis of Seismic Responses among Different Testing Cases

Figure 23 graphically shows the trends in responses of acceleration, deflection, bending moment and stress. The structural frequencies were stable amongst different excitation levels of testing, with a mild decreasing rate of 5% from 1.73 Hz to 1.64 Hz. Acceleration, deflection, and moment of bending were linear to the applied PGA, indicating the specimen mostly responded in a linear elastic range.

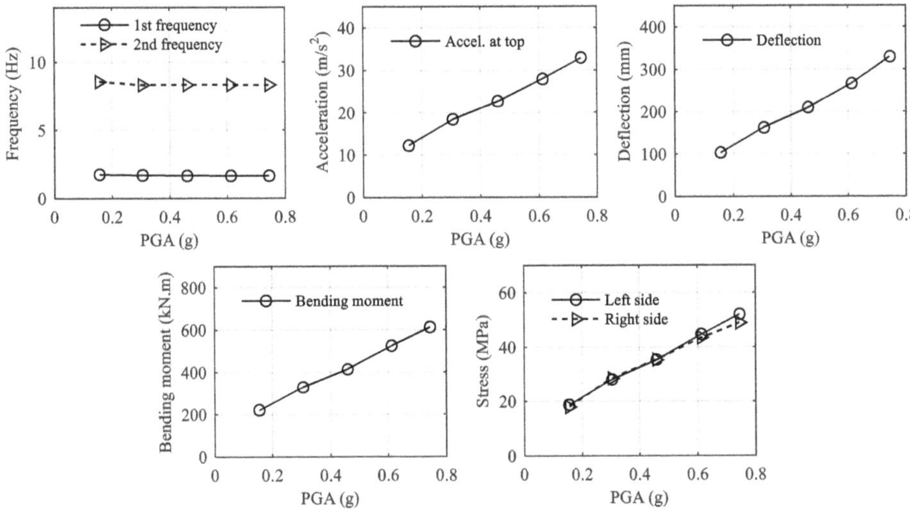

Figure 23. The summary of seismic responses of the specimen with incremental PGAs.

The damping of the specimen can be analyzed by the dynamic parameter identification procedure [51]. The structure was simplified as an SDOF system, described as $m\ddot{u} + c\dot{u} + ku = -m\ddot{u}_g$. The input acceleration at the base and response acceleration at the top were taken as the input and output of the SDOF system, respectively. An equivalent viscous damping ratio was employed to describe the damping effect of the structure, and the model analysis functions in the Matlab Control System Toolbox [52] were used to identify system damping based on the transfer function between the input and the output. As shown in Figure 24a,b, the amplitude of the transfer function was decreased, indicating a significant increase in damping. The damping ratios resulting from curve fitting operations are shown in Figure 24c, increasing from 2% in the 1.4 × 0.1 g PGA test to 6% in the 1.4 × 0.5 g PGA test. The flexural testing showed that the stiffness degradation happened when the bending moment was 580 kN·m. The 593 kN·m in the shaking table test indicated the specimen had just entered a state of mild stiffness degradation, and damage to the specimen was very small, if there was any. It demonstrated that the damping ratio of composite insulators was a sensitive indicator for detecting structural deterioration. This establishes the basis for further investigation in using the change in damping ratio to detect damage to substation equipment with composite insulators.

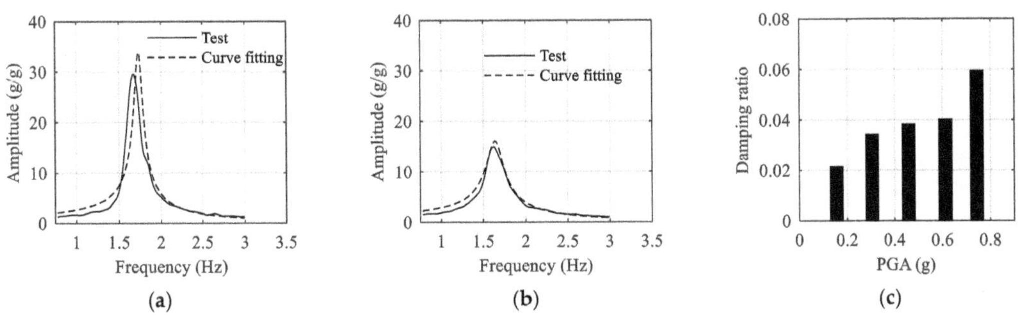

Figure 24. Curve fitting of the transfer function for damping identification. (**a**) 1.4× 0.1 g testing case. (**b**) 1.4 × 0.5 g testing case. (**c**) Damping ratios in different testing cases.

5. Conclusions

This study employed full-scale testing to investigate the seismic performance of the retrofitted composite insulators used in the 1000 kV surge arrester. The conclusions reached are as follows:

(1) The bonded sleeve joint between the GFRP tube and the metal end fitting was a weak point for the bending strength of the composite insulator, which can be retrofitted by introducing ring-shaped shear stiffeners at the joint and replacing the cast iron flange with an aluminium one.
(2) Full-scale cantilever bending tests were carried out to assess the bending strength of the retrofitted composite insulator. The shear stiffener was seen to prevent bonding failure at the joint and increase the specimen's bending strength. The numerical modelling and design optimizations of shear stiffeners are open to further study.
(3) The full-scale shaking table test of the 1000 kV surge arrester with retrofitted insulators demonstrated the equipment's satisfactory performance during an extreme seismic action of 1.4×0.5 g PGA, making it suitable for installation in the UHV substation in regions with high seismic risk.
(4) The responses of the equipment during the shaking table test were mainly in the elastic range, but its damping ratio increased from 2% to 6%, indicating that the damping ratio is a sensitive parameter for the damage detection of equipment with composite insulators.

Author Contributions: Conceptualization, S.L. and Y.C.; methodology, Z.L.; validation, S.L. and H.W.; formal analysis, S.L. and Y.X.; writing—original draft, S.L. and Y.X.; writing—review and editing, S.L. and N.L.; project administration, Z.L. and H.W.; supervision, N.L.; funding acquisition, Y.C. All authors have read and agreed to the published version of the manuscript.

Funding: This research was funded by the Scientific Research Program of State Grid Corporation of China (Grant No. 5200-202156074A-0-0-00) and the State Grid Fujian Electric Power Co., Ltd. (Grant number SGFJJY00BDJS2100019), titled 'Key technology and application research on design and installation reliability assessment for main equipment in the UHV substations'.

Data Availability Statement: Not applicable.

Acknowledgments: The tests in the study were organized by the State Grid Corporation of China and the China Electric Power Research Institute.

Conflicts of Interest: The authors declare no conflict of interest.

References

1. Chalishazar, V.H.; Brekken, T.K.A.; Johnson, D.; Yu, K.; Newell, J.; Chin, K.; Weik, R.; Dierickx, E.; Craven, M.; Sauter, M.; et al. Connecting Risk and Resilience for a Power System Using the Portland Hills Fault Case Study. *Processes* **2020**, *8*, 1200. [CrossRef]
2. Tofani, A.; D'agostino, G.; di Pietro, A.; Giovinazzi, S.; Pollino, M.; Rosato, V.; Alessandroni, S. Operational Resilience Metrics for Complex Inter-Dependent Electrical Networks. *Appl. Sci.* **2021**, *11*, 5842. [CrossRef]
3. Dinh, N.H.; Kim, J.Y.; Lee, S.J.; Choi, K.K. Seismic Vulnerability Assessment of Hybrid Mold Transformer Based on Dynamic Analyses. *Appl. Sci.* **2019**, *9*, 3180. [CrossRef]
4. Yu, Y.; Li, G.; Li, P.; Zhu, Q. Investigation and Analysis of Electric Equipment Damage in Sichuan Power Grid Caused by Wenchuan Earthquake. *Power System Technology* **2008**, *32*, 1–6.
5. Pavel, F.; Vacareanu, R. Seismic Risk Assessment for Elements of the Electric Network in Romania. *Buildings* **2022**, *12*, 244. [CrossRef]
6. Ang, H.S.; Pires, J.A.; Villaverde, R. A Model for the Seismic Reliability Assessment of Electric Power Transmission Systems. *Reliab. Eng. Syst. Saf.* **1996**, *51*, 7–22. [CrossRef]
7. Kazama, M.; Noda, T. Damage Statistics (Summary of the 2011 off the Pacific Coast of Tohoku Earthquake Damage). *Soils Found.* **2012**, *52*, 780–792. [CrossRef]
8. Kwasinski, A.; Eidinger, J.; Tang, A.; Tudo-Bornarel, C. Performance of Electric Power Systems in the 2010-2011 Christchurch, New Zealand, Earthquake Sequence. *Earthq. Spectra* **2014**, *30*, 205–230. [CrossRef]
9. You, H.; Zhao, F. M7.0 Earthquake in Lushan and Damage Cause Analysis of Power Facilities. *Electr. Power Constr.* **2013**, *34*, 100–104. [CrossRef]

10. Salem, A.A.; Lau, K.Y.; Abdul-Malek, Z.; Zhou, W.; Al-Ameri, S.; Al-Gailani, S.A.; Rahman, R.A. Investigation of High Voltage Polymeric Insulators Performance under Wet Pollution. *Polymers* **2022**, *14*, 1236. [CrossRef]
11. Koo, J.H.; Seong, J.K.; Hwang, R.; Lee, B.W. Experimental Assessment on Air Clearance of Multiple Valve Unit Considering Switching Impulse and Dc Superimposed Switching Impulse. *Energies* **2020**, *13*, 2958. [CrossRef]
12. Liu, Z.; Cheng, Y.; Lu, Z. Shake Table Test on UHV Standardization Lightning Arrester Installed with Shear-Type Lead Dampers. *High Volt. Eng.* **2018**, *44*, 2595–2602.
13. Saleem, M.Z.; Akbar, M. Review of the Performance of High-Voltage Composite Insulators. *Polymers* **2022**, *14*, 431. [CrossRef] [PubMed]
14. Papailiou, K.O.; Schmuck, F. *Silicone Composite Insulators: Materials, Design, Applications*; Springer: Berlin/Heidelberg, Germany, 2013. [CrossRef]
15. EPRI. *Survey of Utility Experiences with Composite/Polymer Components in Transmission Class (69-765 KV Class) Substations*; EPRI: Palo Alto, CA, USA, 2004.
16. Cheng, Y.; Li, S.; Lu, Z.; Liu, Z.; Zhu, Z. Seismic Risk Mitigation of Cylindrical Electrical Equipment with a Novel Isolation Device. *Soil Dyn. Earthq. Eng.* **2018**, *111*, 41–52. [CrossRef]
17. Yang, Z.; Xie, Q.; He, C.; Xue, S. Numerical Investigation of the Seismic Response of a UHV Composite Bypass Switch Retrofitted with Wire Rope Isolators. *Earthq. Eng. Eng. Vib.* **2021**, *20*, 275–290. [CrossRef]
18. Fang, H.; Bai, Y.; Liu, W.; Qi, Y.; Wang, J. Connections and Structural Applications of Fibre Reinforced Polymer Composites for Civil Infrastructure in Aggressive Environments. *Compos. B Eng.* **2019**, *164*, 129–143. [CrossRef]
19. Kumosa, M.; Armentrout, D.; Kumosa, L.; Han, Y.; Carpenter, S.H. Analyses of Composite Insulators with Crimped End-Fittings: Part II—Suitable Crimping Conditions. *Compos. Sci. Technol.* **2002**, *62*, 1209–1221. [CrossRef]
20. Kumosa, M.; Han, Y.; Kumosa, L. Analyses of Composite Insulators with Crimped End-Fittings: Part I—Non Linear Finite Element Computations. *Compos. Sci. Technol.* **2002**, *62*, 1191–1207. [CrossRef]
21. Luo, F.J.; Yang, X.; Bai, Y. Member Capacity of Pultruded GFRP Tubular Profile with Bolted Sleeve Joints for Assembly of Latticed Structures. *J. Compos. Constr.* **2015**, *20*, 4015080. [CrossRef]
22. Luo, F.J.; Bai, Y.; Yang, X.; Lu, Y. Bolted Sleeve Joints for Connecting Pultruded FRP Tubular Components. *J. Compos. Constr.* **2016**, *20*, 4015024. [CrossRef]
23. Qiu, C.; Feng, P.; Yang, Y.; Zhu, L.; Bai, Y. Joint Capacity of Bonded Sleeve Connections for Tubular Fibre Reinforced Polymer Members. *Compos. Struct.* **2017**, *163*, 267–279. [CrossRef]
24. Fawzia, S.; Zhao, X.L.; Al-Mahaidi, R. Bond-Slip Models for Double Strap Joints Strengthened by CFRP. *Compos. Struct.* **2010**, *92*, 2137–2145. [CrossRef]
25. Duriatti, D.; Béakou, A.; Levillain, R. Optimisation of the Crimping Process of a Metal End-Fitting onto a Composite Rod. *Compos. Struct.* **2006**, *73*, 278–289. [CrossRef]
26. Prenleloup, A.; Gmür, T.; Botsis, J.; Papailiou, K.O.; Obrist, K. Stress and Failure Analysis of Crimped Metal–Composite Joints Used in Electrical Insulators Subjected to Bending. *Compos. Part A Appl. Sci. Manuf.* **2009**, *40*, 644–652. [CrossRef]
27. Qureshi, J. A Review of Fibre Reinforced Polymer Structures. *Fibers* **2022**, *10*, 27. [CrossRef]
28. Lacey, A.W.; Chen, W.; Hao, H.; Bi, K. Review of Bolted Inter-Module Connections in Modular Steel Buildings. *J. Build. Eng.* **2019**, *23*, 207–219. [CrossRef]
29. Castagnetti, D.; Corigliano, P.; Barone, C.; Crupi, V.; Dragoni, E.; Guglielmino, E. Predicting the Macroscopic Shear Strength of Tightened-Bonded Joints from the Intrinsic High-Pressure Properties of Anaerobic Adhesives. *Metals* **2022**, *12*, 1141. [CrossRef]
30. Rośkowicz, M.; Godzimirski, J.; Komorek, A.; Jasztal, M. The Effect of Adhesive Layer Thickness on Joint Static Strength. *Materials* **2021**, *14*, 1499. [CrossRef]
31. Orefice, A.; Mancusi, G.; Dumont, S.; Lebon, F. An Experimental/Numerical Study on the Interfacial Damage of Bonded Joints for Fibre-Reinforced Polymer Profiles at Service Conditions. *Technologies* **2016**, *4*, 20. [CrossRef]
32. Zhang, Z.; Wu, C.; Nie, X.; Bai, Y.; Zhu, L. Bonded Sleeve Connections for Joining Tubular GFRP Beam to Steel Member: Numerical Investigation with Experimental Validation. *Compos. Struct.* **2016**, *157*, 51–61. [CrossRef]
33. Li, S.; Tsang, H.-H.; Cheng, Y.; Lu, Z. Seismic Testing and Modeling of Cylindrical Electrical Equipment with GFRP Composite Insulators. *Compos. Struct.* **2018**, *194*, 454–467. [CrossRef]
34. Roh, H.; Oliveto, N.D.; Reinhorn, A.M. Experimental Test and Modeling of Hollow-Core Composite Insulators. *Nonlinear Dyn.* **2012**, *69*, 1651–1663. [CrossRef]
35. Epackachi, S.; Dolatshahi, K.M.; Oliveto, N.D.; Reinhorn, A.M. Mechanical Behavior of Electrical Hollow Composite Post Insulators: Experimental and Analytical Study. *Eng. Struct.* **2015**, *93*, 129–141. [CrossRef]
36. Bonhôte, P.; Gmür, T.; Botsis, J.; Papailiou, K.O. Stress and Damage Analysis of Composite–Aluminium Joints Used in Electrical Insulators Subject to Traction and Bending. *Compos. Struct.* **2004**, *64*, 359–367. [CrossRef]
37. Wang, H.; Cheng, Y.; Lu, Z.; Zhu, Z.; Zhang, S. Research on Bending Rigidity at Flange Connections of UHV Composite Electrical Equipment. *Shock Vib.* **2020**, *2020*. [CrossRef]
38. Cheng, Y.; Zhu, Z.; Lu, Z.; Huang, B.; Li, S. *Q/GDW 11391-2015 Technical Code for Seismic Test of High Voltage Pillar Type Electrical Equipment*; Technical Standard; State Grid Corporation of China: Beijing, China, 2015; pp. 1–22.
39. Whittaker, A.S.; Fenves, G.L.; Gilani, A.S.J. Seismic Evaluation and Analysis of High-Voltage Substation Disconnect Switches. *Eng Struct* **2007**, *29*, 3538–3549. [CrossRef]

40. Fahad, M. *Seismic Evaluation and Qualification of Transformer Bushings*; State University of New York at Buffalo: Buffalo, NY, USA, 2013.
41. Alessandri, S.; Giannini, R.; Paolacci, F.; Amoretti, M.; Freddo, A. Seismic Retrofitting of an HV Circuit Breaker Using Base Isolation with Wire Ropes. Part 2: Shaking-Table Test Validation. *Eng. Struct.* **2015**, *98*, 263–274. [CrossRef]
42. He, C.; He, Z.; Xie, Q. Experimental Evaluations on Seismic Performances of Porcelain and GFRP Composite UHV GIS Bushings. *Materials* **2022**, *15*, 4035. [CrossRef]
43. Cheng, Y.; Liu, Z.; Lu, Z.; Zhu, Z.; Li, S.; Liu, H. Experimental Study on Seismic Performance and Structural Characteristics of ±800 KV Composite Post Insulators. *Electr. Power* **2017**, *50*, 89–96. [CrossRef]
44. GB. *Chinese Standard—GB50260 Code for Seismic Design of Electrical Installations*; China Planning Press: Beijing, China, 2013.
45. Zebing, D.; Zhicheng, L.; Yongfeng, C.; Zhubing, Z.; Min, Z.; Fengxin, Z.; Hongbing, Y. *Q/GDW 11132 Technical Specification for Seismic Design of Ultra-High Voltage Porcelain Insulating Equipment and Installation/Maintenance to Energy Dissipation Devices*; Technical Standard; State Grid Corporation of China: Beijing, China, 2013; pp. 1–31.
46. Li, S.; Tsang, H.H.; Cheng, Y.; Lu, Z. Considering Seismic Interaction Effects in Designing Steel Supporting Structure for Surge Arrester. *J. Constr. Steel Res.* **2017**, *132*, 151–163. [CrossRef]
47. IEEE. *IEEE 693 Recommended Practice for Seismic Design of Substations*; IEEE: New York, NY, USA, 2005.
48. Kayhan, A.H.; Demir, A.; Palanci, M. Multi-Functional Solution Model for Spectrum Compatible Ground Motion Record Selection Using Stochastic Harmony Search Algorithm. *Bullet. Earthq. Eng.* **2022**, *20*, 6407–6440. [CrossRef]
49. Cheng, Y.; Zhang, F.; Li, S.; Lu, Z.; Huang, B.; Huang, H.; Zhang, Z.; You, H. *Q/GDW 11594-2016 Seismic Performance Testing Method for Composite Post Insulator*; Technical Standard; State Grid Corporation of China: Beijing, China, 2016; pp. 1–13.
50. Solomon, O.M., Jr. PSD Computations Using Welch's Method. *Sandia Rep.* **1991**, *91*, 1–64. [CrossRef]
51. Gattulli, V.; Potenza, F.; Piccirillo, G. Multiple Tests for Dynamic Identification of a Reinforced Concrete Multi-Span Arch Bridge. *Buildings* **2022**, *12*, 833. [CrossRef]
52. Mathworks Control System Toolbox. Available online: www.mathworks.com (accessed on 20 June 2021).

buildings

MDPI

Article

Study on Optimal Design of Grotto-Eave System with Cable Inerter Viscous Damper for Vibration Control

Jizhong Huang [1], Ruoyu Zhang [1,2], Qingyang Luo [3], Xiuwei Guo [1,2] and Meigen Cao [4,*]

1 Institute for Conservation of Cultural Heritage, Shanghai University, Shanghai 200444, China; hjizhong@163.com (J.H.); 21820359@shu.edu.cn (R.Z.); guoxw@shu.edu.cn (X.G.)
2 School of Mechanics and Engineering Science, Shanghai University, Shanghai 200444, China
3 School of Materials Science and Engineering, Shanghai University, Shanghai 200444, China; lqy4227@shu.edu.cn
4 School of Civil Engineering, North China University of Technology, Beijing 100144, China
* Correspondence: caomeigen@ncut.edu.cn

Abstract: In this paper, the mechanical model of grotto–eave system with cable inerter viscous damper (CIVD) is established, and the vibration control equations are established. Firstly, the stochastic response is carried out, and the optimization design of design parameters of CIVD is carried out for the grotto–eave systems with different connection types. Finally, the vibration mitigation control performance of CIVD under different seismic inputs is analyzed. The research shows that in the optimal design of CIVD, the inerter–mass ratio and damping ratio should be reduced as much as possible to improve the feasibility of the application of CIVD in cultural relics protection engineering under the condition of meeting the target damping ratio. The demand-based optimal method can minimize the cost by enhancing damping element deformation in a small damping ratio, while ensuring that the value of displacement index of grotto–eave system can be reached. Hence, the deformation and damping force of CIVD will increase simultaneously due to the efficient tuning and damping amplification of CIVD. CIVD can enlarge the apparent mass through rotation and damping force through enhancement deformation. Hence, compared with other conventional dampers (such as viscous damper), optimal CIVD has lower damping ratio under the same demand index of grotto–eave system. It can be realized that the lightweight and high efficiency of the damper, and can be applied to the vibration mitigation and reinforcement of the grotto–eave system.

Keywords: grotto–eave system; inerter element; stochastic response; demand-based optimal method; dynamic response

Citation: Huang, J.; Zhang, R.; Luo, Q.; Guo, X.; Cao, M. Study on Optimal Design of Grotto-Eave System with Cable Inerter Viscous Damper for Vibration Control. *Buildings* **2022**, *12*, 661. https://doi.org/10.3390/buildings12050661

Academic Editor: Chiara Bedon

Received: 2 May 2022
Accepted: 13 May 2022
Published: 16 May 2022

1. Introduction

Historic buildings and cultural relics are valuable and must be preserved carefully. However, such structures, such as grotto–eave system, may be destroyed during earthquakes owing to the deterioration of their structural performance over time [1]. Moreover, the seismic resistance of most historic buildings is inferior to that of modern structures because the mechanical performance of the construction materials deteriorates with time [2,3]. Therefore, it is of significance to protect historic buildings and the cultural relics housed in such buildings. There are various types of immovable heritage structures, including wooden, bricked and rammed structures [4–6]. The historical buildings of the most historical and cultural value are mainly wooden structures and bricked structures. In general, non-structural components are often used for decoration, and they may have a certain architectural or utility function. However, in the immovable heritage structures [7], the non-structural components are also regarded as part of the heritage. Their cultural value is higher than the value of structures. It is the main difference between heritage buildings and modern buildings in seismic protection that non-structural components are as important as the main structure.

In order to improve the seismic performance of immovable cultural relics, we can use traditional reinforcement methods that including strengthening the connecting parts of structural members or adding shear walls and supports, and improving the rigidity of the structure. Bento et al. [8] proposed the use of steel or concrete to strengthen the connection of bricked walls, thereby increasing the lateral resistance of the structure. Aty [9] proposed another modification method by using wood support system, and analyzed the seismic performance of X-shaped wood support and K-shaped wood support structure. Witzany et al. [10] conducted experimental research and analysis on the failure mechanism and ultimate compressive bearing capacity of carbon fiber reinforced polymer (CFRP) reinforced bricked walls. Akcay et al. [11] performed reinforcement tests and numerical analyses of historic bricked buildings using different conventional techniques. Although the structural rigidity increases, it also increases the seismic damage to the structure, which heritage buildings may not be able to withstand. These traditional renovation methods all help to improve the seismic performance of heritage buildings. However, the original information of the cultural relics cannot be preserved, and the modern buildings and the cultural relics inside may be damaged during the renovation. Structural control is an efficient approach for suppressing the dynamic response of civil structures under external actions such as those exerted by wind, earthquake, and other hazardous events. Hence, it is necessary to develop a lightweight and high efficiency damper that can control the vibration response of historic buildings and cultural relics simultaneously. An cable inerter viscous damper (CIVD) for seismic response mitigation of typical historic buildings and immovable cultural relics system is proposed in this paper.

The vibration control technologies of inerter-based dampers have been developed based on electromechanical similarity theory [12]. Compared with the TMD, the inerter-based damper can control the inertial force at the two terminals directly. Moreover, the inerter element can effectively enlarge the small apparent mass through converting the translational motion into rotary motion such as ball screw. In 2001, Smith [13] put forward the concept of inerter element and gave the basic forms of ball screw inerter element and rack and pinion inerter element, and designed a hydraulic inerter element in 2013 [14]. Subsequently, shock absorbers were proposed such as tuned viscous mass damper (TVMD) and tuned inerter damper (TID). The design method of the inerter system was also studied. Ikago et al. [15] derived a closed-form formula for TVMD optimization design based on fixed-point theory. Pan et al. [16] considered the natural damping of the original structure and the cost of the inerter-based damper, and make up for the deficiency of fixed-point theory. Then, he proposed the design method of SPIS-II inerter-based damper based on stochastic response mitigation ratio [17]. Hwang et al. [18] proposed a rotation inerter system connected with a toggle brace based on ball screw. It is shown that the system can be effectively used in the structure with small drift. Zhang et al. [19,20] applied the inerter damper system to high-rise structures such as chimneys and wind power towers, and proved the effectiveness of the inerter-based damper in high-rise structures. Gao et al. [21] put forward an optimum design method of viscous inerter damper (VID) based on the feedback control theory. De Domenico [22–32] proposed the optimal design methods of inerter-based TMD systems for seismic response mitigation.Although some scholars have used the inerter-based damper in practical engineering [33], most of the research on the inerter-based damper is still in the stage of theoretical analysis, and only a few scholars have proposed the connection mode and design method of the inerter system applied in building structures [34]. Xie et al. [35,36] put forward a cable-bracing inerter system (CBIS), and shows that it is easy to install and can effectively control displacement and acceleration of structure. Wang et al. [37] put forward a new tuned inerter-negative-stiffness damper (TINSD) based on fixed-point method, which is more effective than the TID, TVMD, and INSD in reducing the dynamic response of structures. It is the satisfactory scheme using inerter-based damper to suppress vibrations of some special structure, such as transmission line [38], tall building [39] and transformer-bushing system [40].

At present, the fixed-point method is mainly used in the parameters design of inerter systems and other vibration control systems, such as TMD, TID and TVMD. A better damping ratio or stiffness ratio can be obtained by fixed-point method, but the optimal inerter-mass ratio (or mass ratio) cannot be obtained directly. In this way, the obtained damping ratio is larger and the inerter element cannot amplify the damping effect as far as possible. At the same time, it is necessary to install lighter dampers to meet the higher seismic requirements for immovable historic buildings and cultural relics such as grottoes and eaves.

In this paper, a cable inerter viscous damper (CIVD) system is proposed. The end of the lightweight inerter viscous damper is directly connected with the elastic cable, which can be quickly installed in various historic buildings and cultural relics. The system can not only realize structural reinforcement and improve the integrity of the structure, but also realize the lightweight of the shock absorber. Firstly, Section 2 introduces the basic principle of CIVD. The motion control equations and frequency response functions of grotto–eave system with CIVD are established. In the Section 3, the parameter analysis is carried out to obtain the minimum additional damping ratio of CIVD under different vibration mitigation ratios. Additionally, the demand-based optimal design method of CIVD is proposed; Finally, it is carried out that the dynamic time history analysis of the grotto–eave system installed with CIVD under the ground motions in Section 4 to verify the vibration control effect and parameter optimization results of CIVD. The research in this paper can provide reference for the design of efficient and lightweight vibration mitigation scheme of immovable historic buildings and cultural relics based on inerter damping system.

2. Theoretical Analysis of Grotto-Eave System with CIVD

2.1. Mechanical Model of Inerter Element and CIVD

Equation (1) shows the output force of inerter element and it has different accelerations in two terminals. Hence, the output of inerter element is proportional to the relative acceleration at two terminals, which can be shown as follows:

$$f_I = m_d(a_2 - a_1) \tag{1}$$

where, f_I is the output force of the inerter element, a_1 and a_2 are the accelerations at terminals, m_d is the inertance, and Figure 1 is the mechanical model of the inerter element.

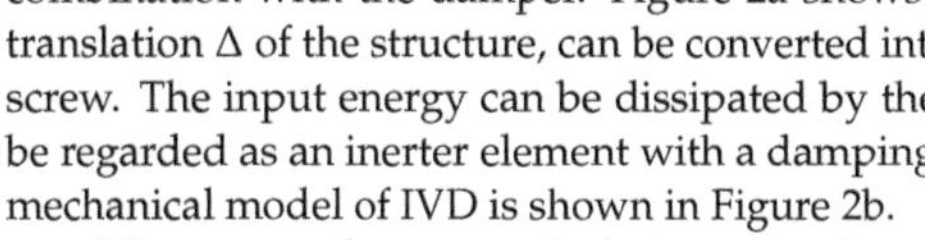

Figure 1. Mechanical model of an inerter element.

The inerter element cannot dissipate energy by itself and it is generally used in combination with the damper. Figure 2a shows the inerter viscous damper (IVD), the translation Δ of the structure, can be converted into rotation φ in the IVD through the ball screw. The input energy can be dissipated by the viscous fluid in the damper. IVD can be regarded as an inerter element with a damping element connected in parallel, and the mechanical model of IVD is shown in Figure 2b.

Where m_d and c_d are equivalent mass and equivalent damping coefficient, corresponding to the translation Δ at both ends of IVD; J and c_{vd} are the apparent mass (moment of inertia) and viscous damping constants. The expressions of m_d and c_d can be obtained, where L is the lead of the ball screw:

$$m_d = \frac{4\pi^2}{L^2}J, c_d = \frac{4\pi^2}{L^2}c_{vd}, \varphi = \frac{2\pi}{L}\Delta \tag{2}$$

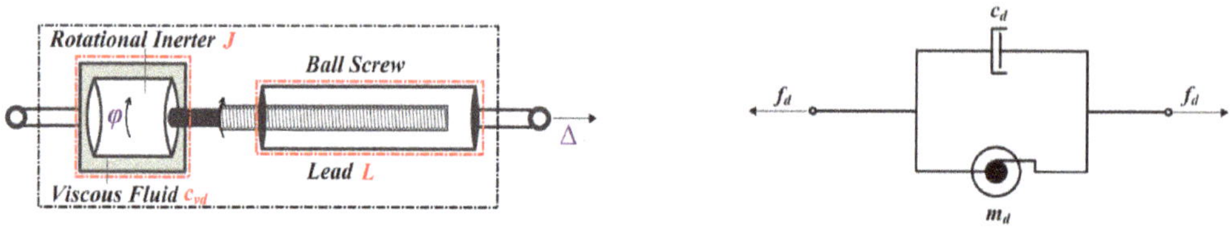

(a) (b)

Figure 2. Structure and mechanical model of an inerter viscous damper (IVD): (**a**) Structure of IVD; and (**b**) mechanical model.

CIVD is IVD connected with an elastic cable, and the cable in this paper is a short-span cable. The reason is that the prestress relaxation of the short cable can be almost ignored. When the angle of cable is an appropriate large constant, the sag of cable and stress relaxation in the short-span structure can be ignored. Compared with the long cable, the short cable has high efficiency of force transmission. The mechanical model is shown in Figure 3. Where k_d is the equivalent stiffness of the elastic cable.

(a) (b)

Figure 3. Mechanical model of a cable inerter viscous damper (CIVD): (**a**) Structure of CIVD; (**b**) and mechanical model.

2.2. Mechanical Model of Grotto-Eave System with CIVD

The grotto–eave system is generally composed of eaves, connection and grotto. Generally, the eaves of the caves are wooden or steel structures. To protect the grottoes, the wooden structures are built in ancient times, while the steel structures are the modern architecture. The height of the structure is generally about 6–20 m, and its overall stiffness is small. The connection is the beam with the eaves lapped on the grotto or inserted into the grotto, so that the two parts are connected into a complete system. The material, type and position of the connection will influence the stiffness of the connection. Similar to the modern frame, the main structure of the grotto is generally an internal hollow structure. The materials of grotto are mostly rock materials such as sandstone, and grotto is generally adjacent to the mountain, the stiffness is relatively large. There is a large difference in stiffness between the eaves and the grottoes. When subjected to external excitation, the incongruity of deformation between the eaves and the grotto body will occur. Many precious cultural heritages are preserved in the grottoes, which have high historical and humanistic value, but are easily damaged by external disturbances; at the same time, since the grottoes are immovable cultural relics, it can only be protected at the original site. It will also lead to the cumulative damage of the cultural relics under frequent external vibration, and resulting in the instantaneous brittle failure of the stone cultural relics.

Under external excitation, it is generally considered to reduce the vibration of the eaves itself to reduce the vibration effect of the eaves on the grottoes. Due to the high demand for vibration mitigation of the system, the bearing capacity of the eave is small and the internal space is limited. Traditional vibration control system (such as TMD) cannot achieve the expected effect. Therefore, it needs to find a lightweight and efficient vibration control system, such as the CIVD, which has the flexible arrangement and the small size damper, and the obvious effect of vibration mitigation. Figure 4 is a layout of CIVD in an eave–grotto system.

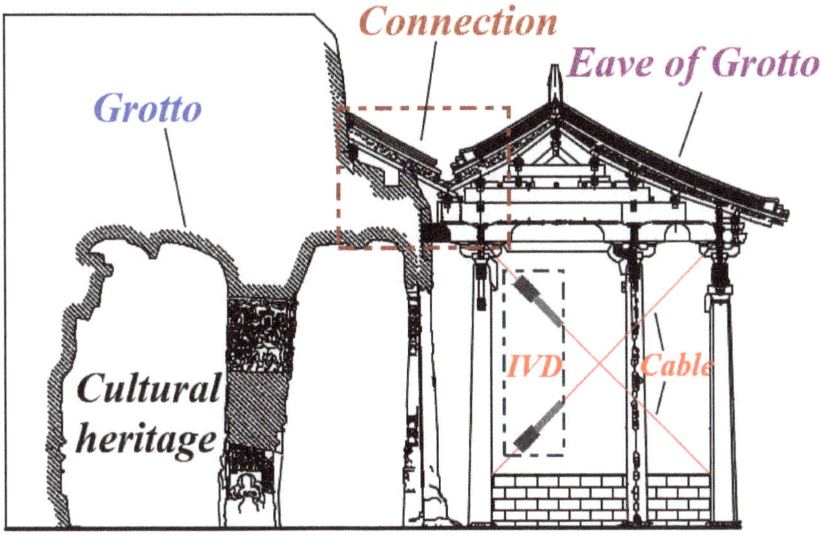

Figure 4. Layout of CIVD in eave–grotto system.

For the grotto–eave system, the shock absorber cannot be directly installed in the grotto, so the CIVD can be installed inside the wooden eaves. Figure 5 shows the mechanical model of the grotto-eave system with CIVD. Where m_e, m_c, m_g and m_d are the mass of the eaves, the connection and the grotto, and the inerter of the CIVD; c_e, c_c, c_g and c_d are the damping coefficients of the eaves, connecting sections, grottoes, and the equivalent damping coefficient of CIVD; k_e, k_c, k_g, and k_d are the stiffnesses of the eaves, connection, grottoes and CIVD.

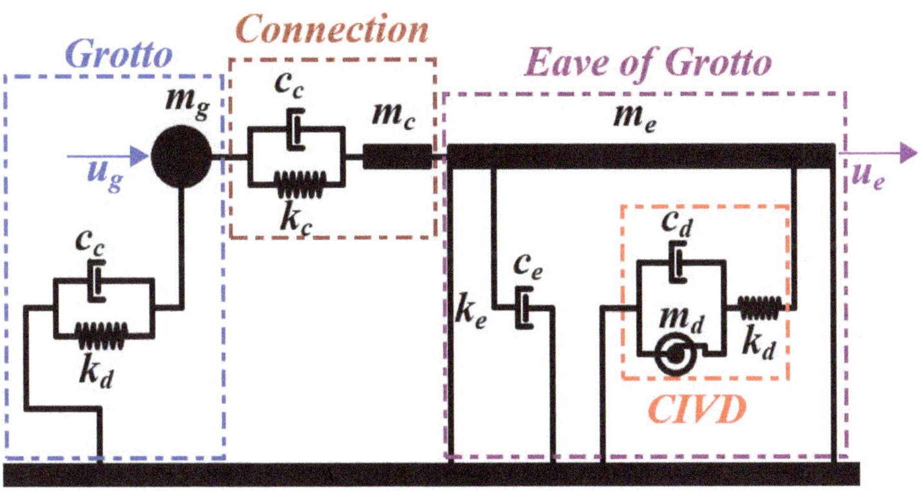

Figure 5. Mechanical model of eave–grotto system with CIVD.

2.3. Motion Control Equation of Grotto-Eave System with CIVD

According to the mechanical model shown in Figure 5, the motion equation of the grotto–eave system with CIVD is established:

$$M\ddot{X} + C\dot{X} + KX = -MIa_g \tag{3}$$

where, $\ddot{X}, \dot{X}, X$ is the acceleration, velocity and displacement vector of the grotto-eave system, a_g is the acceleration of ground motion; M, C, K are the mass, damping and stiffness matrices, and I is the ground motion excitation vector. Therefore, M, C, K, I and motion vector can be expressed as:

$$M = \begin{bmatrix} m_e + m_c & & \\ & m_d & \\ & & m_g \end{bmatrix}, C = \begin{bmatrix} c_e + c_c & & -c_c \\ & c_d & \\ -c_c & & c_g + c_c \end{bmatrix}, K = \begin{bmatrix} k_e + k_c + k_d & -k_d & -k_c \\ -k_d & k_d & \\ -k_c & & k_g + k_c \end{bmatrix} \tag{4}$$

$$X = \begin{Bmatrix} u_e \\ u_d \\ u_g \end{Bmatrix}, \dot{X} = \begin{Bmatrix} \dot{u}_e \\ \dot{u}_d \\ \dot{u}_g \end{Bmatrix}, \ddot{X} = \begin{Bmatrix} \ddot{u}_e \\ \ddot{u}_d \\ \ddot{u}_g \end{Bmatrix}, I = \begin{Bmatrix} 1 \\ 0 \\ 1 \end{Bmatrix} \tag{5}$$

where $u, \dot{u}, \ddot{u}$ is the vectors of displacement, velocity and acceleration; the subscripts e, d, g represents the eaves, CIVD and grottoes. For the convenience of parameter analysis, we define the following dimensionless parameters. Where Equation (6) is the natural vibration frequency and damping ratio parameters of the eaves, connection and grottoes; Equation (7) is parameters of CIVD; Equation (8) is the parameters of relative mass.

$$\omega_e = \sqrt{\frac{k_e}{m_e}};\ \zeta_e = \frac{c_e}{2m_e\omega_e};\ \omega_c = \sqrt{\frac{k_c}{m_c}};\ \zeta_c = \frac{c_c}{2m_c\omega_c};\ \omega_g = \sqrt{\frac{k_g}{m_g}};\ \zeta_g = \frac{c_g}{2m_g\omega_g} \tag{6}$$

$$\zeta_d = \frac{c_d}{2m_e\omega_e};\ \kappa_d = \frac{k_d}{k_e} \tag{7}$$

$$\mu_d = \frac{m_d}{m_e};\ \mu_{ce} = \frac{m_c}{m_e};\ \mu_{cg} = \frac{m_c}{m_g} \tag{8}$$

where ω_e is the natural circular frequency of the eaves, ζe is the damping ratio of the eaves; ω_c is the natural circular frequency of the connection, and ζ_c is the damping ratio of the connection; ω_g is the natural circular frequency of the grotto, and ζ_g is the damping ratio of the grotto; ζ_d is the damping ratio of CIVD, and κ_d is the stiffness ratio of CIVD; μ_d is the inerter–mass ratio, μ_{ce} is the mass ratio of connection and eave, and μ_{cg} is the mass ratio of connection and grotto.

From Equations (5)–(8), Laplace transform is applied to Equation (3), so the equation of motion (9) and (10) of the CIVD grotto–eave system in frequency domain can be obtained, where $\hat{M}, \hat{C}, \hat{K}, \hat{I}$ is the dimensionless mass, damping, stiffness matrix and external excitation vector of the system.

$$\left(\hat{M}s^2 + \hat{C}s + \hat{K}\right)\hat{U} = \hat{I}A_g;\ \hat{U} = \begin{Bmatrix} U_e \\ U_d \\ U_g \end{Bmatrix};\ \hat{I} = \begin{Bmatrix} -1 - \mu_{ce} \\ 0 \\ -1 \end{Bmatrix} \tag{9}$$

$$\hat{M} = \begin{bmatrix} 1 + \mu_{ce} & & \\ & \mu_d & \\ & & 1 \end{bmatrix};\ \hat{C} = \begin{bmatrix} 2\omega_e\zeta_e + 2\omega_c\zeta_c\mu_{ce} & & -2\omega_c\zeta_c\mu_{ce} \\ & 2\omega_e\zeta_d & \\ -2\omega_c\zeta_c\mu_{cg} & & 2\omega_g\zeta_g + 2\omega_c\zeta_c\mu_{cg} \end{bmatrix};\ \hat{K} = \begin{bmatrix} \omega_e^2(1+\kappa) + \omega_c^2\mu_{ce} & -\omega_e^2\kappa_d & -\omega_c^2\mu_{ce} \\ -\omega_e^2\kappa_d & \omega_e^2\kappa_d & \\ -\omega_c^2\mu_{cg} & & \omega_g^2 + \omega_c^2\mu_{cg} \end{bmatrix} \tag{10}$$

where s is the Laplace operator, $s = i\Omega$, Ω is the external excitation frequency, U_e, U_d, U_g and A_g are the Laplace transforms of u_e, u_d, u_g and a_g. The linear matrix Equations (9) and (10) can be solved to obtain U_e, U_d and U_g, at the same time, the transfer functions $H_{Ue}(s)$, $H_{Ud}(s)$ and $H_{Ug}(s)$ of the grotto–eave system with CIVD can be obtained:

$$H_{U_e}(s) = \frac{U_e(s)}{A_g(s)}, H_{U_d}(s) = \frac{U_d(s)}{A_g(s)}, H_{U_g}(s) = \frac{U_g(s)}{A_g(s)} \tag{11}$$

Meanwhile, the transfer function of output force of CIVD $H_{Fd}(s)$ can be expressed as follows:

$$H_{F_d}(s) = \frac{U_d(s)}{A_g(s)}\left(\mu_d s^2 + 2\zeta_d \omega_e s\right) \tag{12}$$

3. Parameter Analysis

According to Parseval's theorem, the root mean square (RMS) response σ of the system excited by white noise is obtained:

$$\sigma = \sqrt{\int_{-\infty}^{+\infty} |H(i\Omega)|^2 S_0 \mathrm{d}\Omega} \tag{13}$$

where S_0 is the power spectrum of white noise. The primary index is the γ_{Ur}, the relative displacement root mean square response ratio of the eaves and the grottoes before and after vibration mitigation by CIVD. It can be compared to evaluate the interaction between the eaves and the grottoes. At the same time, the γ_U, the displacement root mean square response ratio of the eaves before and after the vibration mitigation by CIVD, can directly measure the effect of the shock absorber, it can be regarded as an additional index. The smaller the γ_U and γ_{Ur} are, the better the vibration mitigation effect of the grotto–eave system is; that γ_U and γ_{Ur} are greater than 1 indicate that the CIVD has a displacement amplification effect on the system. The relative displacement ratio γ_{Ur} and displacement ratio γ_U are expressed as:

$$\gamma_{U_r}(\zeta_d, \kappa_d, \mu_d) = \frac{\left|\sigma_{U_e} - \sigma_{U_g}\right|}{\left|\sigma_{U_{e,0}} - \sigma_{U_{g,0}}\right|} \cdot \gamma_U(\zeta_d, \kappa_d, \mu_d) = \frac{\sigma_{U_e}}{\sigma_{U_{e,0}}} \tag{14}$$

where $\sigma_{Ue,0}$ and $\sigma_{Ug,0}$ are the displacement root mean square responses of the eaves and grottoes in the uncontrolled state. The CIVD parameter analysis selects the inerter–mass ratio μ_d, the stiffness ratio κ_d and the damping ratio ζ_d; and the parameter analysis index is the relative displacement ratio γ_{Ur}. Similarly, using Equations (12) and (13), the RMS response of output force of CIVD σ_{Fd} can be determined. In parameter analysis, the value range of parameters of CIVD is 0.01 to 1.

In addition to the CIVD parameters, it is also necessary to analyze the influence of the stiffness of the connection between grotto and eave. Defining the parameters of the Benchmark model of the grotto–eave system: $\omega_g/\omega_e = 5$, $\zeta_c = 0.015$, $\zeta_e = \zeta_g = 0.02$, $\mu_{ce} = 0.2$, $\mu_{cg} = 0.04$, and subsequent analyses are carried out according to this Benchmark model; then defining the frequency ratio of the connection: $\beta = \omega_c/\omega_e$, β can reflect the stiffness of the connection, and the smaller the β, the smaller the stiffness of the connection. Table 1 shows the frequency ratio β of different types of connection.

Table 1. Frequency ratio of different types of connection.

Types of Connection	Soft Connection	Equal-Stiffness	Hard Connection
Symbol	*SC*	*EC*	*HC*
β	0.2	1	5

3.1. Parameter Analysis of CIVD

Based on the above analysis indexes, the parameters of the grotto–eave Benchmark model and different types of connection, the influence study of design parameters of CIVD, including inerter–mass ratio μ_d, stiffness ratio κ_d and damping ratio ζ_d.

Firstly, the influence of types of connection on grotto-eave system under different inerter-mass ratio μ_d is analyzed, and Figures 6–8 show three-dimensional contour plot of γ_{Ur}. When the stiffness ratio is 0.01 to 0.3 and the damping ratio is 0.1 to 1, it can be seen from figures that as the stiffness ratio and damping ratio increase, the relative displacement between the grotto and the eaves will decrease; when the damping ratio is 0.01 to 0.1, the value of the minimum γ_{Ur} changes with the increase of the inerter–mass ratio, the larger the inerter-mass ratio, the greater requirement of additional stiffness of CIVD. However, in the case of low damping ratio and low stiffness ratio, the hard connection system has displacement amplification, which should be avoided in parameters selection. The maximum γ_{Ur} of three types of connection is 0.2, but the stiffness ratio and damping ratio required by CIVD should be as large as possible. Figures 6–8 show the analytical solutions of the equivalent mathematical problems, which cannot be realized in engineering applications. Generally, it is more appropriate to control the damping ratio within 0.2. Hence, further analysis of other parameters and indexes should be carried out.

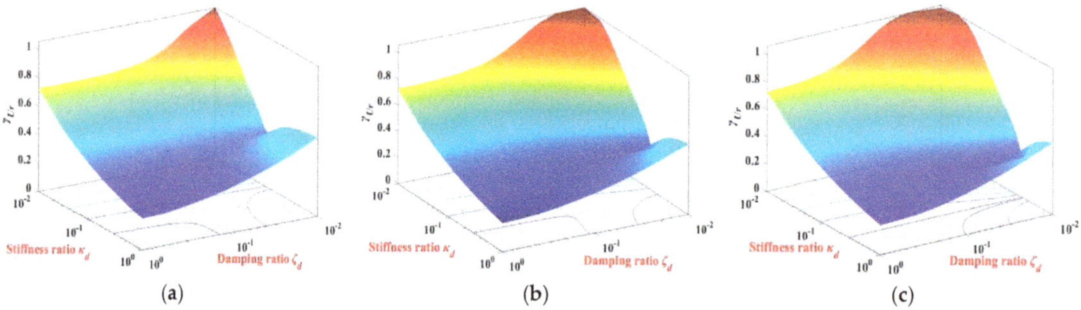

Figure 6. Contour plot of γ_{Ut} of SC system: (**a**) μ_d = 0.1; (**b**) μ_d = 0.2; and (**c**) μ_d = 0.3.

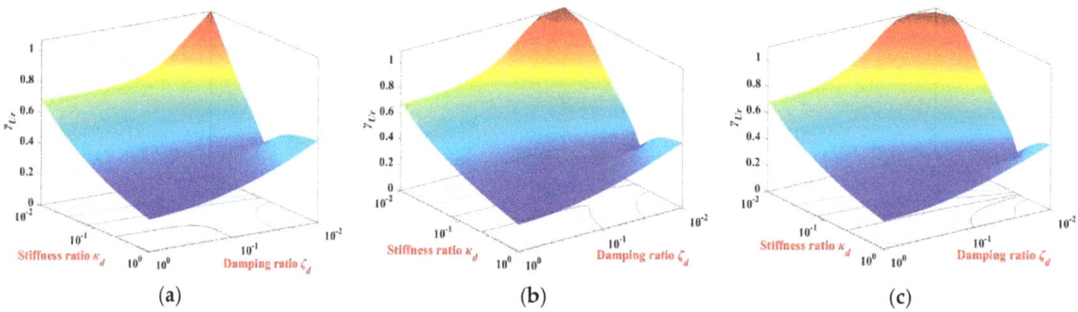

Figure 7. Contour plot of γ_{Ut} of EC system: (**a**) μ_d = 0.1; (**b**) μ_d = 0.2; and (**c**) μ_d = 0.3.

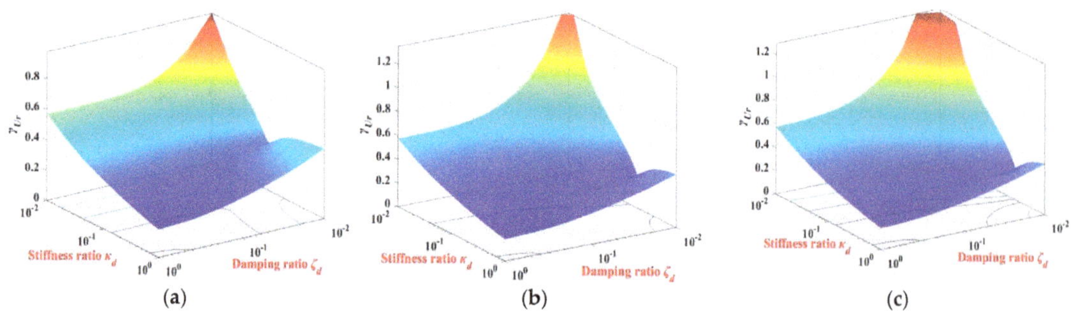

Figure 8. Contour plot of γ_{Ut} of HC system: (**a**) μ_d = 0.1; (**b**) μ_d = 0.2; and (**c**) μ_d = 0.3.

Taking γ_{Ur} as the index and fixing the damping ratio ζ_d to be 0.05, 0.1, and 0.15, from contour plot of γ_{Ur} with κ_d-μ_d space, we can get the variation trend of γ_{Ur} with ζ_d. Figures 9–11 show the 2D contour plot of γ_{Ur} in κ_d-μ_d space under different types of connection, in which the numbers in the figure are contour values, representing the value of γ_{Ur}, the variation range is 0 to 1, and the interval value is 0.1. When the damping ratio ζ_d varies from 0.05 to 0.15, the value of the contour line of the soft connection system is 0.2 to 1, the equal stiffness connection system is 0.4 to 1, and the hard connection system is 0.5 to 1. Furthermore the smaller the damping ratio, the higher the contour line value, and the worse effect of vibration mitigation of relative displacement. At the same time, as the stiffness of the connection between the grotto and the eaves increases, CIVD needs greater the damping ratio to obtain the same relative displacement ratio γ_{Ur}. We can also draw a conclusion from the number of contour lines intuitively, the more "hard" the connection (means the stiffness of connection is large), the less the number of contour lines. CIVD has a better control effect on the relative displacement index γ_{Ur}, and it can indicate that the arrangement of CIVD can reduce the impact of the eaves on the grotto. However, at the same time, considering the vibration mitigation effect of the eaves themselves, if the displacement of the eaves themselves are too large, the CIVD cannot be used as the best vibration control device. To this end, we should continue to study another index γ_U with CIVD parameters.

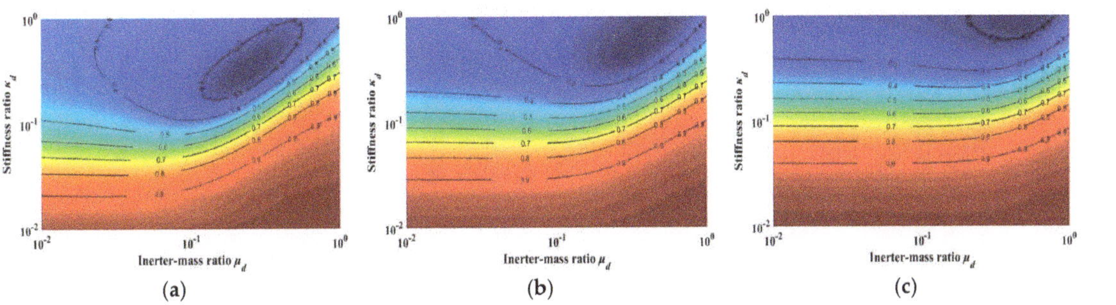

Figure 9. 2D Contour plot of γ_{Ur} of SC system: (**a**) ζ_d = 0.05; (**b**) ζ_d = 0.10; and (**c**) ζ_d = 0.15.

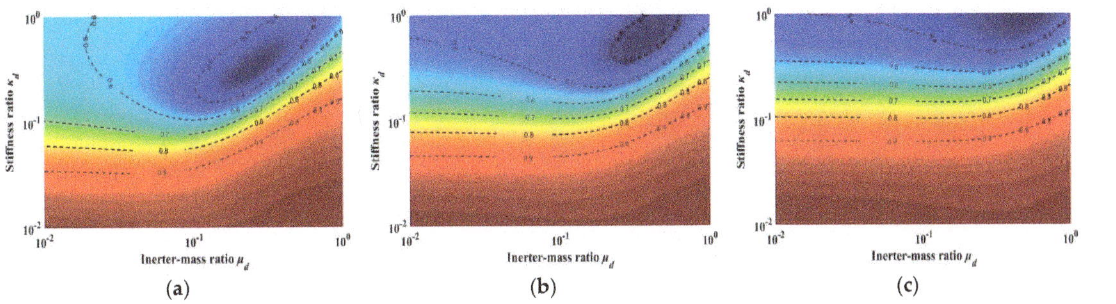

Figure 10. 2D Contour plot of γ_{Ur} of EC system: (**a**) ζ_d = 0.05; (**b**) ζ_d = 0.10; and (**c**) ζ_d = 0.15.

The γ_U is also controlled by three parameters. For the convenience of research, a three-dimensional μ_d-ζ_d-κ_d space is established. The γ_U is any point in the parameter space, and it is represented by a specified color and a value corresponding to the color. We make slice plots on some specific parameter planes, and the two-dimensional spaces under the specified parameters are displayed. In this paper, we fixed μ_d to 0.01, 0.03, 0.1, 0.3 and 1, and it is used to study the response trend of the eaves under different types of connection based on the Benchmark model. Figure 12 is the γ_U slice plot of different connection systems. It can be seen from the figure that the displacement ratio of the SC system is the smallest, and the minimum value is 0.2. As the stiffness of the connection increasing, the minimum value of γ_U is larger, and the displacement mitigation of single eaves is worse. Figure 12a shows the displacement ratio of the eaves without connection, and it is similar to EC system. If the target displacement ratio of the eaves is specified as 0.5 ($\gamma_U = 0.5$), it can be seen that when μ_d is in the range of 0.03 to 0.3, the enclosed area of target contour γ_U is larger (the blue area), and the enclosed area tends to increase firstly and then decrease. When μ_d is in the range of 0.01 to 0.03, the γ_U is almost constant, so the range of μ_d can be determined between 0.03 and 0.3, and further discussions of influence of stiffness are continued.

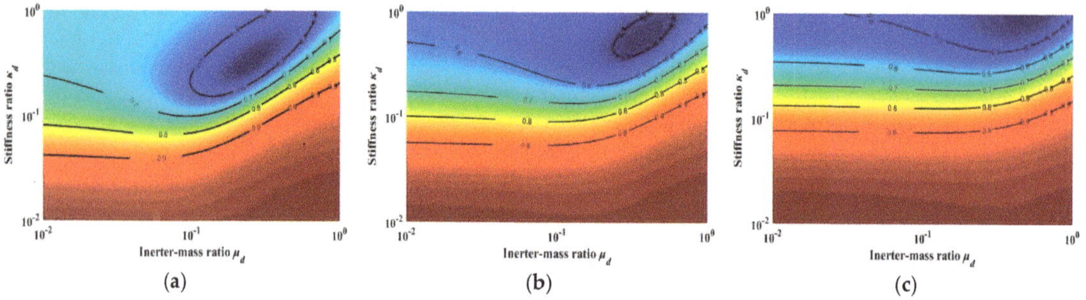

Figure 11. 2D Contour plot of γ_{Ur} of HC system: (**a**) $\zeta_d = 0.05$; (**b**) $\zeta_d = 0.10$; and (**c**) $\zeta_d = 0.15$.

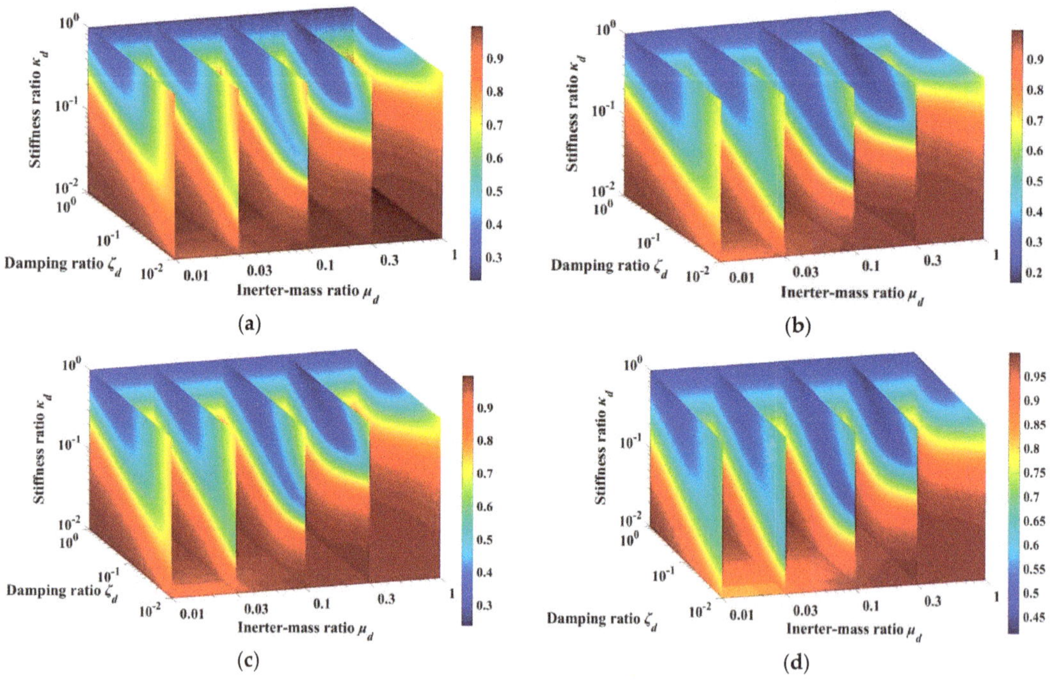

Figure 12. Slice plot of γ_U: (**a**) $\beta = 0$; (**b**) $\beta = 0.2$; (**c**) $\beta = 1$; and (**d**) $\beta = 5$.

Since the large stiffness of the cable is easier to achieve in practical, it can be assumed that the stiffness of the cable can be taken as any value. The γ_U of SC system under different inerter–mass ratios are shown in Figure 13. It can be seen from the figure that a large inerter–mass ratio requires a large stiffness to achieve the same displacement ratio of single eaves. At the same time, when ζ_d is less than 0.05, γ_U decreases firstly and then increases under a small damping ratio. Additionally, as the inerter–mass ratio increasing, this phenomenon becomes more obvious. Hence, the robustness of CIVD with small damping ratio is low. When ζ_d is greater than 0.1, increasing the stiffness of the cable cannot reduce the displacement ratio. In addition, a larger damping ratio (ζ_d is greater than 0.3) does not continue to reduce γ_U, but the stability of the curve and the trend of γ_U are better. However, in practical engineering, increasing damping ratio will lead to a higher cost and a larger volume of the damper. Therefore, while ensuring the index, further parameter optimization design is required to reduce the damping ratio.

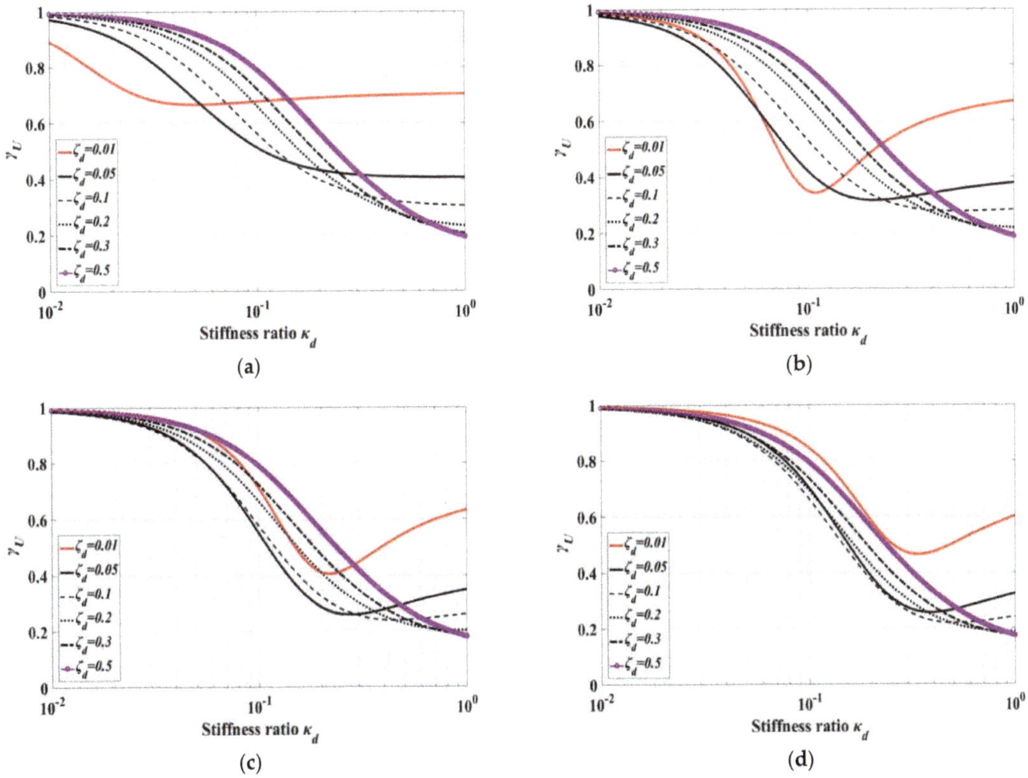

Figure 13. γ_U of SC system: (**a**) μ_d = 0.03; (**b**) μ_d = 0.1; (**c**) μ_d = 0.2; and (**d**) μ_d = 0.3.

3.2. Demand-Based Optimal Design of CIVD

The parameters of CIVD and other inerter-based shock absorbers can be determined by the fixed-point method [41]. Ikago et al. [16] has proposed corresponding inerter system design methods for single degree of freedom (SDOF) structures and multi-degrees of freedom structures. For the SDOF system of the eaves structure with CIVD, if the arrangement angle of the cables is not considered, the stiffness ratio and damping ratio can be obtained by the following formulas after the inerter–mass ratio μ_d is determined:

$$\kappa_d = \frac{\mu_d}{1-\mu_d}, \zeta_d = \frac{\mu_d}{2}\sqrt{\frac{3\mu_d}{(1-\mu_d)(2-\mu_d)}} \tag{15}$$

However, the fixed-point theory does not consider the inherent damping ratio, external excitation characteristics and performance demands of main structure. For example, it cannot reflect the key index such as the relative displacement response of grottoes and eaves. Zhang et al. [42] consider the main performance of the structure based on closed-form solution and use extreme conditions to determine parameters of CIVD. Compared with fixed-point method, better structural performance index can be obtained by this methods. In addition to performance index, output force of CIVD can represent the cost index of shock absorber. The contour plot of σ_{Ue} and σ_{Fd} of SC system are shown in Figure 13.

It can be seen from Figure 14 that the minimum value of σ_{Ue} is located at the upper right of the contour plot (white dot), but at this time, it corresponds to a large stiffness ratio and damping ratio. That means the output value of CIVD is also large, which is difficult to achieve in practical. However, compared with Figure 14a,b, it can be seen that there exists a parameter combination of a small stiffness ratio and a damping ratio when the inerter–mass ratio is determined, which can make the displacement and output smaller simultaneously (dark yellow dot). It means that the parameter combination can meet both the performance index (displacement ratio) and the cost index (output force of CIVD). If the main performance index of grottoes and eaves is determined, the optimization problem can be transformed from multi-objectives optimization problem to single-objective optimization problem, as shown in the following formula 16. It can be seen that the objective function of single-objective optimization problem is σ_{Fd}.

$$\begin{aligned} &\text{minimize} \quad \sigma_{Fd}(\zeta_d, \kappa_d, \mu_d) \\ &\textit{subject to} \begin{cases} \gamma_{Ur}(\zeta_d, \kappa_d, \mu_d) = \gamma_{Ur,t} \\ \zeta_d \in (0, 0.2) \\ \kappa_d \in (0, 1) \\ \mu_d \in (0, 1) \end{cases} \end{aligned} \tag{16}$$

where, $\gamma_{Ur,t}$ is the relative displacement ratio of the grotto and the eaves. It is difficult to realize in practical engineering with large damping ratio. For this reason, the damping ratio is set within 0.2.

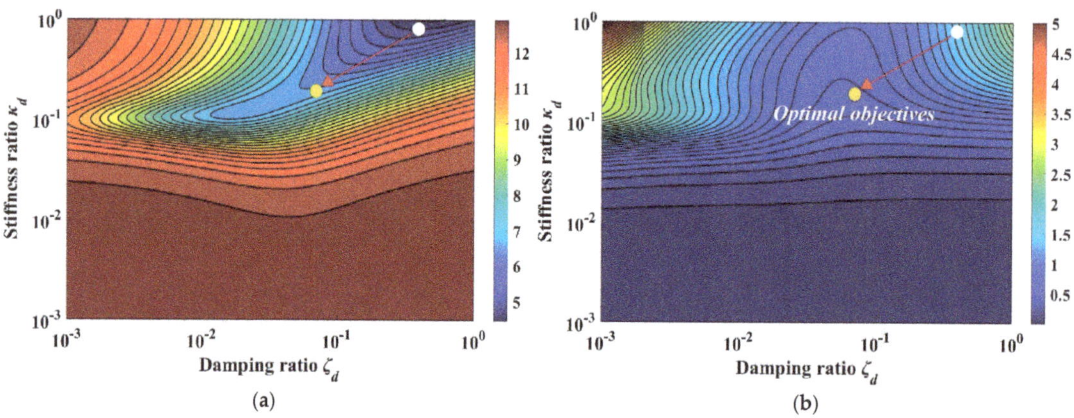

Figure 14. 2D Contour plot of σ_{Ue} and σ_{Fd} of SC system (μ_d = 0.1): (**a**) σ_{Ue}; and (**b**) σ_{Fd}.

Moreover, compared with the conventional viscous damper (VD), the different topological connection forms of the inerter element can enhance the deformation of the damping element and improve the capacity of energy dissipation [43]. Therefore, a new index of damping effects of CIVD is proposed by comparing VD, namely, inerter-enhanced energy dissipation coefficient η:

$$\eta = \frac{\sigma_{U0}(\zeta_0) - \sigma_{U_e}(\zeta_0, \zeta_d, \kappa_d, \mu_d)}{\sigma_{U0}(\zeta_0) - \sigma_{U_{eVD}}(\zeta_0, \zeta_d)} \tag{17}$$

where σ_{UeVD} are the RMS displacement responses of the eave with VD. Additionally, η means the damping element with inerter element has a higher deformation effect under the same additional damping ratio. When η is greater than 1, it shows that the inerter can strengthen the energy dissipation of the structure, and it can be used as an index to evaluate the robustness of the CIVD. According to massive numerical case studies, the recommended range of η is [1,2] and the η can be also a constraint condition of formula 16. In this paper, η is fixed as 1.5 and $\gamma_{Ur,t}$ is fixed as 0.5, 0.6 and 0.7. Table 2 shows the results of parameters of CIVD after optimization. It can be seen from Table 2 that the value of parameters' combination of HC system are larger, but the values of CIVD are generally small, especially the damping ratio (less than 0.05).

Table 2. Optimal parameters of CIVD.

γ_{Ur}	Parameters	$\beta = 0$	SC (β = 0.2)	EC (β = 1)	HC (β = 5)
0.7	μ_d	0.0121	0.0163	0.0277	0.0391
	κ_d	0.0213	0.0254	0.0442	0.0782
	ζ_d	0.0054	0.0061	0.0114	0.0183
0.6	μ_d	0.0221	0.0225	0.0485	0.0584
	κ_d	0.0332	0.0455	0.0782	0.1327
	ζ_d	0.0091	0.0111	0.0216	0.0341
0.5	μ_d	0.0344	0.0483	0.0692	0.0994
	κ_d	0.0671	0.0755	0.1301	0.2031
	ζ_d	0.0164	0.0192	0.0366	0.0485

Figure 15 shows the transfer function of relative displacement of the grotto–eaves system under different types of connection. Where, the modulus of transfer function of the relative displacement H_{Ur} of the grotto–eaves is:

$$|H_{U_r}| = \left|H_{U_e} - H_{U_g}\right| \tag{18}$$

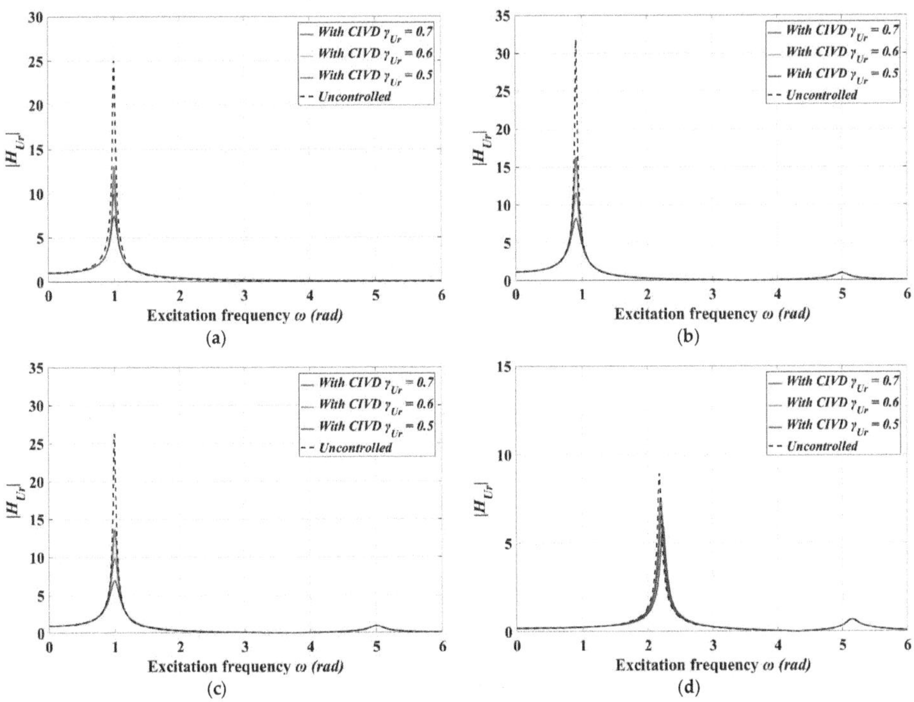

Figure 15. Transfer function curves of relative displacement of grotto-eaves system: (**a**) β = 0; (**b**) β = 0.2; (**c**) β = 1; and (**d**) β = 5.

It can be seen from Figure 15 that peak relative displacements of the grotto–eaves system are reduced under harmonic excitation, which controls the narrow-band resonance response of the system significantly. However, the HC system needs a smaller relative displacement ratio to control the resonance response. In order to better illustrate the advantages of this optimal method, the response of eaves and the deformation of damping element by using the classic fixed-point method should be compared under the same target relative displacement ratio.

Figures 16 and 17 show the displacement amplification factor of eaves and transfer function of deformation of damping element of CIVD in SC system with different optimal design methods. It can be seen from Figures 16 and 17 that the maximum displacement of demand-based optimal method is less than it of the fixed-point method. Meanwhile, the damping element deformation enhancement of the inerter system can be brought into full play by using demand-based optimal method. Hence, the demand-based optimal method can minimize the cost by enhancing damping element deformation in a small damping ratio.

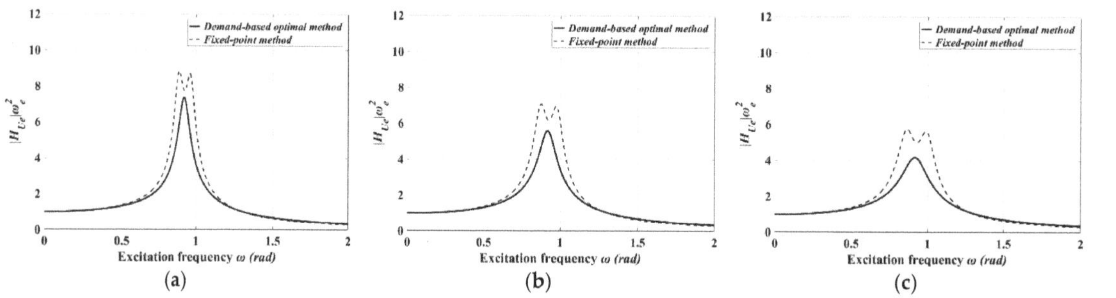

Figure 16. Displacement amplification factor of eaves in SC system: (**a**) γ_{Ur} = 0.7; (**b**) γ_{Ur} = 0.6; and (**c**) γ_{Ur} = 0.5.

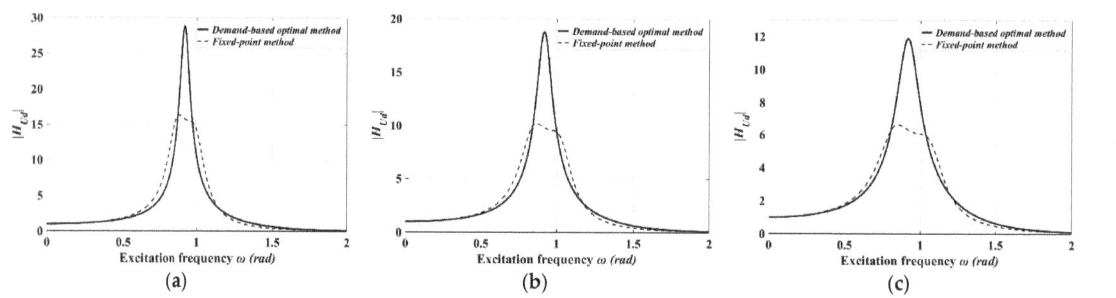

Figure 17. Transfer function of deformation of damping element of CIVD in SC system: (**a**) γ_{Ur} = 0.7; (**b**) γ_{Ur} = 0.6; and (**c**) γ_{Ur} = 0.5.

4. Dynamic Response

In the previous study, parameter analyses in the frequency domain are carried out based on the performance index. In order to further verify the vibration mitigation effect of CIVD and the optimal parameters. Dynamic response analyses of the grotto–eave system with CIVD are carried out under the non-stationary ground motions in this section. The structural parameters of the grotto–eaves system refer to the Benchmark model in Section 2.1, and the parameters of CIVD are shown in Table 2. The ground motion records EL Centro record, Taft record, Chi-chi record and Kobe record are selected. The predominant frequencies of the four records are all different. Figure 18 shows the acceleration response spectrum of the four ground motion records.

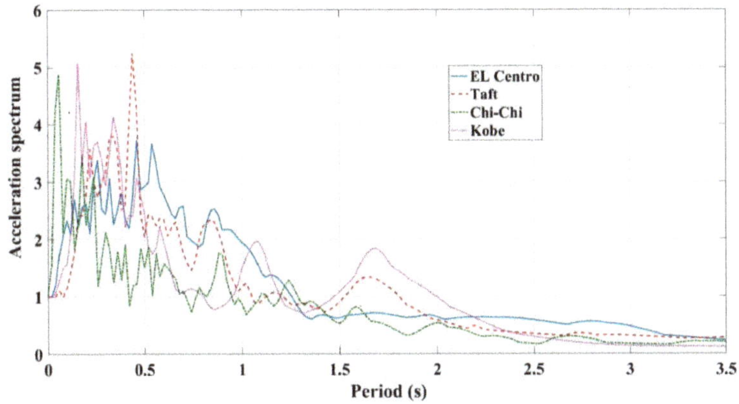

Figure 18. Normalized acceleration spectra of ground motion records.

Figure 19 shows the relative displacement time history response of SC eave–grotto system. It can be seen that CIVD has better vibration control performance intuitively. Additionally, the continuous vibration mitigation of CIVD is good on the grotto–eave system, and the controlled system can be stable and have small displacements quickly. As the seismic response progresses, the vibration control effects of CIVD in the later stage are better, indicating that a stable vibration state of eave–grotto system can be quickly got under control of CIVD.

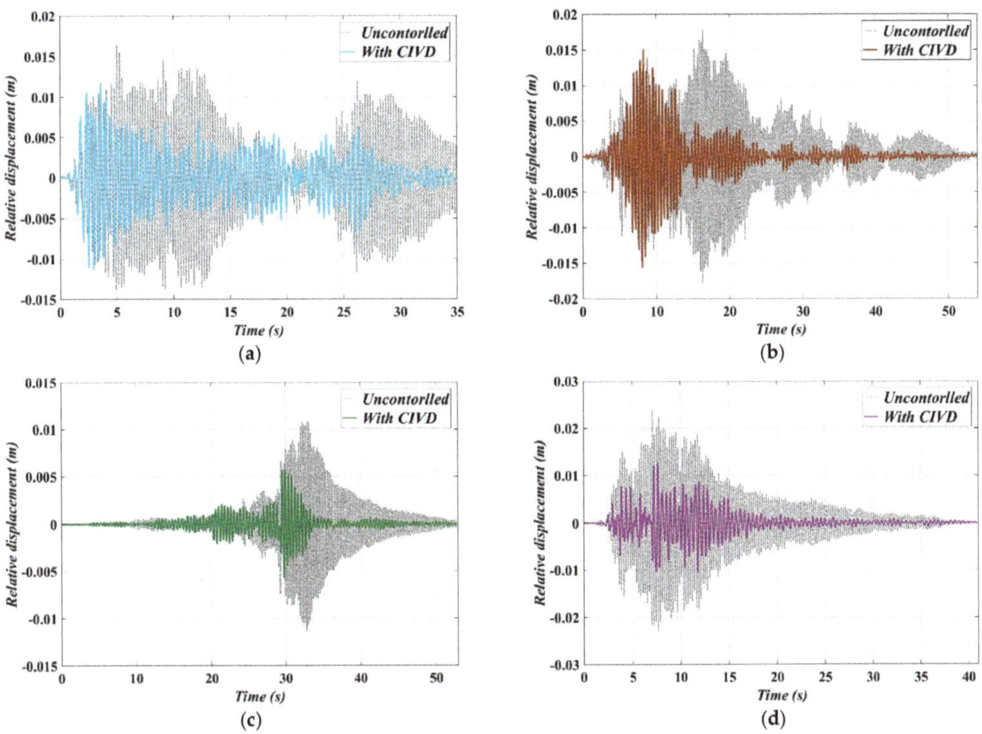

Figure 19. Relative displacement responses of SC eave-grotto system (γ_{Ur} = 0.6): (**a**) EL Centro; (**b**) Taft; (**c**) Chi-Chi; and (**d**) Kobe.

Figure 20 shows that the decline of RMS displacement of the eave is between 25% and 60%. At the same time, it can be seen that the deformation of damping element in CIVD is always greater than displacement of main structure (eave). Compared with the hysteretic loops of CIVD and VD under the same damping coefficient as shown in Figure 21, it can be seen that the deformation and damping force of CIVD will increase simultaneously due to the efficient tuning and damping amplification of CIVD.

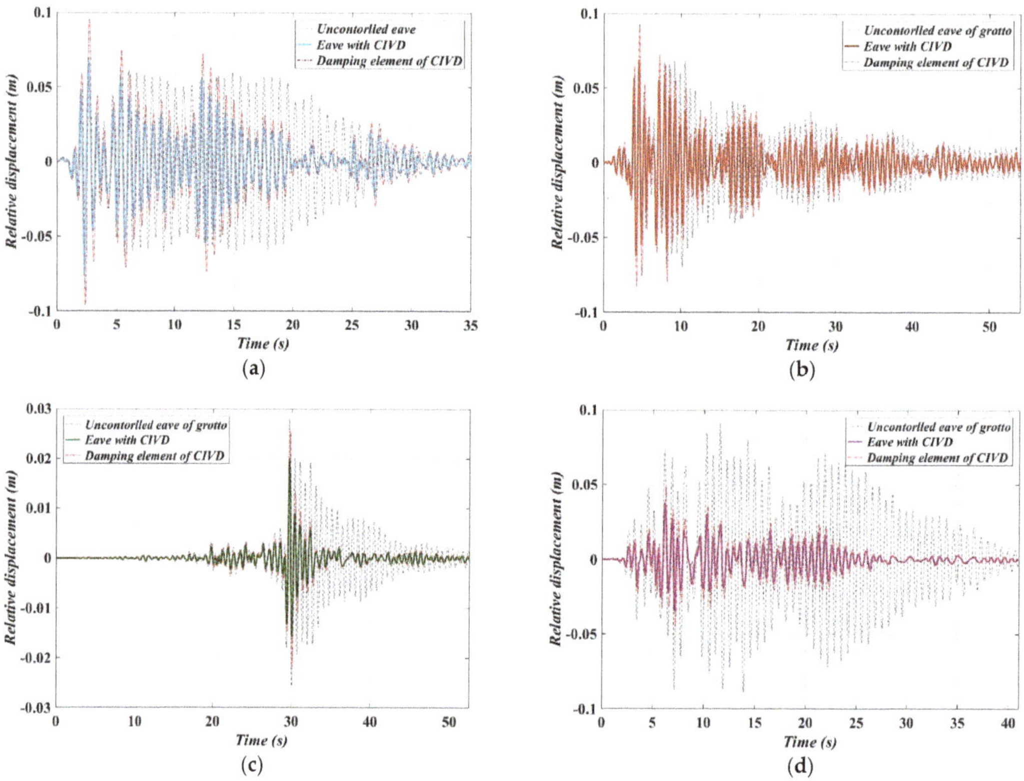

Figure 20. Displacement responses of SC eave and damping element of CIVD (γ_{Ur} = 0.6): **(a)** EL Centro; **(b)** Taft; **(c)** Chi-Chi; and **(d)** Kobe.

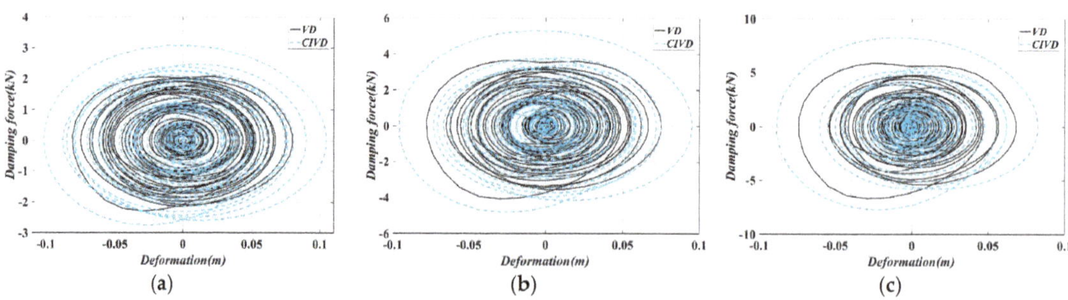

Figure 21. Hysteretic loops damping element of CIVD in SC eave under EL Centro record: **(a)** γ_{Ur} = 0.7; **(b)** γ_{Ur} = 0.6; and **(c)** γ_{Ur} = 0.5.

Hence, if the damping ratio ζ_d is a fixed, the damping force will be greater as the deformation of the damping element of CIVD is greater. Additionally, the damping effect of CIVD will be improved. Then, the deformation enhancement coefficient of damping element γ_{Ud} can be defined, as shown in Equation (19):

$$\gamma_{U_d}(\zeta_d, \kappa_d, \mu_d) = \frac{\sigma_{U_d}}{\sigma_{U_e}} \tag{19}$$

where, σ_{Ud} is the displacement root mean square responses of damping element of CIVD. The larger γ_{Ud} means the better deformation amplification effect of the inerter on the damping element, corresponding the smaller displacement mitigation ratio γ_U. Additionally, the deformation enhancement coefficient of damping element of VD $\gamma_{Ud,VD}$ is equal to 1, indicating there is no deformation enhancement effect of VD. A large damping ratio is required in VD system in order to achieve the same displacement mitigation ratio γ_U as CIVD. That means the VD will produce large damping force even under small displacement the eave, which will have a great impact on historical buildings. Due to a small optimal damping ratio, CIVD have a smaller damping force under small displacement the eave, but can still control the eave through better tuning capacity.

Figure 22 shows the contour plot of γ_U and γ_{Ud} of eave with optimal μ_d of CIVD, it can be seen that under the same damping ratio, the deformation of damping element of CIVD is about 2 times of VD, which can obtain better energy dissipation effect and reduce the energy transmitted to the historical buildings' underground motions. Meanwhile, about five times the damping coefficient of CIVD are required for VD to achieve the same vibration mitigation effect as CIVD. Thus, compared with VD, the damping force of CIVD transmitted to the historical buildings will be much lower.

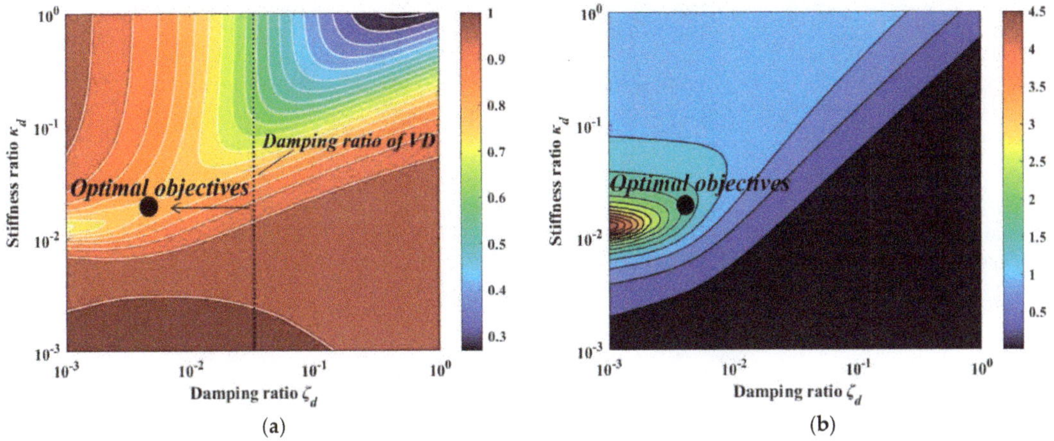

Figure 22. 2D Contour plot of γ_U and γ_{Ud} of eave with optimal μ_d ($\beta = 0$, $\gamma_U = 0.7$): (**a**) γ_U; and (**b**) γ_{Ud}.

5. Conclusions

In this paper, the motion equation of grotto–eave systems with cable inerter viscous damper (CIVD) is established. Then, the stochastic analysis and demand-based parameters optimization of CIVD are carried out. Finally, the vibration control performance of CIVD is analyzed under different seismic inputs. The main conclusions are as follows:

1. The CIVD is easy to install, and can quickly improve the seismic performance of the structure. Therefore, it can be used in immovable cultural relics such as eaves and grottoes where the damper installation conditions are harsh.
2. In the parameters optimization design of CIVD, damping ratio should be reduced as much as possible under the condition of satisfying the target vibration mitigation ratio based on performance demand. The optimal design of a CIVD in grotto–eave

system should be a balance process between response of different types of grotto–eave system and the cost of CIVD. It can improve the feasibility of the application of CIVD in cultural relics protection projects.
3. The proposed demand-based optimal method can minimize the cost by enhancing damping element deformation in a small damping ratio, while ensuring that the value of displacement index of grotto–eave system can be reached. Moreover, the inerter–mass ratio of the CIVD can be also determined by displacement index of grotto–eave system when using fixed-point method.
4. Considering the lower damping ratio and inerter–mass ratio of CIVD, applications of the CIVD designed by demand-based optimal method can be extended and its installation made more flexible in specific structures like eaves. The corresponding verification experiment should be conducted in the near future.

Author Contributions: J.H., supervision, funding acquisition, methodology, investigation, validation, writing—review and editing. R.Z., methodology, conceptualization, software, investigation, writing—original draft, writing—review and editing; Q.L., validation, writing—review and editing. X.G., validation, writing—review and editing. M.C., conceptualization, writing—review and editing. All authors have read and agreed to the published version of the manuscript.

Funding: This work is supported in part by National Key R&D Program of China (Grant no. 2019YFC1520500, 2020YFC1523004).

Institutional Review Board Statement: Not applicable.

Informed Consent Statement: Not applicable.

Data Availability Statement: The data used to support the findings of this study are included within the article.

Conflicts of Interest: The authors declare that there is no conflict of interest regarding the publication of this paper.

References

1. Mosoarca, M.; Onescu, I.; Onescu, E.; Azap, B.; Chieffo, N.; Sirbu, M.S. Seismic vulnerability assessment for the historical areas of the Timisoara city, Romania. *Eng. Fail. Anal.* **2019**, *101*, 86–112. [CrossRef]
2. Cardoso, R.; Lopes, M.; Bento, R. Seismic evaluation of old masonry buildings. Part I: Method description and application to a case-study. *Eng. Struct.* **2005**, *27*, 2024–2035. [CrossRef]
3. Betti, M.; Vignoli, A. Modelling and analysis of a Romanesque church under earthquake loading: Assessment of seismic resistance. *Eng. Struct.* **2008**, *30*, 352–367. [CrossRef]
4. Zhao, X.B.; Zhang, F.L.; Xue, J.Y.; Ma, L.L. Shaking table tests on seismic behavior of ancient timber structure reinforced with CFRP sheet. *Eng. Struct.* **2019**, *197*, 109405. [CrossRef]
5. Sayin, B.; Yildizlar, B.; Akcay, C.; Gunes, B. The retrofitting of historical masonry buildings with insufficient seismic resistance using conventional and non-conventional techniques. *Eng. Fail. Anal.* **2019**, *97*, 454–463. [CrossRef]
6. Silva, R.A.; Mendes, N.; Oliveira, D.V.; Romanazzi, A.; Martínez, O.D.; Miranda, T. Evaluating the seismic behaviour of rammed earth buildings from Portugal: From simple tools to advanced approaches. *Eng. Struct.* **2018**, *157*, 144–156. [CrossRef]
7. Zhang, R.; Wu, M.; Lu, W.; Li, X.; Lu, X. Seismic retrofitting of a historic building by using an isolation system with a weak restoring force. *Soil Dyn. Earthq. Eng.* **2021**, *148*, 106836. [CrossRef]
8. Bento, R.; Lopes, M.; Cardoso, R. Seismic evaluation of old masonry buildings. Part II: Analysis of strengthening solutions for a case study. *Eng. Struct.* **2005**, *14*, 2014–2023. [CrossRef]
9. Aty, Y.Y.A. Proposals for seismic retrofitting of timber roofs to enhance their in-plane stiffness and diaphragm action at historical masonry buildings in Cairo. *J. Cult. Herit.* **2018**, *32*, 73–83.
10. Witzany, J.; Zigler, R.; Kroftová, K. Strengthening of compressed brick masonry walls with carbon composites. *Constr. Build. Mater.* **2016**, *112*, 1066–1079. [CrossRef]
11. Akcay, C.; Bozkurt, T.S.; Sayin, B.; Yildizlar, B. Seismic retrofitting of the historical masonry structures using numerical approach. *Constr. Build. Mater.* **2016**, *113*, 752–763. [CrossRef]
12. Lee, S.H.; Min, K.W.; Hwang, J.S.; Kim, J. Evaluation of equivalent damping ratio of a structure with added dampers. *Eng. Struct.* **2004**, *26*, 335–346. [CrossRef]
13. Smith, M.C.; Wang, F.C. Performance benefits in passive vehicle suspensions employing inerters. *Veh. Syst. Dyn.* **2004**, *42*, 235–257. [CrossRef]
14. Smith, M.C. Synthesis of mechanical networks: The inerter. *IEEE Trans. Autom. Control* **2002**, *47*, 1648–1662. [CrossRef]

15. Ikago, K.; Saito, K.; Inoue, N. Seismic control of single-degree-of-freedom structure using tuned viscous mass damper. *Earthq. Eng. Struct. Dyn.* **2012**, *41*, 453–474. [CrossRef]
16. Pan, C.; Zhang, R.; Luo, H.; Li, C.; Shen, H. Demand-based optimal design of oscillator with parallel-layout viscous inerter damper. *Struct. Control Health Monit.* **2018**, *25*, e2051. [CrossRef]
17. Pan, C.; Zhang, R. Design of structure with inerter system based on stochastic response mitigation ratio. *Struct. Control Health Monit.* **2018**, *25*, e2169. [CrossRef]
18. Hwang, J.S.; Kim, J.; Kim, Y.M. Rotational inertia dampers with toggle bracing for vibration control of a building structure. *Eng. Struct.* **2007**, *29*, 1201–1208. [CrossRef]
19. Zhang, R.F.; Zhao, Z.P.; Dai, K.S. Seismic response mitigation of a wind turbine tower using a tuned parallel inerter mass system. *Eng. Struct.* **2019**, *180*, 29–39. [CrossRef]
20. Zhao, Z.P.; Zhang, R.F.; Wierschem, N.E.; Jiang, Y.Y.; Pan, C. Displacement mitigationoriented design and mechanism for inerter-based isolation system. *J. Vib. Control* **2020**, *27*, 1991–2003. [CrossRef]
21. Gao, H.; Wang, H.; Li, J.; Wang, Z.; Liang, R.; Xu, Z.; Ni, Y. Optimum design of viscous inerter damper targeting multi-mode vibration mitigation of stay cables. *Eng. Struct.* **2021**, *226*, 111375. [CrossRef]
22. De Domenico, D.; Ricciardi, G. An enhanced base isolation system equipped with optimal tuned mass damper inerter (TMDI). *Earthq. Eng. Struct. Dyn.* **2018**, *47*, 1169–1192. [CrossRef]
23. De Domenico, D.; Impollonia, N.; Ricciardi, G. Soil-dependent optimum design of a new passive vibration control system combining seismic base isolation with tuned inerter damper. *Soil Dynam. Earthq. Eng.* **2018**, *105*, 37–53. [CrossRef]
24. De Domenico, D.; Ricciardi, G. Optimal design and seismic performance of tuned mass damper inerter (TMDI) for structures with nonlinear base isolation systems. *Earthq. Eng. Struct. Dyn.* **2018**, *47*, 2539–2560. [CrossRef]
25. Zhao, Z.P.; Zhang, R.F.; Jiang, Y.Y.; Pan, C. Seismic response mitigation of structures with a friction pendulum inerter system. *Eng. Struct.* **2019**, *193*, 110–120. [CrossRef]
26. Nakamura, Y.; Fukukita, A.; Tamura, K.; Yamazaki, I.; Matsuoka, T.; Hiramoto, K.; Sunakoda, K. Seismic response control using electromagnetic inertial mass dampers. *Earthq. Eng. Struct. Dyn.* **2014**, *43*, 507–527. [CrossRef]
27. Wang, F.C.; Chen, C.W.; Liao, M.K.; Hong, M.F. Performance analyses of building suspension control with inerters, in: Decision and Control. In Proceedings of the 2007 IEEE Conference on Decision and Control, New Orleans, LA, USA, 12–14 December 2007; pp. 3786–3791.
28. Lazar, I.F.; Wagg, D.J.; Neild, S.A. An inerter vibration isolation system for the control of seismically excited structures. In Proceedings of the 10th International Conference on Urban Earthquake Engineering, Tokyo, Japan, 1–2 March 2013.
29. Luo, H.; Zhang, R.F.; Weng, D.G. Mitigation of liquid sloshing in storage tanks by using a hybrid control method. *Soil Dyn. Earthq. Eng.* **2016**, *90*, 183–195. [CrossRef]
30. Marian, L.; Giaralis, A. Optimal design of a novel tuned mass-dampereinerter (TMDI) passive vibration control configuration for stochastically support-excited structural systems. *Probabilistic Eng. Mech.* **2014**, *38*, 156–164. [CrossRef]
31. Hashimoto, T.; Fujita, K.; Tsuji, M.; Takewaki, I. Innovative base-isolated building with large mass-ratio TMD at basement for greater earthquake resilience. *Future Cities Environ.* **2015**, *1*, 9. [CrossRef]
32. Pietrosanti, D.; Angelis, M.D.; Basili, M. Optimal design and performance evaluation of systems with tuned mass damper inerter (TMDI). *Earthq. Eng. Struct. Dyn.* **2017**, *46*, 1367–1388. [CrossRef]
33. Sugimura, Y.; Goto, W.; Tanizawa, H.; Saito, K.; Nimomiya, T. Response control effect of steel building structure using tuned viscous mass damper. In Proceedings of the 15th World Conference on Earthquake Engineering, Lisbon, Portugal, 24 September 2012; Volume 9, pp. 24–28.
34. Kurata, M.; Leon, R.T.; Desroches, R. Rapid seismic rehabilitation strategy: Concept and testing of cable bracing with couples resisting damper. *J. Struct. Eng.* **2012**, *138*, 354–362. [CrossRef]
35. Xie, L.; Ban, X.; Xue, S.; Ikago, K.; Kang, J.; Tang, H. Theoretical Study on a Cable-Bracing Inerter System for Seismic Mitigation. *Appl. Sci.* **2019**, *9*, 4096. [CrossRef]
36. Xue, S.; Kang, J.; Xie, L.; Zhang, R.; Ban, X. Cross-Layer Installed Cable-Bracing Inerter System for MDOF Structure Seismic Response Control. *Appl. Sci.* **2020**, *10*, 5914. [CrossRef]
37. Wang, H.; Gao, H.; Li, J.; Wang, Z.; Ni, Y.; Liang, R. Optimum design and performance evaluation of the tuned inerter-negative-stiffness damper for seismic protection of single-degree-of-freedom structures. *Int. J. Mech. Sci.* **2021**, *212*, 106805. [CrossRef]
38. Liu, X.; Yang, Y.; Sun, Y.; Zhong, Y.; Zhou, L.; Li, S.; Wu, C. Tuned-Mass-Damper-Inerter Performance Evaluation and Optimal Design for Transmission Line under Harmonic Excitation. *Buildings* **2022**, *12*, 435. [CrossRef]
39. Weber, F.; Huber, P.; Borchsenius, F.; Braun, C. Performance of TMDI for Tall Building Damping. *Actuators* **2020**, *9*, 139. [CrossRef]
40. Zhang, R.; Cao, M.; Huang, J. Study on Seismic Response and Parameter Influence in a Transformer-Bushing with Inerter Isolation System. *Buildings* **2022**, *12*, 530. [CrossRef]
41. Den Hartog, J.P. *Mechanical Vibrations*, 4th ed.; Dover: New York, NY, USA, 1956.
42. Zhang, R.; Cao, M. Study on Vibration Control and Parameters Influence of Cable Inerter Viscous Damping System. *Shock Vib.* **2022**, *2022*, 2983700. [CrossRef]
43. Zhang, R.; Huang, J.; Cao, M.; Luo, Q.; Guo, X. Study on Parameters' Influence and Optimal Design of Tuned Inerter Dampers for Seismic Response Mitigation. *Buildings* **2022**, *12*, 558. [CrossRef]

buildings

MDPI

Article

Investigations on the Dynamic Response of Adjacent Buildings Connected by Viscous Dampers

Peng Chen [1,2,*] and Xiaobin Wu [2]

1 Department of Civil Engineering, Sichuan University, Chengdu 610225, China
2 China Southwest Architectural Design and Research Institute Co., Ltd., Chengdu 610041, China
* Correspondence: pengchen@scu.edu.cn

Abstract: This paper investigates the seismic performance of two adjacent buildings connected by viscous dampers. Three types of damper placement are discussed, including installing dampers within a single building, connecting two buildings at the same floor level, and connecting two buildings at the inter-story level. Analytical models are established to consider various dynamic properties of the adjacent buildings, and the theoretical solutions are obtained, including the transmissibility curves, additional modal damping, and input energy under the seismic design spectrum. Time history analyses of an engineering project are performed with different damper placements. Different numerical models are compared for frequently and rarely occurred earthquakes. The seismic mitigation effect is discussed with regard to the story drift reduction rate and dynamic energy. Theoretical and numerical results demonstrate that the connecting dampers provide added modal damping while causing the coupled response. As a result, it is less efficient than traditional ways of placement within a building. Furthermore, the connecting dampers significantly increase the reaction of the floors without installed dampers. When designing dampers to connect the adjacent buildings, careful engineering calculations should be made.

Keywords: adjacent building; viscous damper; seismic response; transmissibility; time history analysis

Citation: Chen, P.; Wu, X. Investigations on the Dynamic Response of Adjacent Buildings Connected by Viscous Dampers. *Buildings* **2022**, *12*, 1480. https://doi.org/10.3390/buildings12091480

Academic Editors: Liqiang Jiang, Jihong Ye and Wei Guo

Received: 31 August 2022
Accepted: 15 September 2022
Published: 17 September 2022

Publisher's Note: MDPI stays neutral with regard to jurisdictional claims in published maps and institutional affiliations.

1. Introduction

Metropolitan buildings are constructed more closely with the growing population and limited land resources. Many commercial buildings are composed of a low-rise podium along with tall towers. For some, podiums and towers are built as a whole structure, but for many others, they are built independently with construction joints. Joints also separate buildings with large or irregular plan sizes. Considering these reasons, researchers and engineers proposed connecting the separated structures using energy dissipative devices to lower seismic damage, the possibility of pounding, or wind-induced vibration [1].

Traditional passive dampers are studied as connectors, which are viscous dampers in the early stage. Luco et al. [2] studied the dynamic response of two adjacent buildings with different heights connected by viscous dampers. The optimal damping values are discussed regarding the different properties of the two buildings. The controlling strategies have been verified by Xu et al. [3] and Zhang et al. [4] using analytical and numerical approaches. Tubaldi et al. [5] presented a performance-based methodology for the seismic assessment of two steel-type buildings connected by viscous dampers and found out that the seismic performance is sensitive to the viscous properties of the dampers. Yang and Lam [6] studied the dynamic response of two buildings connected by viscoelastic dampers under bidirectional excitations. It is concluded that the bidirectional input can increase the reactions of coupled asymmetric buildings, and the installation dampers may induce a sudden change in the lateral stiffness of the taller building.

For experiment verifications, several small-scaled shaking table tests were performed to validate the controlling effect by Xu et al. [7]. Yang et al. [8] performed comparative experiments to verify the seismic mitigation effect using viscous dampers to connect a scaled five-story and a six-story steel frame. A significant increment in modal damping is observed while the natural frequencies remain almost unchanged. Wu et al. [9] conducted shaking table tests for two adjacent isolated steel frames connected by viscous dampers. Test results proved that the dampers could help reduce the overturning effect and maximum bearing deformation for adjacent high-rise isolated structures. In addition to viscous dampers, other passive energy dissipative devices are studied, including viscoelastic dampers [10], friction dampers [11], tuned mass dampers [12], inertial mass dampers [13], negative stiffness devices [14], etc.

Researchers focus on the optimal placement of dampers for passive devices to connect adjacent buildings. Bigdeli et al. [15] simplify the optimization process as a bi-level optimization problem and compares five methods to solve the problem. Theoretical and numerical studies are also performed to verify the optimizing results [16,17]. Tubaldi et al. [18] used the linearized reduced-order model of the coupled system to establish a simplified design method for both linear and nonlinear dampers between adjacent buildings. The simplified design method is verified by time history analysis. Palermo and Silvestri [19] studied the optimal additional damping for connecting viscous dampers and the trends of damping reduction factors concerning the main dynamic parameters of the coupled systems.

Other strategies to connect the adjacent buildings adopt active or semi-active technologies. Ou and Li [20] investigated the characteristics of forces in active control systems connected to adjacent buildings. The design approach for all control devices is proposed based on the results, which can provide an expected reduction in seismic response. Xu et al. [13] proposed using active tuned mass dampers on the adjacent buildings' top floors combined with viscous dampers and numerically verified the vibration controlling effect. Uz and Hadi [21] studied the adoption of Magnetorheological (MR) dampers and the corresponding optimal design process. It has been proved that the controlling technology can enhance seismic performance economically. Guenidi et al. [22] numerically investigated using shared tuned mass damper (TMD) and MR dampers and concluded that a shared TMD could provide adequate response reduction compared to that obtained using two TMDs separately. Al-Fahdawi et al. [23,24] compared the controlling benefits by connecting two adjacent buildings using passive viscous and MR dampers with properly designed control algorithms. The reduction in story drift and absolute acceleration are compared, and it is concluded that the active control method (MR dampers) is more effective than the traditional passive device.

The above literature focuses on connecting the two adjacent buildings to reduce seismic damage. In addition, the pounding prevention and wind-induced vibration controlling effect are discussed and verified by refs. [25–31], in which many energy dissipative devices are also included. Nevertheless, most of the present studies did not consider the soil–structure interaction. Further research is required to include the SSI effect on the seismic response of connected adjacent buildings, but not for lightweight structures [32–34].

Although many studies have explored reducing seismic action between two adjacent buildings using energy dissipation devices, most of them discussed the most likely benefit of the dampers. The complex situations are not thoroughly studied, including the adjacent buildings' different dynamic properties and nonlinear seismic behavior. The main contribution of this paper is the detailed analytical and numerical investigations to study the seismic performance of connecting adjacent buildings with dampers. The possible adverse effect caused by the additional dampers is discussed. In addition, a new connecting type is proposed by installing dampers between the inter-story levels, and the controlling efficiency is compared. Discrete simplified analytical models are established, and the transmissibility law is compared considering the different properties of the adjacent buildings. Nonlinear time history (NLTH) analyses are also performed to validate the theoretical results based on an actual engineering project.

2. Problem Statement

2.1. Analytical Model

Adjacent buildings are idealized as linear shear-type models. The mass is concentrated at the centers of the floors. For the comparison of damper placement, three analytical models are established as multi-degree of freedom (MDOF) systems, as shown in Figure 1:

(1) Model a: The tower with dampers installed within the building;
(2) Model b: Two adjacent buildings connected by dampers at the same story level;
(3) Model c: Two adjacent buildings with inter-story damper placement.

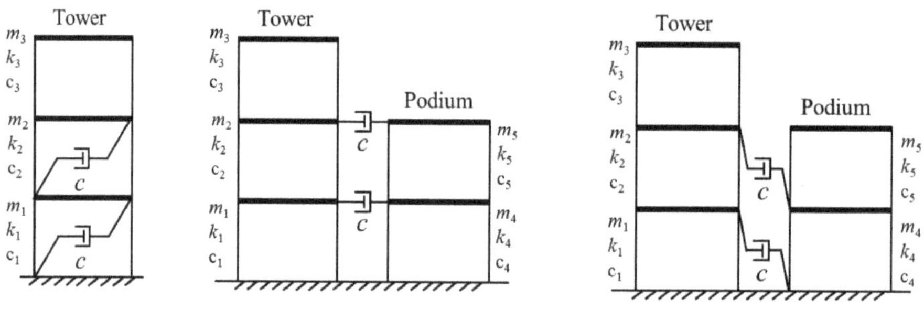

Figure 1. MODF analytical model with different damper layouts: (**a**) Model a: with single building placement; (**b**) Model b: with same level placement; (**c**) Model c: with inter-story placement.

2.2. Equations of Motion

For the different analytical models in Section 2.1, the equation of motion can be described as,

$$M\ddot{X} + (C + c)\dot{X} + KX = -MI\ddot{u}_g(t), \tag{1}$$

where $\boldsymbol{M}$, $\boldsymbol{C}$, and $\boldsymbol{K}$ are the mass, damping, and stiffness matrices, respectively. c is the additional damping coefficient provided by a single viscous damper element. $\ddot{X}$, $\dot{X}$, and X are the relative acceleration, velocity, and displacement matrices of the system, respectively. $\boldsymbol{I}$ is the unit vector with all elements equal to 1, and $\ddot{u}_g(t)$ is the time history of the ground motion.

The mass, damping, and stiffness matrices of Model a–Model c are given as follows,

$$M_a = \begin{bmatrix} m_1 & 0 & 0 \\ 0 & m_2 & 0 \\ 0 & 0 & m_3 \end{bmatrix}, \tag{2}$$

$$M_b, M_c = \begin{bmatrix} m_1 & 0 & 0 & 0 & 0 \\ 0 & m_2 & 0 & 0 & 0 \\ 0 & 0 & m_3 & 0 & 0 \\ 0 & 0 & 0 & m_4 & 0 \\ 0 & 0 & 0 & 0 & m_5 \end{bmatrix}, \tag{3}$$

$$C_a + c = \begin{bmatrix} c_1 + c_2 + 2c & -c_2 - c & 0 \\ -c_2 - c & c_2 + c_3 + c & -c_3 \\ 0 & -c_3 & c_3 \end{bmatrix}, \tag{4}$$

$$C_b + c = \begin{bmatrix} c_1 + c_2 + c & -c_2 & 0 & -c & 0 \\ -c_2 & c_2 + c_3 + c & -c_3 & 0 & -c \\ 0 & -c_3 & c_3 & 0 & 0 \\ -c & 0 & 0 & c_4 + c_5 + c & -c_5 \\ 0 & -c & 0 & -c_5 & c_5 + c \end{bmatrix}, \tag{5}$$

$$C_c + c = \begin{bmatrix} c_1 + c_2 + c & -c_2 & 0 & 0 & 0 \\ -c_2 & c_2 + c_3 + c & -c_3 & 0 & 0 \\ 0 & -c_3 & c_3 & 0 & 0 \\ 0 & 0 & 0 & c_4 + c_5 + c & -c_5 \\ 0 & 0 & 0 & -c_5 & c_5 + c \end{bmatrix}, \tag{6}$$

$$K_a = \begin{bmatrix} k_1 + k_2 & -k_2 & 0 \\ -k_2 & k_2 + k_3 & -k_3 \\ 0 & -k_3 & k_3 \end{bmatrix}, \tag{7}$$

$$K_b, K_c = \begin{bmatrix} k_1 & -k_2 & 0 & 0 & 0 \\ -k_2 & k_2 + k_3 & -k_3 & 0 & 0 \\ 0 & -k_3 & k_3 & 0 & 0 \\ 0 & 0 & 0 & k_4 + k_5 & -k_5 \\ 0 & 0 & 0 & -k_5 & k_5 \end{bmatrix}, \tag{8}$$

where M_a, M_b, and M_c are the mass matrices, $C_a + c$, $C_b + c$, and $C_c + c$ are the damping matrices with added viscous dampers, and K_a, K_b, and K_c are the stiffness matrices of Model a–Model c, respectively. c_1–c_5, and k_1–k_5 represent the story stiffness value and the damping coefficient for the 1st–5th story, respectively.

2.3. Transmissibility in Frequency Domain

For the motion of equations given in Equation (1), it can be transformed into the frequency domain using Fourier transformation for both sides of the equation as

$$\left[-\omega^2 M + i\omega(C + c) + K\right] U(\omega) = -MI\ddot{U}_g(\omega), \tag{9}$$

where $U(\omega) = \mathcal{F}\{X(t)\}$, $\ddot{U}_g(\omega) = \mathcal{F}\{\ddot{u}_g(t)\}$.

The displacement transmissibility between the i-th floor and the base can be obtained,

$$H_i(\omega) = \frac{U(\omega)}{\ddot{U}(\omega)} = \frac{-MI}{-\omega^2 + i\omega(C + c) + K}, \tag{10}$$

where $H_i(\omega)$ denotes the displacement transmissibility between the ground and the i-th degree of freedom.

The acceleration transmissibility of the i-th floor is

$$H_{ai}(\omega) = -\omega^2 H_i(\omega), \tag{11}$$

3. Analytical Results

3.1. Model Parameters

The 5-DOF analytical models were established to simulate the main tower building constructed with a bottom podium. Different engineering situations where the adjacent tower and podium usually have different or similar architectural heights and dynamic properties were included. Different dynamic properties of the adjacent buildings could lead to different seismic performances of the connecting dampers. Cases No. 1, No. 2, and No. 3 were used to reflect the differences between the two adjacent structures by changing the mass and stiffness of the degree of freedom. The story mass and stiffness are listed in Table 1, and inherent structural damping matrices are simplified using Rayleigh damping matrices by assuming the first two-order modal damping ratio to be 5%. The natural periods of the 5-DOF model in different cases are given in Table 2, and the normalized mode shapes are presented in Figure 2. The first natural period of the tower for different cases varies from 2.11 to 3.17 s, with slight differences in the second and third periods and the normalized modal shapes.

Table 1. Structural parameters of the 5-DOF models.

Case No. 1		Case No. 2		Case No. 3	
Mass (ton)	Stiffness (kN/m)	Mass (ton)	Stiffness (kN/m)	Mass (ton)	Stiffness (kN/m)
m_1 = 400	k_1 = 16,000	m_1 = 400	k_1 = 16,000	m_1 = 400	k_1 = 16,000
m_2 = 400	k_2 = 16,000	m_2 = 400	k_2 = 16,000	m_2 = 400	k_2 = 16,000
m_3 = 300	k_3 = 16,000	m_3 = 300	k_3 = 16,000	m_3 = 1200	k_3 = 22,000
m_4 = 400	k_4 = 16,000	m_4 = 400	k_4 = 36,000	m_4 = 400	k_4 = 16,000
m_5 = 300	k_5 = 16,000	m_5 = 300	k_5 = 36,000	m_5 = 300	k_5 = 16,000

Table 2. Natural periods of the 5-DOF models with different parameter cases (without dampers).

Mode	Podium (Case No. 1 & No. 3)	Tower (Case No. 1 & No. 2)	Podium (Case No. 2)	Tower (Case No. 3)
1st	1.45 s (0.69 Hz)	2.08 s (0.48 Hz)	0.97 s (1.03 Hz)	3.17 s (0.32 Hz)
2nd	0.59 s (1.70 Hz)	0.76 s (1.32 Hz)	0.39 s (2.56 Hz)	0.84 s (1.19 Hz)
3rd	—	0.54 s (1.85 Hz)	—	0.54 s (1.84 Hz)

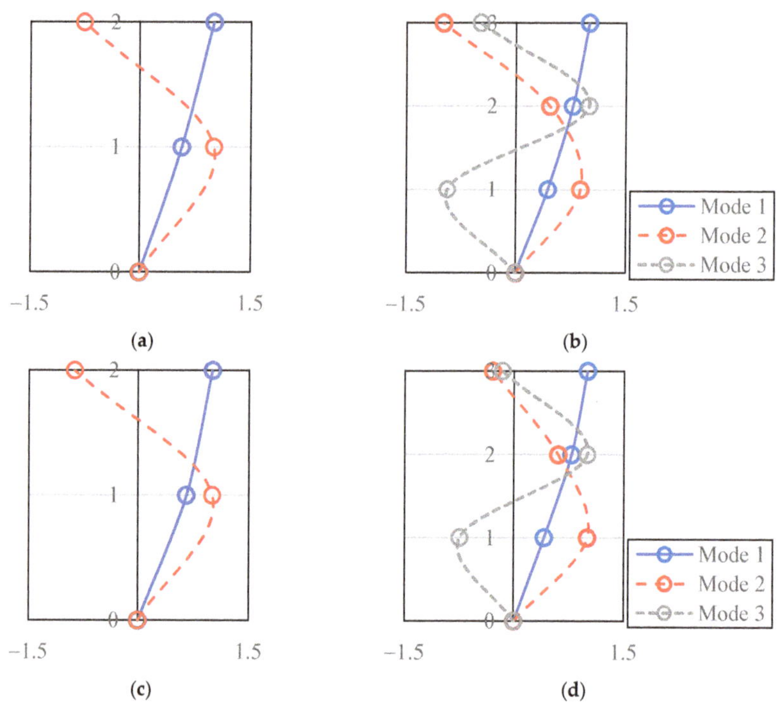

Figure 2. The normalized mode shapes: (**a**) Podium of Case No. 1 and No. 3; (**b**) Tower of Case No. 1 and No. 2; (**c**) Podium of case No. 2; (**d**) Tower of Case No. 3.

For simplicity, it is assumed that the viscous linear dampers can provide a 6% additional damping ratio for the podium. The corresponding damping coefficients are calculated from the modal damping matrices of the podium as c = 216 kN·s^2/m.

3.2. Transimissibility Comparison

Based on the transmission equations in Equations (10) and (11), the top floor acceleration transmissibility is presented in Figure 3a. It has been proved that the damper can reduce dynamic response by providing additional modal damping for all modes. However, due to the installation of connecting dampers, the dynamic response is coupled, and the coupled response can be significantly observed in the frequency domain. The podium is influenced more obviously than the tower because of the difference in structural size. Due to the contribution of the coupled response, the dynamic motion may be amplified under earthquakes.

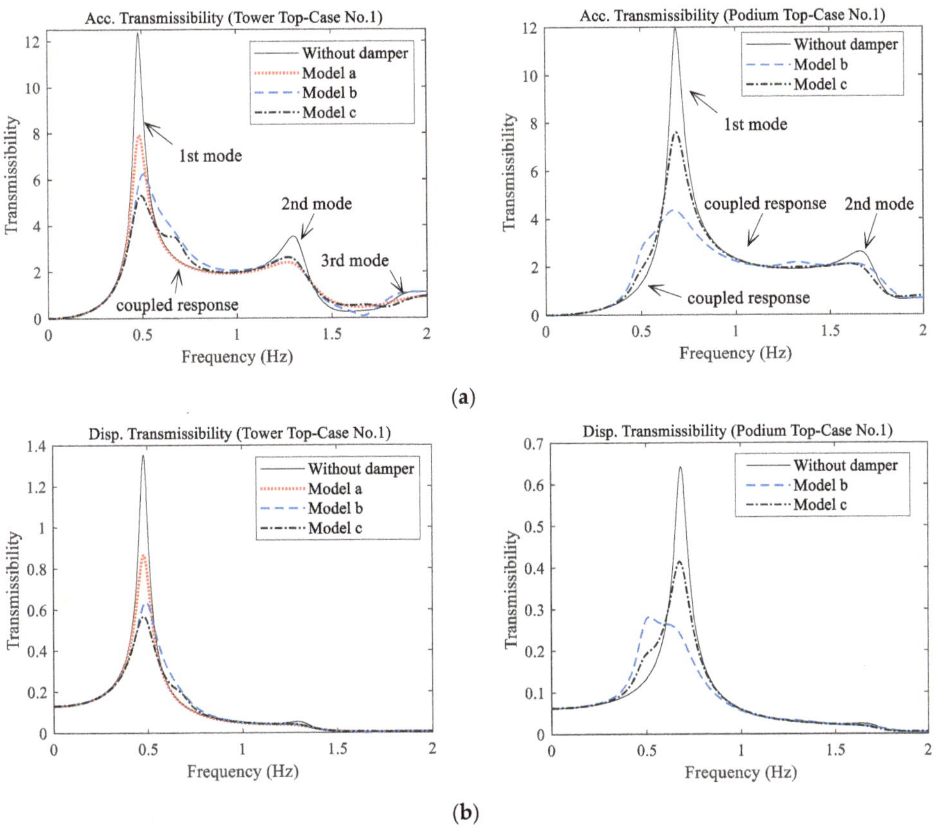

Figure 3. Transmissibility of the tower and podium (Case No. 1): (**a**) absolute top acceleration; (**b**) absolute top displacement.

The main factors induced by the connecting dampers that can affect the seismic response of the tower and the podium are: (1) additional modal damping by the energy dissipation of dampers; and (2) the coupled dynamic response, which is essentially caused by the phase difference between the two ends of the damper under motion. Additional modal damping is always beneficial for reducing the seismic response; however, the amplification phenomena due to the coupling increase the complexity of the connecting system. Figure 3b presents the top displacement transmissibility for the tower and podium. Compared with the acceleration response, the displacement response mainly depends on the first mode contribution, and the higher mode response is minor. At the same time, the coupled response is also significant for the podium structure.

3.3. Quantification of Additional Modal Damping

The Levy Method [35] is a widely accepted system identification algorithm that adopts a mathematic transmission model to fit the frequency data in a least-squares sense. This method can calculate the modal damping from the absolute acceleration transfer function. The modal damping ratios of the all modes are obtained and compared herein. The additional modal damping ratio is obtained by the identified damping ratio minus the original modal damping. The results are discussed considering different structural parameters in Figure 4. The main conclusions are:

(a) The damper layout in Model a (installed within the tower building) provides more significant modal damping in higher modes, while the dampers in Model b and Model c that connect buildings can dissipate more energy in the first mode response;
(b) Dampers in Model b and Model c have similar performance for providing additional modal damping;
(c) Under specific circumstances, dampers can decrease the modal damping ratio (minus the modal damping ratio). That is because if the coupled response is near the modal frequency, the transmissibility value will increase, and the identified modal damping using the Levey Method will decrease accordingly.

Figure 4a,b present the additional damping ratio for Case No. 1 and No. 2. For these two cases, the bottom podium of Case No. 2 is more rigid. The natural period of the podium in Case No. 1 and No. 2 is 1.45 s and 0.97 s, respectively. Results illustrate that the modal damping in higher modes of tower building increase owing to the rigid podium. For the podium, the modal damping significantly increases in Model b but decreases in Model c.

Figure 4a,c show the additional damping ratio for Case No. 1 and No. 3. For these two cases, the bottom podium of Case No. 3 is softer to represent a high-rise tower building. The natural period of the tower in Case No. 1 and No. 3 is 2.08 s and 3.17 s respectively. The damping ratio for the tower decreases in Case No. 3 because the tower size becomes significantly larger. The third mode additional damping ratio is a minus value, which amplifies the third mode response. The podium damping ratio obtains an increment in modal damping, primarily for Model b.

3.4. Seismic Input Energy

Though the additional modal damping ratio has been compared in Section 3.3, evaluating the damping benefits with different damper layouts for the whole structural system is challenging. Even though a higher first mode damping ratio is obtained using connecting dampers, a significant coupled response is observed (Figure 3), which can amplify the dynamic response.

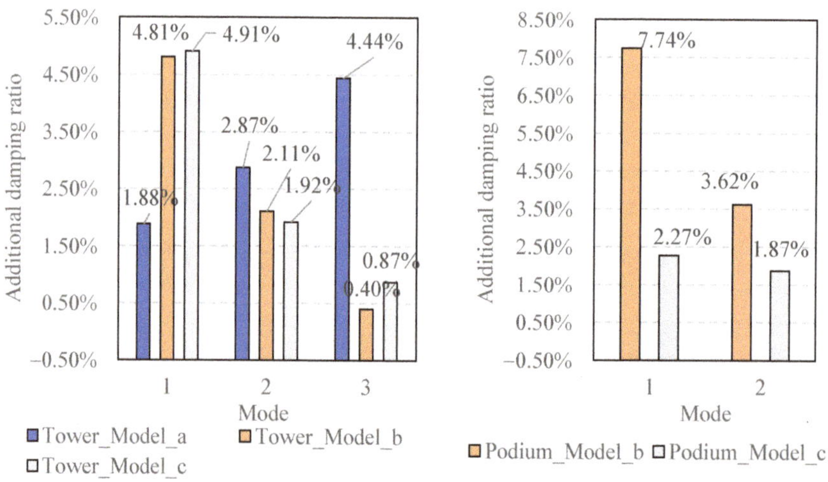

(a)

Figure 4. *Cont.*

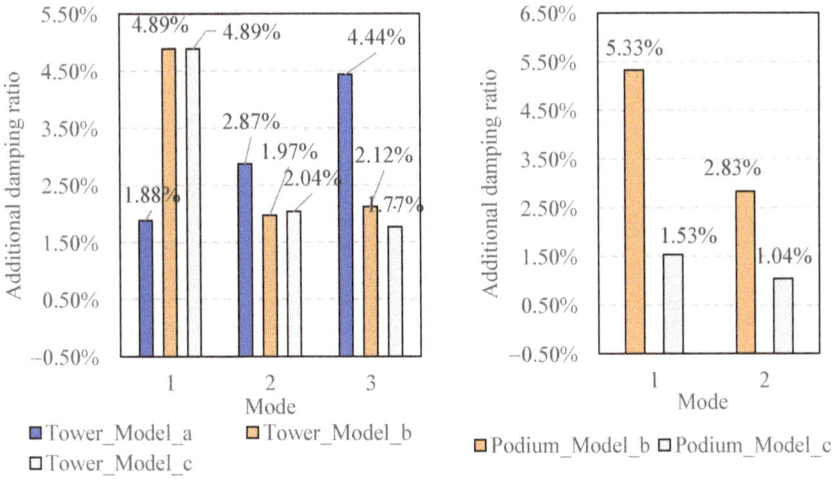

(b)

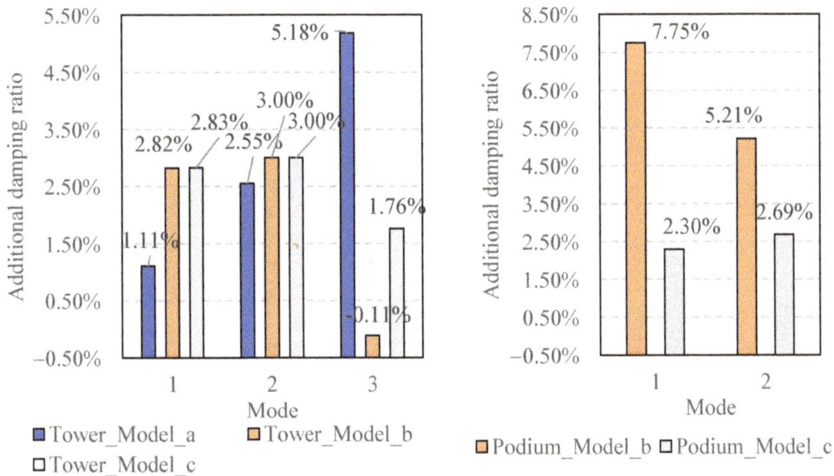

(c)

Figure 4. Additional damping ratio under different damper placement with 6% damping level: **(a)** Case No. 1; **(b)** Case No. 2; **(c)** Case No. 3.

The seismic input energy is calculated and compared to appropriately evaluate the seismic mitigation effect. The ordinary approach to obtaining input energy is to conduct a time history analysis, though the energy can be calculated more briefly in the frequency domain,

$$E_I(t) = \int_0^{\infty} \left|\ddot{U}_g(\omega, t)\right|^2 F(\omega) d\omega, \tag{12}$$

where $\ddot{U}_g(\omega, t)$ is the Fourier Transform of the seismic wave in the time range between 0 and t. $F(\omega)$ is the energy transfer function given in the following form,

$$\begin{aligned} F(\omega) &= \mathrm{Re}\left[\frac{i\omega M^T M}{-\omega^2 M + i\omega C + KI}\right] / \pi, \\ &= \mathrm{Re}\left[-i\omega M^T \cdot H_i(\omega)\right] / \pi, \end{aligned} \tag{13}$$

The design spectrum of the Chinese Code [36] is adopted and transformed into the power spectrum as

$$S_g(\omega) = \frac{\xi}{\pi\omega} S_a^2(\omega, \xi) / \ln\left[\frac{-\pi}{\omega T} \ln(1 - P)\right], \tag{14}$$

where $S_g(\omega) = \left|\ddot{U}_g(\omega)\right|^2$ is the power spectrum of the seismic input. $S_a(\omega, \xi)$ is the design acceleration spectrum with designated damping ratio ξ. T is the duration of the earthquake and P is the exceedance probability (T = 30 s, P = 0.85 in this paper). It is assumed that the phase angle of the seismic wave is random. Thus, the code-based artificial wave can be formed by the trigonometric series. The selected design spectrum and generated artificial spectrum after iterative correction are given in Figures 5 and 6. The characteristic period of the selected spectrum is 0.40 s, and the damping ratio is 5%. The peak acceleration value is 70 cm/s^2, corresponding to the frequent earthquakes of fortification intensity of eight in the Chinese Code [36].

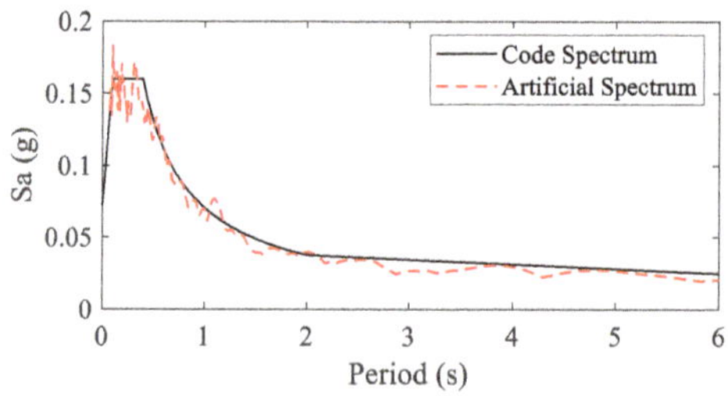

Figure 5. Seismic design spectrum and artificial spectrum.

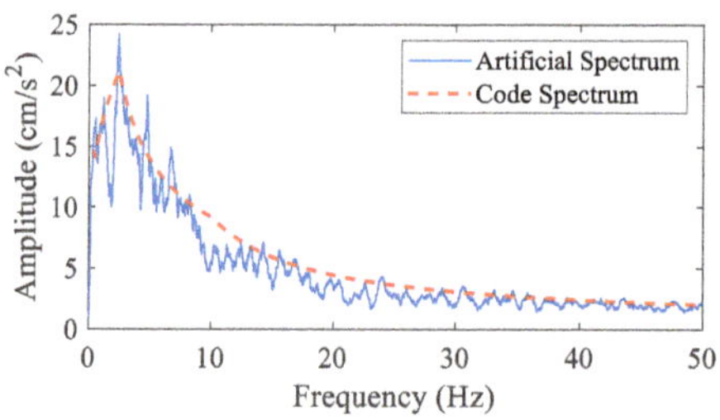

Figure 6. Fourier spectrum generated by the Code and the artificial spectrum.

Energy comparison in time history is calculated by Equation (12). The energy is calculated using the frequency domain method, while the step-by-step approach is used in the time domain to obtain incremental results. An example is given in Figure 7. It presents the seismic input energy for Case No. 1 of each degree of freedom in Figure 1. The input energy can accurately capture the seismic response of each floor, especially including both the effect of energy dissipation of dampers and the coupled response. It is concluded that installing dampers on the lower floors of the tower will increase the seismic response on the upper floors (3-DOF). The connecting dampers in Model b are likely to increase the seismic response of the tower but with a significant seismic reduction effect for the podium.

In comparison, Model b can reduce the dynamic response of the tower but amplify the podium response. According to Figure 4a, the additional modal damping of Model b and Model c are similar for the tower but are higher for the podium of Model b, leading to a significant seismic reduction effect. The energy results for the tower can be explained in Figure 3: For Model b, the coupled response is remarkable in the frequency domain between 0.5 and 1.0 Hz, which is caused by the first mode response of the podium. Figure 6 also presents that the seismic energy is high in the same frequency range as the Fourier spectrum generated by the Code.

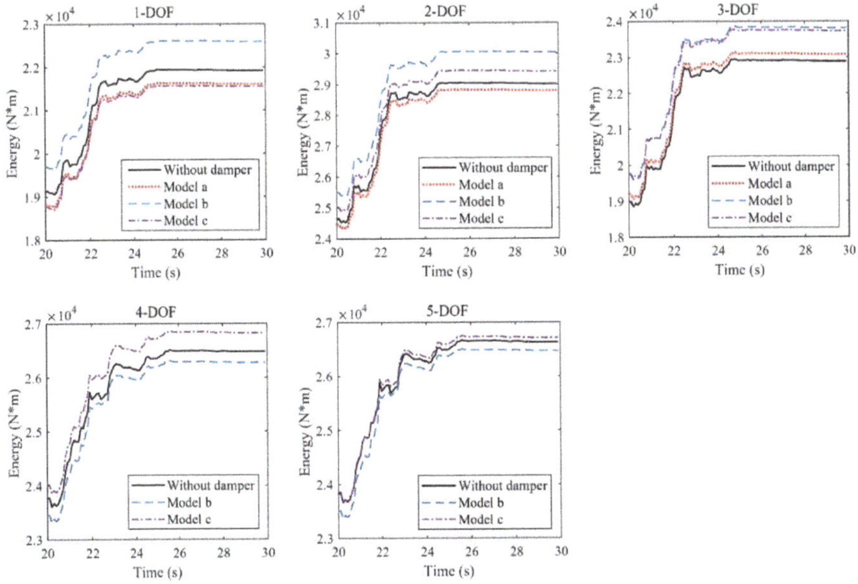

Figure 7. Seismic input energy for each degree of freedom (DOF) of tower and podium (Case No. 1).

The seismic input energy reduction ratio R_E is defined to evaluate the damping effect provided by different layouts of dampers,

$$R_E = (EI_d - EI)/EI, \tag{15}$$

where EI is the seismic input energy (total dissipated energy at the final time step) of the structures without dampers. EI_d is the seismic input energy of the damped structure.

The energy reduction ratio is summarized in Figures 8–10, including different adjacent building cases and damper layout models. The main conclusions that can be made from the comparison results are: (a) All the damper layouts will increase the response on the upper floor of towers, especially for the dampers connecting two adjacent buildings. (b) For the traditional damper layout in the single tower in Model a, the energy dissipation effect is guaranteed for the tower. While the dynamic response will increase in many cases considered in this study due to the coupled response of the adjacent buildings. (c) The damping effect caused by the connecting dampers is significantly influenced by the dynamic properties of the two adjacent buildings. From Case No. 1 to Case No. 2, the damper is more beneficial for the tower, with the podium becoming rigid. In addition, comparing Case No. 1 and Case No. 3, the dampers are more effective for both the tower and podium with a higher tower building. The response energy results conclude that the dampers connecting two buildings are not always beneficial for the seismic response. If dampers are designed to connect two adjacent buildings, a detailed analysis should be made to fully consider the possible adverse effect.

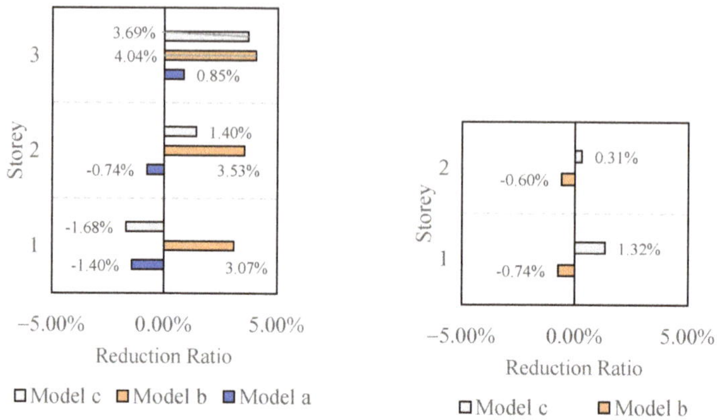

(**a**) (**b**)

Figure 8. Energy reduction ratio of Case No. 1: (**a**) Tower; (**b**) Podium.

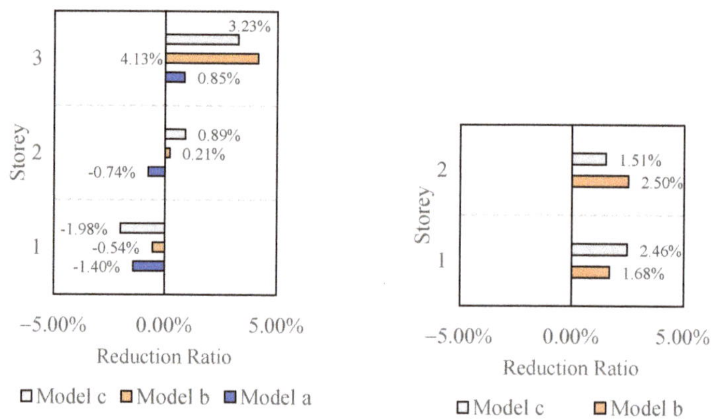

(**a**) (**b**)

Figure 9. Energy reduction ratio of Case No. 2: (**a**) Tower; (**b**) Podium.

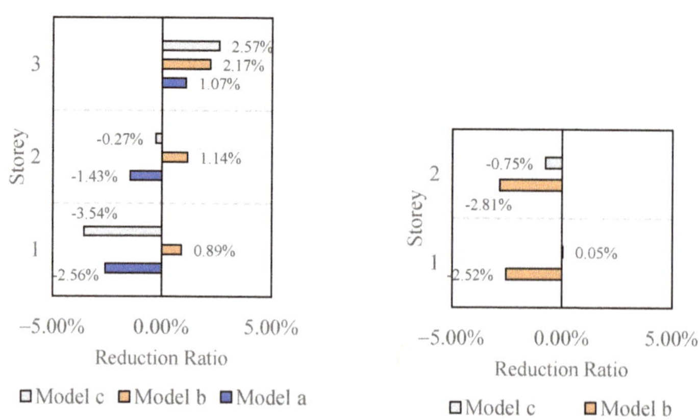

(**a**) (**b**)

Figure 10. Energy reduction ratio of Case No. 3: (**a**) Tower; (**b**) Podium.

4. Engineering Verification

The different damper layouts in adjacent building models have been discussed thoroughly in analytical solutions regarding energy in Section 3. In general, results show that the dampers that connect two adjacent buildings may increase the dynamic response in some circumstances instead of dissipating the motion energy. Numerical studies examine the seismic response of the two adjacent buildings with dampers.

4.1. Numerical Model

Figure 11 presents a 23-story reinforced concrete (RC) frame shear wall building with an adjacent 10-story steel moment frame as the podium. In between is a construction joint, which divides the two buildings. The numerical model is based on an actual project in Yunnan Province, China, designed by the China Southwest Architectural Design and Research Institute. The basic dynamic properties are listed in Table 3. Due to the architectural function, the maximum number of dampers in the global x-direction is limited. Linear viscous dampers are installed within the tower and across the construction joint in the x-direction to satisfy the seismic mitigation requirement of the RC tower. After the preliminary design, the viscous damping coefficient of viscous dampers is taken as 36,300 kN·s/m, providing a 50-ton maximum damping force.

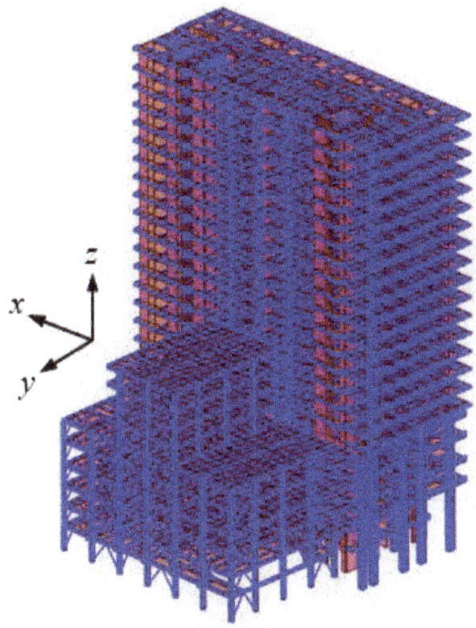

Figure 11. The numerical model of adjacent buildings.

Table 3. Dynamic properties of adjacent tower and podium.

	Tower	Podium	Direction
Mass	76,250 ton	15,000 ton	-
1st Period	2.35 s	1.29 s	y
2nd Period	2.12 s	1.28 s	x
3rd Period	1.98 s	0.85 s	torsional

The SAP2000 v22.0 software was adopted to implement the seismic performance simulation of the case study. For the nonlinear behavior of columns and beams, the P-M2-M3 and M3 hinges were defined, respectively. The steel moment frame podium braces were modelled using the P hinges. The shear walls were defined using the nonlinear layered shells, in which the concrete and rebar were defined separately using layers with designated thickness and positions. The section size and rebars defined in the model are based on the actual engineering design. For the viscous dampers, the link element represents the Maxwell model.

Based on the Chinese Design Code [36], two field-recorded seismic waves and one code-based artificial wave (AW) were selected to evaluate the building performance. The process of generating the AW wave is described in Section 3.4. The detailed information is given in Table 4, the response spectrum is compared with the code spectrum in Figure 12, and the normalized acceleration time history excitation is given in Figure 13. To include the damper performance under different seismic magnitudes, seismic performance analysis with a peak acceleration of 70 gal and 400 gal represents frequently and rarely occurred earthquakes.

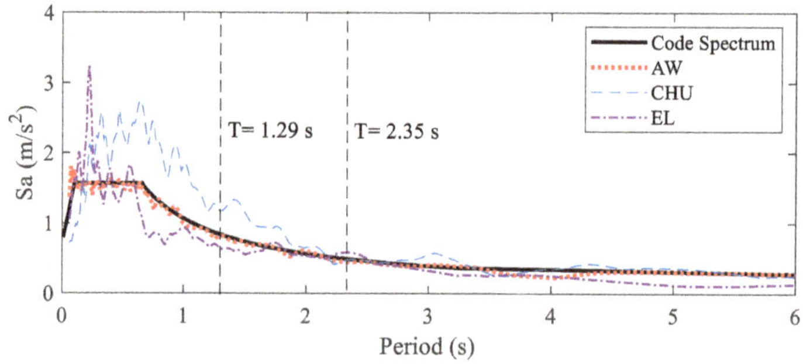

Figure 12. Spectrum comparison between the seismic waves and the Code spectrum.

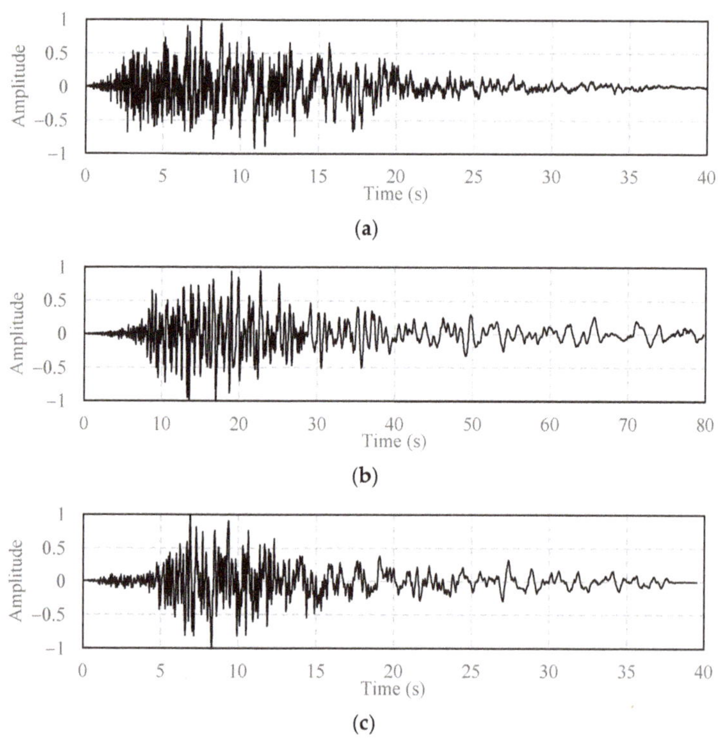

Figure 13. Normalized acceleration time history of the seismic records: **(a)** AW; **(b)** CHE; **(c)** EL.

Table 4. The information of seismic waves.

Event	Location	Station	Year
AW	Artificial Wave	—	—
CHE	Chuetsu-oki, Japan	NIG014	2007
EL	Imperial Valley, US	El Centro Array	1979

4.2. Models with Different Damper Layout

Models with different damper layouts are established to compare the seismic controlling effect via time history analysis:

(1) Original model: RC tower without viscous dampers;
(2) Model A: Few viscous dampers installed within the tower;
(3) Model B: Additional dampers installed on the same level across the construction joint (based on Model A);
(4) Model C: Additional dampers in the inter-story layout (based on Model A).

Figure 14 presents an illustrative sketch of the dampers designed within the tower in a typical story. Dampers in the x-direction are installed at positions ① and ②, while positions ③–⑤ for dampers in the y-direction. In positions ① and ③, dampers are placed from the second floor to the top floor (the 23rd floor) and from the fourth floor to the top floor for other positions. In total, 42 and 62 dampers are used separately in the x and y-direction.

The dampers that connect the two buildings are installed differently in Model B and Model C. The additional dampers are all considered to be used in the x-direction to investigate whether they can provide an approximately identical supplemental damping as in the y-direction. The placement of dampers is presented in Figure 15.

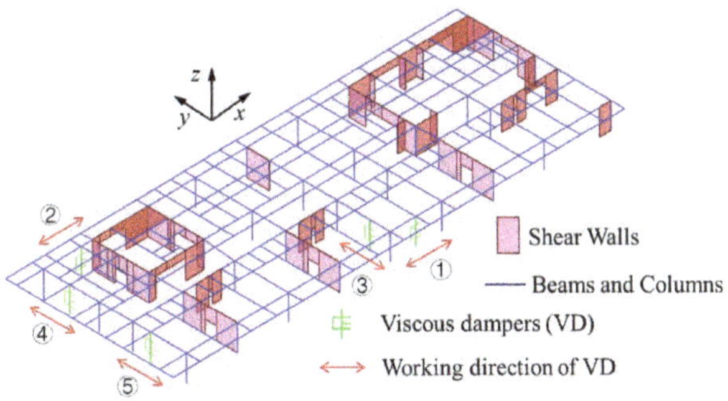

Figure 14. Typical placement of dampers within the towers.

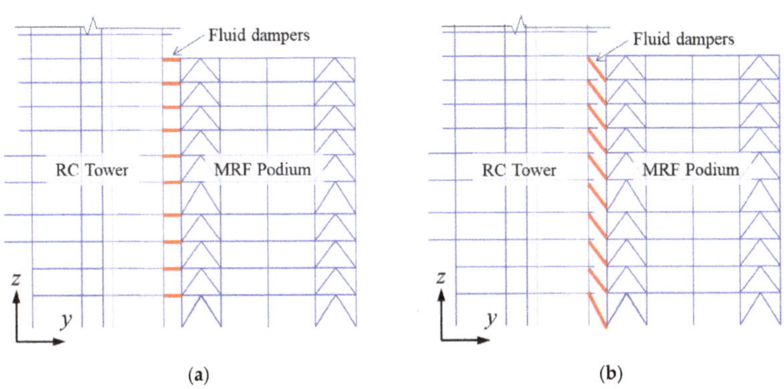

Figure 15. Connecting damper placement for the two adjacent buildings: (**a**) Model B; (**b**) Model C.

4.3. Seismic Response under Frequent Earthquakes

The dynamic seismic response of different models in the x-direction is calculated via time history analysis, and the maximum story drift is obtained. The Maxwell model is adopted to model the viscous dampers, and the Fast Nonlinear Analysis (FNA) method is used to calculate the time history response of the elastic model with limited nonlinear components (viscous damper elements). The drift reduction rate R_d is calculated accordingly to compare the damping effect. Figure 16 presents the drift reduction results from the original model to Model A. It is concluded that the viscous dampers can significantly reduce the maximum story drift under frequent earthquakes at an average rate of about 20%.

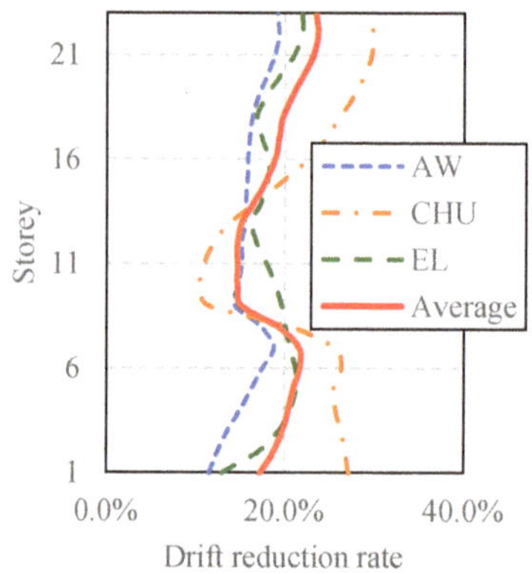

Figure 16. Drift reduction rate between original RC tower and Model A.

The drift reduction rate between Model A and Model B is presented in Figure 17. According to the drift comparison, the seismic response is amplified on the upper floors after installing the dampers to connect the two adjacent buildings. In contrast, the seismic mitigation effect is observed on the bottom floors with connecting dampers. Based on the average value of the three seismic waves, the amplification and reduction rates are between 0 and 5%. The results correspond to the analytical seismic energy comparison in Figures 8–10.

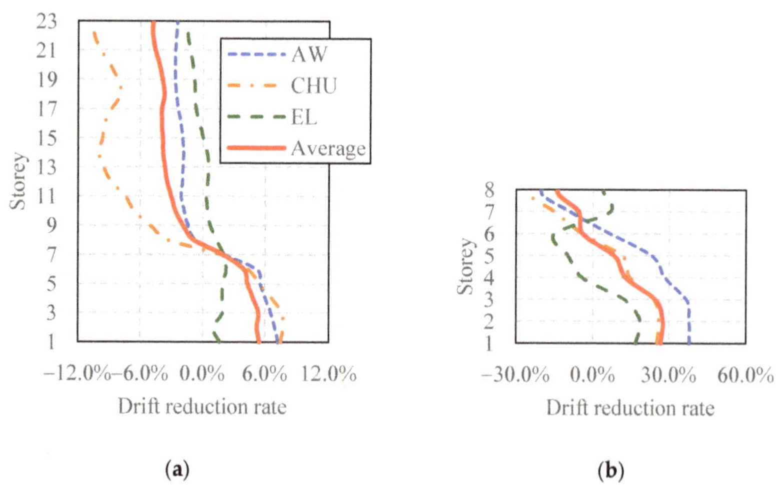

Figure 17. Drift reduction rate between Model A and Model B: (**a**) RC tower; (**b**) Steel podium.

The drift reduction rate between Model B and Model C is presented in Figure 18. The damper placement is different (see Figure 15). According to the seismic reduction rate, it is concluded that the dampers in Model C can provide a better controlling effect for both the RC tower and the steel podium based on average meaning. However, the dampers still cause adverse effects on the upper floors of the RC tower.

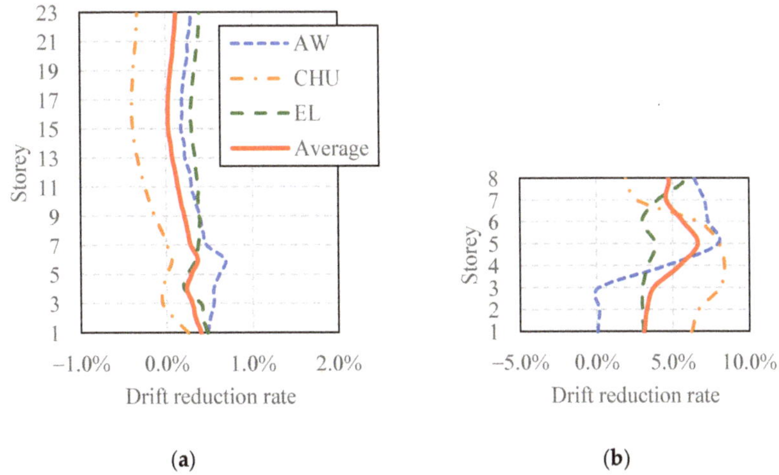

Figure 18. Drift reduction rate between Model B and Model C: (**a**) RC tower; (**b**) Steel podium.

4.4. Nonlinear Response under Rare Earthquake

Although multiple theoretical and numerical analyses have been compared to study the behavior of the dampers connecting adjacent buildings in this paper and previous studies, most of which utilized the elastic model. The nonlinear response of connected buildings under rarely occurred earthquakes should be investigated in detail. The time history response is calculated using the step-by-step method. The details of structural members, including the rebars, are included based on actual engineering designs to reflect the nonlinear response more accurately. Due to the computing time, only the result of the AW wave is calculated in this section.

The story drift reduction rate of Models A, B, and C compared to the original model are given in Figure 19. The drift of Model A is slightly amplified on floors 1–5 because there are no dampers on the first floor and fewer dampers on the second and third floors. It is concluded that the maximum story drift of upper floors can be reduced between 0% and 10% with additional viscous dampers. The drift reduction rate under rare earthquakes significantly differs from that of frequent earthquakes (Figure 16). The overall seismic mitigation efficiency is lower than cases under frequent earthquakes due to the nonlinear behavior of structural members. Under maximum considered earthquakes, the controlling effect increases with the increment of the building height. For the additional effect caused by the dampers across the architectural joint, the seismic response decreases on the floors, but the response is amplified on the upper floors for both Model B and Model C. The primary influence caused by dampers in Model B and Model C is similar.

To discuss the difference between the damper placement in the same and inter-story levels, the drift reduction rate of Model B and Model C compared to Model A is given in Figure 20. The response of both the RC tower and steel podium are compared. For the RC tower with connecting dampers under maximum earthquakes, results indicate that the dampers installed at the inter-story levels are less effective in reducing the seismic response in the floors with additional dampers. However, they cause a less adverse effect on the upper floors without connecting dampers. For the steel podium, the influence is more significant, the drift of the bottom floors is reduced between 0% and 10%, but the maximum drift is amplified by at most 65% on the upper floors. Previous analytical studies have proven that the connecting damper can potentially cause an adverse effect on buildings, which has been verified herein.

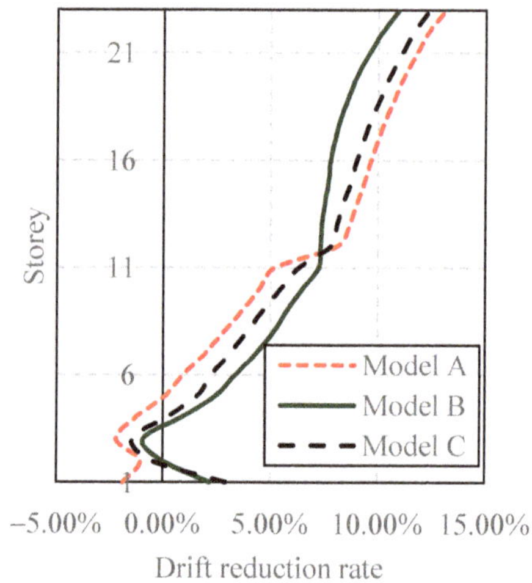

Figure 19. Drift reduction rate of Model A, B, and C compared to the original model.

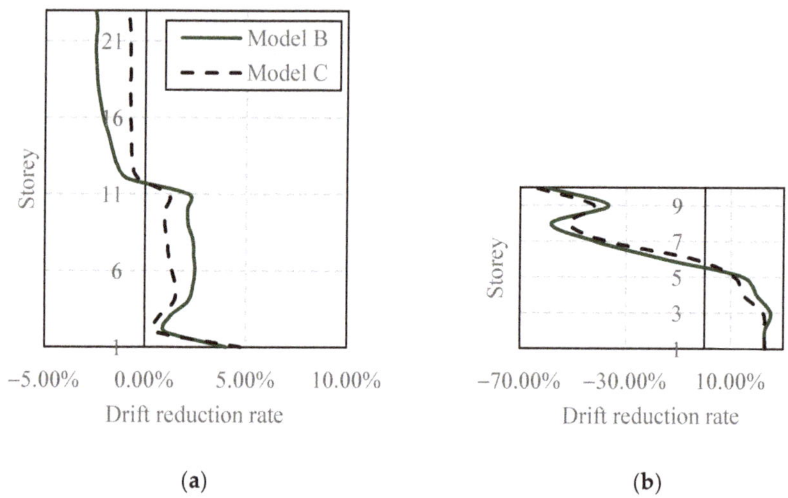

Figure 20. Drift reduction rate of Model B and Model C compared to Model A: (**a**) RC tower; (**b**) Steel podium.

For different models under maximum considered earthquakes, the seismic input energy, kinetic energy, potential energy, modal damping energy dissipated by structural hysteretic loops (hinges and nonlinear layered shells), and energy dissipated by viscous dampers are calculated and compared. An example of the time history of seismic energy is given as an example in Figure 21a. The hinge state of a typical frame of the concrete tower is presented in Figure 21b. All the hinges occurred at the beam end within the LS (life safety) and IO (immediate occupancy) states. The columns and walls stay at the elastic stage, which presents an optimal seismic performance.

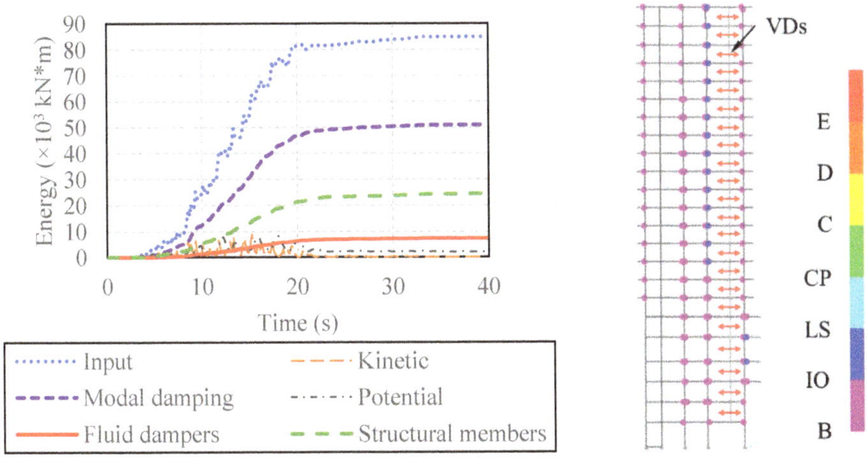

(a) (b)

Figure 21. Nonlinear response of Model A: (**a**) Energy comparison; (**b**) Hinge state.

The maximum value of each energy component from different models is displayed in Figure 22. The viscous dampers dissipate no energy for the original RC tower. For Model A, the dampers are installed within the tower building, and additional connecting dampers are considered in Model B and Model C. The model damping ratio is set as 5% during the nonlinear analysis, and then the overall damping ratio provided by viscous dampers is calculated by the proportion of the dissipated energy. For Model A, Model B, and Model C, the additional damping ratio is 0.50%, 0.63%, and 0.72%, respectively.

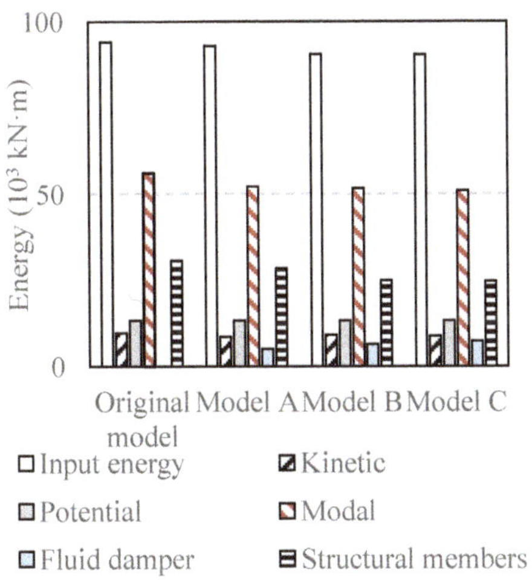

Figure 22. Total energy comparison of different models.

The total energy dissipated by the dampers and the structural members can reflect the damper performance and the structural damage level. The total energy values are compared in Figure 23. Comparing Model A with the Original Model, the dampers within the tower can have a significant effect in dissipating motion energy and mitigating structural damage. Comparing Model B/C with Model A, in the aspect of overall energy, the connecting dampers significantly decrease the energy component of the structural members. It is also concluded that connecting dampers in Model C are slightly more efficient than in Model B.

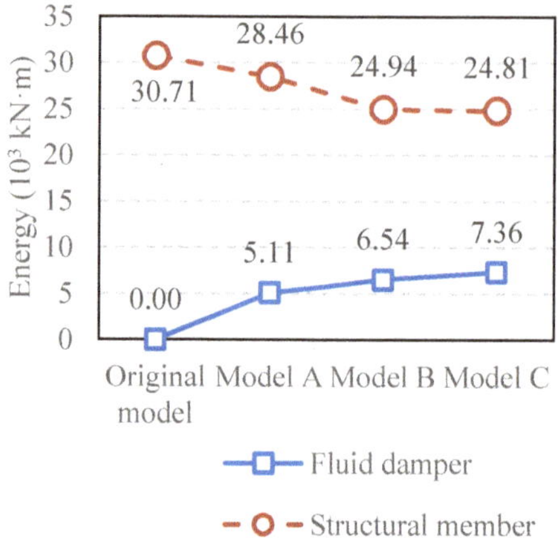

Figure 23. Energy comparison of viscous dampers and structural members.

5. Conclusions

Analytical and numerical studies have been performed in this paper to compare the seismic mitigation effect of viscous dampers to connect two adjacent buildings. The dynamic properties of two adjacent buildings and different damper placements are discussed. In addition, Nonlinear time history (NLTH) numerical analyses for an actual engineering project under frequent and rare earthquakes are performed. The main conclusions are summarized as follows:

(1) From the comparison of transmissibility curves, the connecting dampers can increase the modal damping ratio as the traditional placement of dampers within a single building. In contrast, the dampers can cause the coupled response of the adjacent building for both the podium and tower, which can potentially amplify the seismic response.

(2) The analytical and numerical studies prove that the connecting dampers can significantly amplify the seismic response in the upper stories of the tower. The seismic mitigation effect is proved in the lower stories on the floors where the VDs are installed. Careful consideration is required when designing dampers to connect two buildings. The seismic performance of the upper floors should especially be guaranteed.

(3) For the considered cases in this study, traditional damper placement can be more effective for seismic controlling of a single building. If dampers are designed to connect adjacent buildings due to architectural requirements or to avoid pounding, connecting the two ends of the dampers on different levels should obtain a better seismic mitigation effect under seismic input.

Author Contributions: Conceptualization, X.W.; methodology, P.C.; software, P.C.; validation, X.W.; formal analysis, P.C.; investigation, P.C.; resources, X.W.; data curation, P.C.; writing—original draft preparation, P.C.; writing—review and editing, P.C.; visualization, P.C.; supervision, X.W.; project administration, X.W.; funding acquisition, X.W. All authors have read and agreed to the published version of the manuscript.

Funding: The authors appreciate the support of the National Key R&D Program of China (2019YFC1509305) and the Fundamental Research Funds for the Central Universities (Grant No. 2022SCU12083).

Data Availability Statement: Not applicable.

Conflicts of Interest: The authors declare no conflict of interest.

References

1. Zhai, C.; Jiang, S.; Li, S.; Xie, L. Dimensional analysis of earthquake-induced pounding between adjacent inelastic MDOF buildings. *Earthq. Eng. Eng. Vib.* **2015**, *14*, 295–313. [CrossRef]
2. Luco, J.E.; De Barros, F.C.P. Optimal damping between two adjacent elastic structures. *Earthq. Eng. Struct. D* **1998**, *27*, 649–659. [CrossRef]
3. Xu, Y.L.; He, Q.; Ko, J.M. Dynamic response of damper-connected adjacent buildings under earthquake excitation. *Eng. Struct.* **1999**, *21*, 135–148. [CrossRef]
4. Zhang, W.S.; Xu, Y.L. Dynamic characteristics and seismic response of adjacent buildings linked by discrete dampers. *Earthq. Eng. Struct. D* **1999**, *28*, 1163–1185. [CrossRef]
5. Tubaldi, E.; Barbato, M.; Dall'Asta, A. Performance-based seismic risk assessment for buildings equipped with linear and nonlinear viscous dampers. *Eng. Struct.* **2014**, *78*, 90–99. [CrossRef]
6. Yang, Z.; Lam, E.S.S. Dynamic responses of two buildings connected by viscoelastic dampers under bidirectional earthquake excitations. *Earthq. Eng. Eng. Vib.* **2014**, *13*, 137–150. [CrossRef]
7. Xu, Y.L.; Zhan, S.; Ko, J.M.; Zhang, W.S. Experimental investigation of adjacent buildings connected by viscous damper. *Earthq. Eng. Struct. D* **1999**, *28*, 609–631. [CrossRef]
8. Yang, Z.; Xu, Y.L.; Lu, X.L. Experimental seismic study of adjacent buildings with viscous dampers. *J. Struct. Eng.* **2003**, *129*, 197–205. [CrossRef]
9. Wu, Q.; Yan, H.; Zhu, H.; Bai, X. Shaking table test study on the seismic isolation effect of a hybrid passive control system. *Measurement* **2020**, *164*, 108125. [CrossRef]
10. Tubaldi, E. Dynamic behavior of adjacent buildings connected by linear viscous/viscoelastic dampers. *Struct. Control. Health Monit.* **2015**, *22*, 1086–1102. [CrossRef]
11. Patel, C.C. Random response analysis of adjacent structures connected with friction damper. *Asian J. Civ. Eng.* **2021**, *22*, 1115–1129. [CrossRef]
12. Xu, L.; Cui, Y.; Wang, Z. Active tuned mass damper based vibration control for seismic excited adjacent buildings under actuator saturation. *Soil Dyn. Earthq. Eng.* **2020**, *135*, 106181. [CrossRef]
13. De Souza Pippi, A.; Avila, S.M.; Doz, G. A review on the use of the inerter device in the structural coupling technique for adjacent building vibration control. *Structures* **2022**, *42*, 480–501. [CrossRef]
14. Zhao, Z.; Wang, Y.; Hu, X.; Weng, D. Seismic performance upgrading of containment structures using a negative-stiffness amplification system. *Eng. Struct.* **2022**, *262*, 114394. [CrossRef]
15. Bigdeli, K.; Hare, W.; Tesfamariam, S. Configuration optimization of dampers for adjacent buildings under seismic excitations. *Eng. Optimiz.* **2012**, *44*, 1491–1509. [CrossRef]
16. Bigdeli, K.; Hare, W.; Nutini, J.; Tesfamariam, S. Optimizing damper connectors for adjacent buildings. *Optim. Eng.* **2016**, *17*, 47–75. [CrossRef]
17. Bigdeli, K.; Hare, W.; Tesfamariam, S. Optimal design of viscous damper connectors for adjacent structures using genetic algorithm and Nelder-Mead algorithm. In Proceedings of the Active and Passive Smart Structures and Integrated Systems, San Diego, CA, USA, 27 March 2012.
18. Tubaldi, E.; Gioiella, L.; Scozzese, F.; Ragni, L.; Dall'Asta, A. A design method for viscous dampers connecting adjacent structures. *Front. Built Environ.* **2020**, *6*, 25. [CrossRef]
19. Palermo, M.; Silvestri, S. Damping reduction factors for adjacent buildings connected by viscous-viscous dampers. *Soil Dyn. Earthq. Eng.* **2020**, *138*, 106323. [CrossRef]
20. Ou, J.; Li, H. Design approaches for active, semi-active and passive control systems based on analysis of characteristics of active control force. *Earthq. Eng. Eng. Vib.* **2009**, *8*, 493–506. [CrossRef]
21. Uz, M.E.; Hadi, M.N.S. Optimal design of semi active control for adjacent buildings connected by MR damper based on integrated fuzzy logic and multi-objective genetic algorithm. *Eng. Struct.* **2014**, *69*, 135–148. [CrossRef]
22. Guenidi, Z.; Abdeddaim, M.; Ounis, A.; Shrimali, M.K.; Datta, T.K. Control of adjacent buildings using shared tuned mass damper. *Procedia Eng.* **2017**, *199*, 1568–1573. [CrossRef]
23. Al-Fahdawi, O.A.S.; Barroso, L.R.; Soares, R.W. Simple adaptive control method for mitigating the seismic responses of coupled adjacent buildings considering parameter variations. *Eng. Struct.* **2019**, *186*, 369–381. [CrossRef]
24. Al-Fahdawei, O.F.S.; Barroso, L.R. Adaptive neuro fuzzy and simple adaptive control methods for full three-dimensional coupled buildings subjected to bi-directional seismic excitations. *Eng. Struct.* **2021**, *231*, 111798. [CrossRef]
25. Karabork, T.; Aydin, E. Optimum design of viscous dampers to prevent pounding of adjacent structures. *Earthq. Struct.* **2019**, *16*, 437–453.
26. Kazemi, F.; Miari, M.; Jankowski, R. Investigating the effects of structural pounding on the seismic performance of adjacent RC and steel MRFs. *Bull. Earthq. Eng.* **2021**, *19*, 317–343. [CrossRef]
27. Jankowski, R.; Mahmoud, S. Linking of adjacent three-storey buildings for mitigation of structural pounding during earthquakes. *Bull Earthq. Eng.* **2016**, *14*, 3075–3097. [CrossRef]
28. Kandemir-Mazanoglu, E.C.; Mazanoglu, K. An optimization study for viscous dampers between adjacent buildings. *Mech. Syst. Signal Process.* **2017**, *89*, 88–96. [CrossRef]

29. Zhu, Z.; Lei, W.; Wang, Q.; Tiwari, N.; Hazra, B. Study on wind-induced vibration control of linked high-rise buildings by using TMDI. *J. Wind Eng. Ind. Aerodyn.* **2020**, *205*, 104306. [CrossRef]
30. Wang, Q.; Tian, H.; Qiao, H.; Tiwari, N.D.; Wang, Q. Wind-induced vibration control and parametric optimization of connected high-rise buildings with tuned liquid-column-damper–inerter. *Earthq. Struct.* **2021**, *226*, 111352. [CrossRef]
31. Ricciardelli, F.; Occhiuzzi, A.; Clemente, P. Semi-active tuned mass damper control strategy for wind-excited structures. *J. Wind Eng. Ind. Aerod.* **2000**, *88*, 57–74. [CrossRef]
32. Pilarska, D.; Maleska, T. Numerical analysis of steel geodesic dome under seismic excitations. *Materials* **2021**, *14*, 4493. [CrossRef] [PubMed]
33. Nair, D.; Ichihashi, K.; Terazawa, Y.; Sitler, B.; Takeuchi, T. Higer mode effects of multistorey substructures on the seismic response of double-layered steel gridshell domes. *Eng. Struct.* **2021**, *243*, 112677. [CrossRef]
34. Fan, F.; Zhi, X.; Li, W. Analysis of the acceleration response spectra of single-layer spherical reticulated shell structures. *Appl. Sci.* **2022**, *12*, 2116. [CrossRef]
35. Zhou, Y.; Chen, P. Investigations on a vertical isolation system with quasi-zero stiffness property. *Smart Struct. Syst.* **2020**, *25*, 543–557.
36. *GB 50011-2016*; Chinese Standard, Code for Seismic Design of Buildings. Chinese Architecture and Building Press: Beijing, China, 2018.

buildings

MDPI

Article

Numerical Investigations on Axial Compressive Behavior of Opening Cold-Formed Thin-Wall C-Steel Combined Double-Limb Column

Chang He, Yong Cai * and Haijun Chen

Department of Structural Engineering, School of Civil Engineering, Central South University, Changsha 410075, China
* Correspondence: caiyong@csu.edu.cn

Abstract: To investigate the axial compressive behavior of an opening cold-formed thin-wall C-steel combined double-limb column, C-steel combined I-section columns were modeled in this paper, and the models were validated by experiments on axial-compressed combined columns. Parametric analyses were carried out on the combined columns. The effects of slenderness ratio, height to thickness ratio, width to thickness ratio, bolt spacing, and opening in the web on the ultimate compressive bearing capacity of the combined columns were investigated. It was observed that the slenderness ratio had the most significant effects on the combined column. Furthermore, the formulas predicting the compressive bearing capacity in the Chinese and AISI standards were compared, and the accuracy of the formulas was studied. Afterward, the formulas with higher accuracy and applicability for the ultimate compressive bearing capacity for the C-steel combined I-section column were proposed. The compression stability factor and reduction factor were fitted in this paper. The proposed formulas and factors can predict the ultimate compressive bearing capacity of the C-steel combined I-section column.

Keywords: combined column; numerical analyses; ultimate compressive bearing capacity; slenderness ratio; opening ratio

Citation: He, C.; Cai, Y.; Chen, H. Numerical Investigations on Axial Compressive Behavior of Opening Cold-Formed Thin-Wall C-Steel Combined Double-Limb Column. *Buildings* **2022**, *12*, 1378. https://doi.org/10.3390/buildings12091378

Academic Editor: Francisco López Almansa

Received: 8 August 2022
Accepted: 31 August 2022
Published: 4 September 2022

1. Introduction

Steel structures and composite structures have been widely used in civil and industrial engineering [1–3]. Due to its advantages of light weight, high strength, energy saving, environmental protection, and rapid construction [4,5], the cold-formed thin-walled steel structure has been widely used in engineering structures [6–9]. In applications of the columns, two or more U-shaped and C-shaped cold-formed thin-walled steel members are usually connected with self-tapping screws, rivets, or bolts to form combined columns with better mechanical properties. The combined components are suitable for multistorey buildings [10–13]. Compared with a single section, a cold-formed thin-walled steel combined column has the advantages of greatly improving the bearing capacity, bending and torsional stiffness, and convenient manufacturing and construction [14,15]. Among all kinds of combined cross-sections, the I-shaped cross-section with two C-shaped cross-sections assembled back-to-back is the most widely used, and its structural schematic diagram is shown in Figure 1. In this paper, this type of combined column is called a C-steel combined I-section column.

In recent decades, much research has been conducted on cold-form thin-walled combined columns. Bae et al. carried out experiments on combined columns with U-shaped, 2U-shaped, and 2U + C-shaped combined cross sections, and proved that steel columns with combined cross sections have higher compressive bearing capacity [16]. Whittle and Ramseyer conducted experimental studies on a C-shaped thin-walled steel combined box section with double limb, and found that the reasonable spacing of connectors can effectively avoid the instability of members [17]. Zhou et al. carried out numerical analyses on the cold-formed thin-walled combined columns to investigate the influence of the sectional form, size, and slenderness ratio on the combination effects, and suggested that the stable bearing capacity of the combined column should be set as 0.7 times that of the overall sectional stability bearing capacity to ensure the reliability of the connector [18]. Chen et al. investigated the influence of the thickness of the backing plate and the spacing of the backing plate on the mechanical properties of the combined column through experiments and numerical studies, and proposed a formula for calculating the bearing capacity of the columns with the effective width method [19–21]. Li et al. proposed a formula for calculating the elastic distortion buckling stress of the combined box-section columns [22]. Dabaon and Ellobody studied the failure mode and deformation behavior of combined columns through experiments, and found that the European code and Australian code were not conservative for the destruction of the combined columns with local buckling [23,24]. Abbasi and Khezri analyzed the elastic buckling of the combined columns, and proposed an element with adjustable stiffness characteristics to simulate connectors between the combined columns [25]. Recently, Rahnavard also carried out numerical analyses on the box-section combined columns, and investigated the influence of the form and spacing of the batten plate on the buckling behavior of combined members [26]. Zhou et al. carried out axial compression tests on long columns with C-shaped cross-section and box-shaped cross-section combined with the C-steel. It was found that the failure mode of the specimens was mainly global bending [27–30].

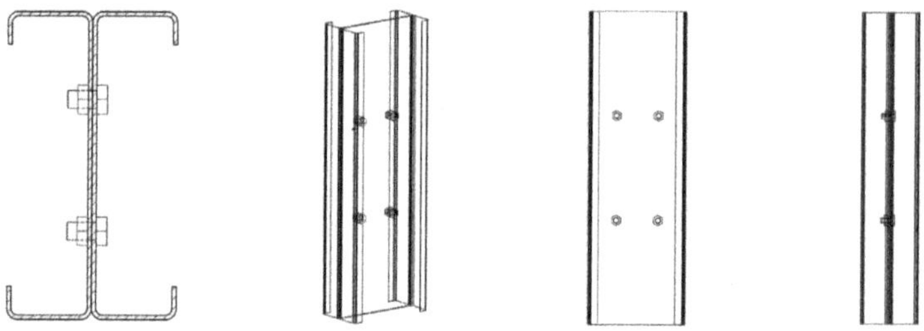

(a) (b) (c)

Figure 1. Diagram of the C-steel combined I-section column (**a**) Cross section; (**b**) 3D diagram; (**c**) Side-view and front view diagram.

Since the I-shaped cross section column has been widely adopted in building engineering, many studies have been conducted on the C-steel combined I-section column with double limb. Stone and Laboube analyzed the buckling form and ultimate bearing capacity of 32 assembled I-section members connected by self-drilling screws under axial compression, and proved that the ultimate bearing capacity calculated by the American code was conservative [31]. Yao studied the effects of the slenderness ratio and screw spacing on the mechanical properties of I-section steel columns. The experiments showed that the slenderness ratio had a great influence on the compressive bearing capacity of I-section steel columns [32]. Fratamicoa carried out buckling and failure tests and numerical analyses on I-shaped cross-section columns with double-limb open-ended grooved steel. It was found that local–global mutual buckling and bending–torsional buckling are common failure modes of open-ended assembled specimens [33,34]. In addition, through Southwell's method [35] to obtain the end stiffness, the design method based on direct strength was proposed. Roy and Ting focused on the influence of the member thickness and slenderness ratio on the compression performance of the combined I-shaped members. The results showed that the thickness and slenderness ratio were important factors affecting the bearing capacity of the composite column [36,37]. In previous research, it was observed that the lateral torsional buckling occurred at the beam or column [38–41]. The lateral torsional buckling of the C-steel combined I-section column should be investigated.

In summary, in previous research, the connection form of cold-formed thin-walled steel I-section columns was mostly screw connection. There are few studies on the bolt-connected combined I-section column. Furthermore, the research was mostly concentrated on the combined I-section column with complete section form. There are few studies on I-section combined column with opening web. The openings in the web may affect the compressive bearing capacity of the I-section combined column. To investigate the axial compressive behavior of the C-steel combined I-section column, the finite element models of the columns were established in this paper. The finite element models were validated by experiments. In addition, parametric analyses were conducted on the I-section combined columns with complete and opening webs to investigate the mechanical behaviors. Moreover, the design formulas of the I-section combined column were proposed, and the applicability and accuracy of the proposed formulas were verified in this paper.

2. Numerical Simulations on C-Steel Combined I-Section Columns

2.1. Finite Element Models

To investigate the compressive performance of the cold-formed thin-walled C-steel combined I-section columns with double limbs and the factors affecting their buckling mode, the C-steel combined I-section columns were simulated in Ansys [42]. In the combined column, the C-steel and terminal pad were simulated by Shell 181 elements, and the bolts in the web were modeled by Solid 185 elements. Since the two C-shaped steels were combined, the interactions between the two C-shaped steels were set as Targel 70 and Contal 174, respectively. Furthermore, the interactions between the C-shaped steels and terminal pads were surface-to-line interaction. Thus, the interactions between the terminal pad and C-shaped steels were set as Targel 70 and Contal 175, which can obtain better simulation effects in surface-to-line interaction [42]. In the finite element models, the pre-tension was set as 125 kN. The combined specimens in [43] were simulated in this paper, and the parameters of the specimens are listed in Table 1. The meanings of the notations in Table 1 are shown in Figure 2. Moreover, δ_0, λ_y, and A denote the initial geometric defect, slenderness ratio of a single limb component in the weak axis direction, and sectional area of the combined column, respectively. The naming rules of the specimens are shown in Figure 3. In Figure 3, the prefixes S, M, and L denote the short, medium, and long columns, respectively.

Table 1. Parameters of the C-steel combined I-section columns in Ref [43].

	L/mm	h/mm	b/mm	c/mm	t/mm	δ_0/mm	λ_y	A/mm^2
SC-90-A1	270	92.8	41.8	14.8	1.19	0.412	8.45	466.5
SC-90-A2	270	93.8	41.8	14.7	1.19	0.521	8.47	468.4
MC-90-A1	1533	93.5	41.5	15.0	1.18	0.565	96.51	464.0
MC-90-A2	1531	92.5	42.0	14.5	1.20	0.433	95.73	469.0
LC-90-A1	3033	91.8	43.2	14.4	1.19	0.521	184.56	468.9
LC-90-A2	3038	92.8	40.2	15.0	1.18	0.374	197.15	456.2
SC-140-A1	451	142.7	42.9	15.1	1.47	0.646	14.47	724.3
SC-140-A2	451	144.2	42.8	14.8	1.48	0.535	14.59	731.1
MC-140-A1	1532	142.0	42.5	15.5	1.48	0.652	98.82	726.9
MC-140-A2	1533	142.0	42.0	15.0	1.48	0.443	100.66	721.0
LC-140-A1	3034	140.8	42.0	15.5	1.49	0.656	197.96	725.0
LC-140-A2	3033	141.5	42.3	16.0	1.47	0.661	195.37	722.5

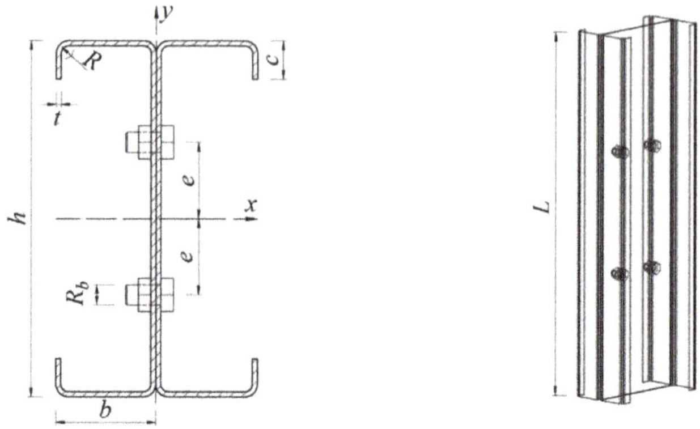

Figure 2. Parameters in the combined column.

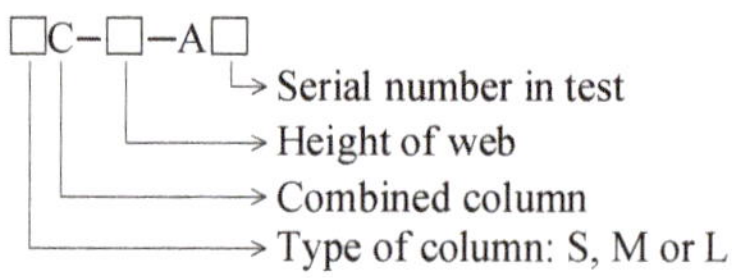

Figure 3. Naming rules of the combined column.

According to the material tests in [43], the elastic modulus, Poisson's ratio and yield strength of the steel were set as 2.05×10^5 N/mm^2, 0.303 and 305.4 N/mm^2, respectively. In the finite element simulations, the von Mises yield rule was used. To simulate the boundary conditions, in the finite element models, the translation displacements u_x, u_y, and u_z of the upper terminal pad were constrain, and the u_x and u_y displacements of the bottom terminal pad were fixed. The finite element model of the C-steel combined I-section column is shown in Figure 4.

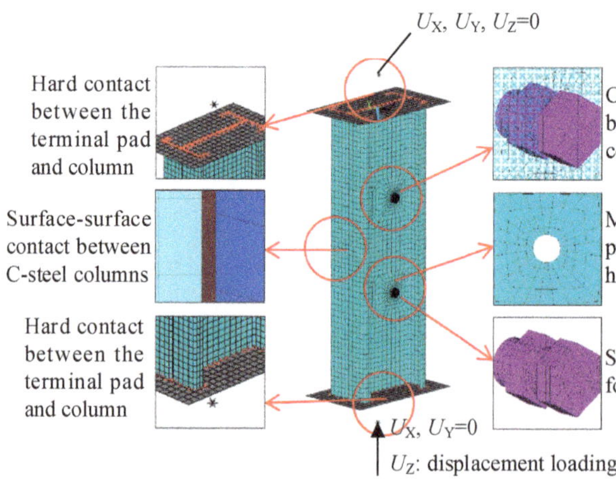

Figure 4. Finite element models of the C-steel combined I-section columns.

2.2. Finite Element Model Validations

The ultimate compressive bearing capacities of the combined columns in experiments and simulations are listed in Table 2. In Table 2, N_{FE} and N_{TE} denote the compressive bearing capacities obtained by the finite element analyses and experiments. According to

Table 2, the compressive bearing capacities obtained by the simulations and experiments were similar, and the average error was 2.6%. Furthermore, the failure mode obtained by the simulations were the same as those observed in the tests. Figure 5 shows the load–vertical displacement (F-Δ) relationship curves of the specimens. For the short, medium, and long combined columns, the curves obtained by tests and simulations were similar. Moreover, the failure modes of the C-steel combined I-section columns are compared in Figure 6. The failure modes observed in the tests and predicted in the numerical analyses were the same. Therefore, the finite element modeling method was validated and can be adopted to investigate the ultimate compressive bearing capacity of the C-steel combined I-section columns.

Table 2. Comparisons between the simulated and tested compressive bearing capacities.

	Simulation		Experiment		N_{FE}/N_{TE}	Error/%
	N_{FE}/kN	Failure Mode	N_{TE}/kN	Failure Mode		
SC-90-A1	133.6	Local buckling	127.7	Local buckling	1.046	4.62
SC-90-A2	133.5	Local buckling	132.8	Local buckling	1.005	0.53
MC-90-A1	102.3	Local + overall buckling + distortion	97.7	Local + overall buckling + distortion	1.047	4.71
MC-90-A2	106.7	Local + overall buckling + distortion	103.0	Local + overall buckling + distortion	1.036	3.59
LC-90-A1	44.5	Overall buckling	42.6	Overall buckling	1.045	4.46
LC-90-A2	44.9	Overall buckling	39.9	Overall buckling	1.125	12.53
SC-140-A1	142.7	Local buckling + distortion	130.7	Local buckling + distortion	1.092	9.18
SC-140-A2	143.1	Local buckling + distortion	139.6	Local buckling + distortion	1.025	2.51
MC-140-A1	106.2	Local + overall buckling + distortion	105.8	Local + overall buckling + distortion	0.856	0.38
MC-140-A2	105.1	Local + overall buckling + distortion	101.0	Local + overall buckling + distortion	1.041	4.06
LC-140-A1	48.0	Overall buckling	49.2	Overall buckling	0.976	−2.44
LC-140-A2	47.7	Overall buckling	46.9	Overall buckling	1.017	1.71
Mean value	-	-	-	-	1.026	3.82
Variance	-	-	-	-	0.063	3.98

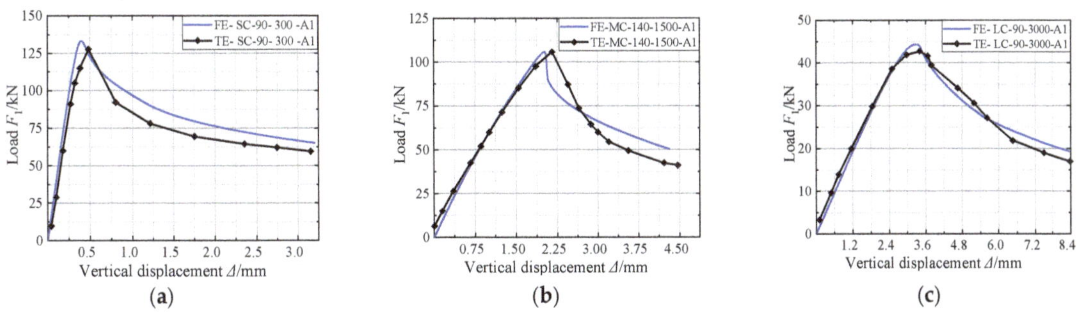

Figure 5. Load–vertical displacement relationship curves of the C-steel combined I-section columns (**a**) SC-90-300-A1; (**b**) MC-140-1500-A1; (**c**) LC-90-3000-A1.

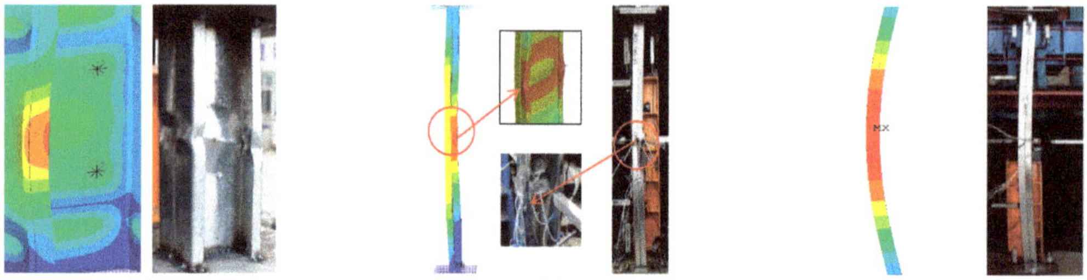

(a) (b) (c)

Figure 6. Comparisons of the failure modes of the combined columns (**a**) SC-90-A2; (**b**) MC-140-A1; (**c**) LC-90-A2.

3. Parametric Analyses on C-Steel Combined I-Section Columns

To investigate the effects of different parameters on the ultimate compressive bearing capacity of the combined columns, 171 combined columns were analyzed in this section. In the analyses, the initial defect mode was set as the first-order buckling mode shape, and the defect value was set as $L/1000$ [44], where L is the length of the column. In this section, the effects of the holes on the web were also studied, and the naming rule of the combined columns in the parametric analyses is shown in Figure 7.

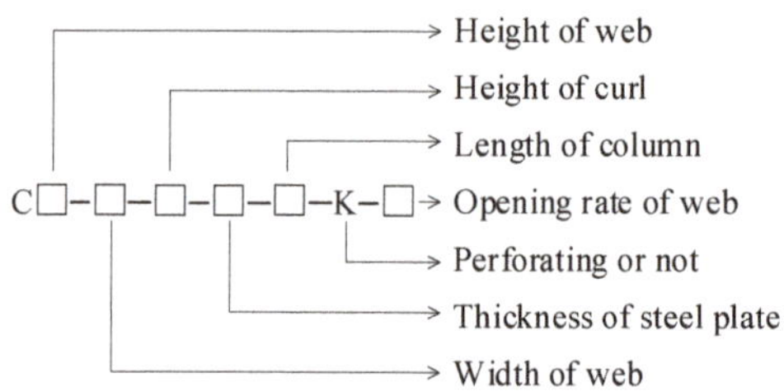

Figure 7. Naming rules of the combined columns in parametric analyses.

3.1. *Effects of Slenderness Ratio*

To investigate the effects of the slenderness ratio on the ultimate compressive bearing capacity, the combined columns, C90-40-15-1.5, C120-40-15-1.5, C140-40-15-1.5, and C160-40-15-1.5, with different lengths (300, 600, 900, 1200, 1500, 1800, and 3000 mm) were modeled in this section. Taking C120-40-1.5 combined columns as examples, with different length, the failure modes of the columns are shown in Figure 8, and the load–vertical displacement relationship curves of C120-40-1.5 combined columns are shown in Figure 9. For the short, medium, and long columns, the failure modes were local, local buckling–distortion, and overall buckling, respectively.

With different slenderness ratios, the ultimate compressive bearing capacities of the four series combined columns are shown in Figure 10. The compressive bearing capacity decreased with the increase in the slenderness ratio. When the slenderness ratios ranged from 10 to 75 and 150 to 210, the compressive bearing capacity decreased slowly. With the slenderness ratios lower than 75, the two C steels in the combined columns constrained with each other, and only local buckling could be observed. When the slenderness ratios were 75~150, the compressive bearing capacity decreased rapidly, caused by the distortion in the combined columns. When the slenderness ratios were greater than 150, overall buckling occurred. Thus, the compressive bearing capacity decreased rapidly.

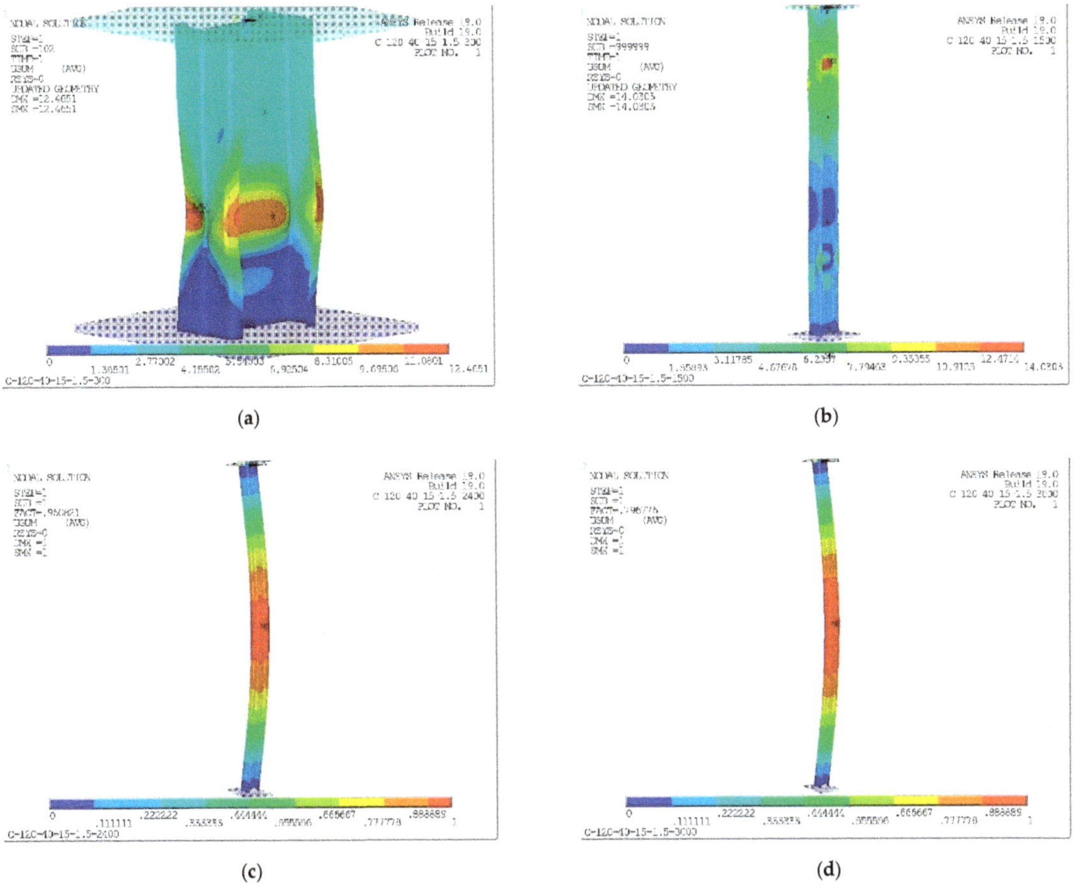

Figure 8. Deformation of the C120-40-15-1.5-*L* combined columns (**a**) C120-40-15-1.5-300; (**b**) C120-40-15-1.5-1500; (**c**) C120-40-15-1.5-2400; (**d**) C120-40-15-1.5-3000.

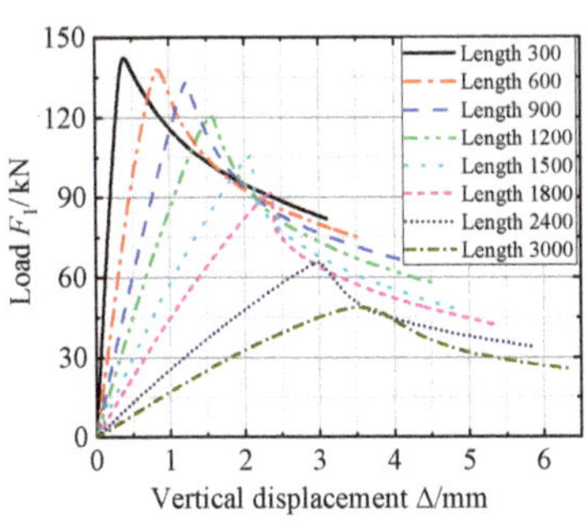

Figure 9. *F*-Δ relationship curves of C120-40-15-1.5-*L*.

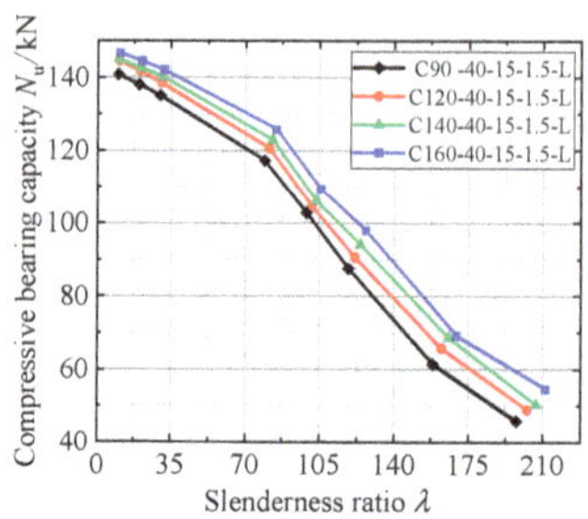

Figure 10. N_u-λ relationship curves of the combined columns.

3.2. Effects of Height to Thickness Ratio of Web

To investigate the effects of height to thickness ratio of the web on the ultimate compressive bearing capacity of the C-steel combined I-section columns, parametric analyses were carried out on the columns with a width of flange, height of the curling, and thickness of the steel of 40 mm, 15 mm, and 1.5 mm, respectively. Furthermore, the columns with length of 300 mm, 900 mm, 1500 mm, and 3000 mm were analyzed. By changing the height of the C steels, the stress distributions of the 1500 mm-length combined columns are shown in Figure 11. With different height of the web, the load–vertical displacement relationship curves are shown in Figure 12. According to Figures 11 and 12, with the increase in the height to thickness ratio, the vertical displacement decreased, and the ultimate bearing capacity increased. The maximum stresses gradually shifted to the web of the component. The ultimate compressive bearing capacity versus height to thickness ratio of the web relationship curves are shown in Figure 13. With the increase in height to thickness ratio of the web, the bearing capacity increased slowly. For the four series of combined columns, the increasing ratio of the height to thickness ratio was 122.2%, and the increasing ratios of the compressive bearing capacity were 7.5%, 8.0%, 11.5%, and 25.8%, respectively. In the models of the combined columns, with the increase in the height to thickness ratio of the web, the cross-sectional areas of the columns increased. Additionally, it further reduced the effect of height to thickness ratio of the web. Moreover, the direction of the web was in the major axis of the combined columns, and the failure of the columns was mainly in the weak axis. Thus, the effects of the height to thickness ratio of the web was not so obvious.

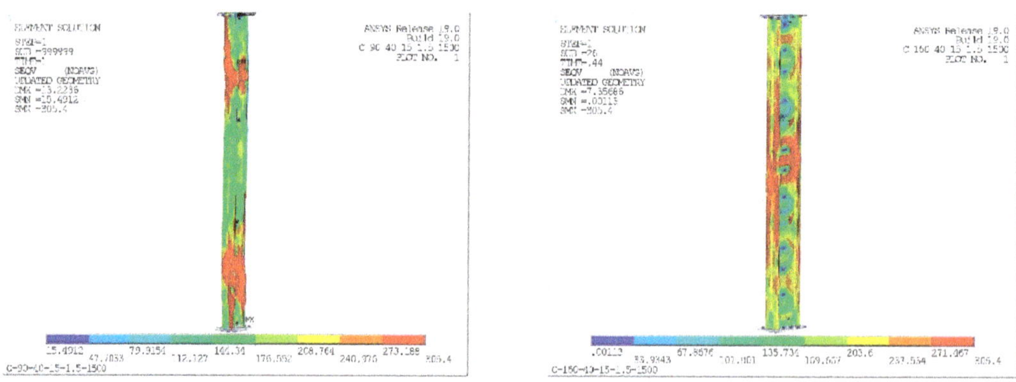

(a) (b)

Figure 11. *Cont.*

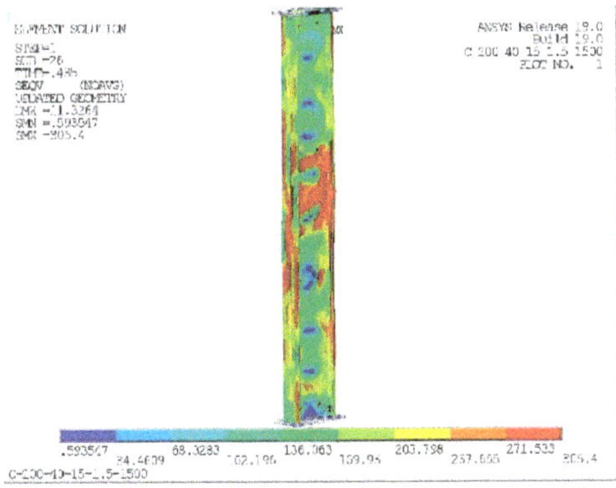

(c)

Figure 11. Stress distortions of the combined columns with different height to thickness ratios of web (**a**) C90-40-15-1.5-1500; (**b**) C160-40-15-1.5-1500; (c) C200-40-15-1.5-1500.

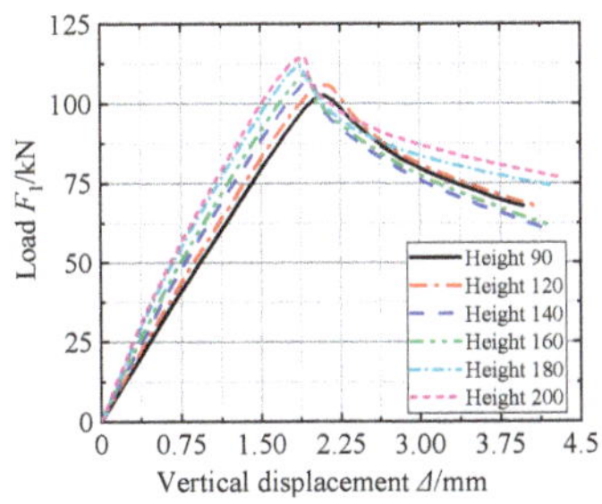

Figure 12. F-Δ relationship curves of CH-40-15-1.5-1500.

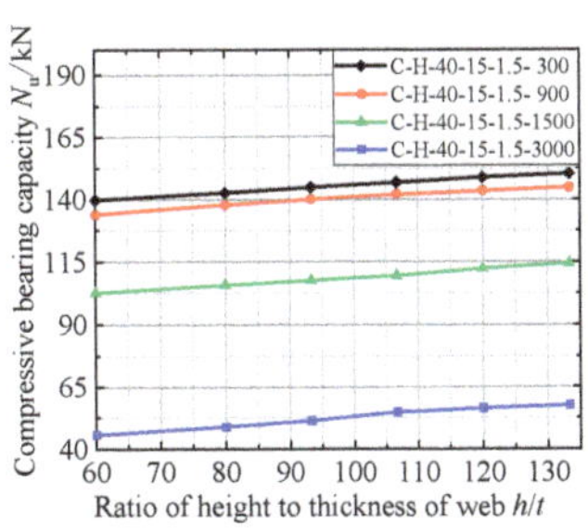

Figure 13. N_u-h/t relationship curves of the combined columns.

3.3. Effects of Width to Thickness Ratio of Flange

In the parametric analyses, the thickness of the thin-walled steel plate was set as 1.5 mm, and the height of the curling and the length of the column were 15 mm and 1500 mm, respectively. The combined columns with a height of 90 mm, 120 mm, 140 mm, and 160 mm were analyzed in this section. For the C90 series combined columns, under the ultimate compressive load, the stress distributions are shown in Figure 14. With the increase in the width to thickness ratio of the flange, the maximum stress gradually distributed on the flanges, which means that the loads bore by the flanges increased. Figure 15 shows the load–vertical displacement relationship curves of C90 series combined columns. The large width to thickness ratio of the flanges constrained the deformation. The ultimate compressive bearing capacity versus width to thickness ratio of the flange are shown in Figure 16. With the increase in the width to thickness ratio, the ultimate compressive bearing capacity increased. Furthermore, when the width to thickness of the flange was in the range of 22.5 to 30, the compressive bearing capacity increased rapidly. The direction of the flange was along the weak axis of the C-steel combined I-section column. Therefore, the effects of the width to thickness ratio of the flange were more obvious than those of the height to thickness ratio of the web. In this section, the parametric analyses were carried out on the medium columns. The large flange constrained the distortion of the columns effectively. It was the reason why the ultimate compressive bearing capacity increased with the increasing width to thickness ratio of the flange.

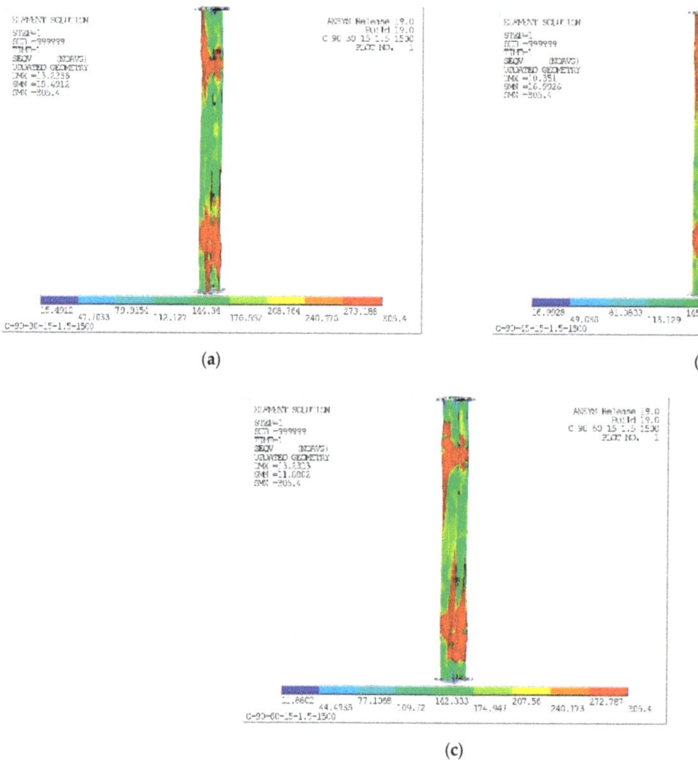

Figure 14. Stress distortions of the combined columns with different width to thickness ratios of flange (**a**) C90-30-15-1.5-1500; (**b**) C90-45-15-1.5-1500; (**c**) C90-60-15-1.5-1500.

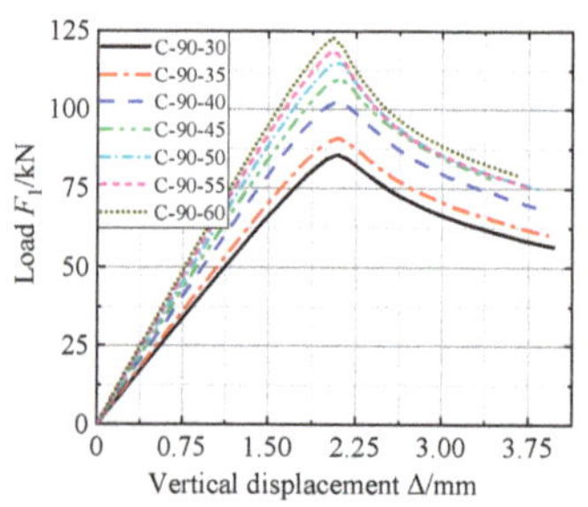

Figure 15. F-Δ relationship curves of C90-B-15-1.5-1500.

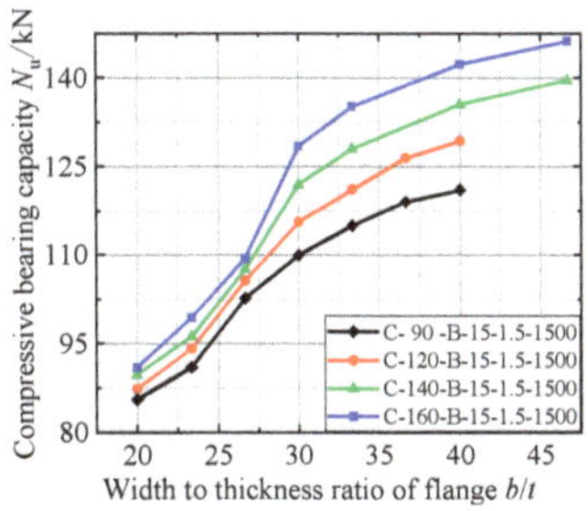

Figure 16. N_u-b/t relationship curves of the combined columns.

3.4. Effects of Bolt Spacing

The two C-steels were combined by the bolts on the webs to form an I-section column. The bolt spacing affects the combined effects. Li et al. recommended that the bolt spacing of the combined column should not be greater than 600 mm, and the number of bolts should be greater than three (when the length of the column is greater than 600 mm) [45]. In this section, parametric analyses were carried out on C140-40-15-1.5 combined columns to investigate the effects of the bolt spacing on the ultimate compressive bearing capacity. The lengths of the columns were set as 300 mm, 900 mm, 1500 mm, and 3000 mm. In the parametric analyses, the bolt spacings of the columns with different length can be found in Table 3. With different bolt spacing, the ultimate compressive bearing capacity of the I-section combined columns are shown in Figure 17. According to Figure 17, the bolt spacing had less influence on the compressive bearing capacity. For the four series of combined columns with different length, the bolt spacing increasing ratios were 275%, 800%, 1100%, and 1100%, respectively. However, the ultimate bearing capacity only increased by 0.6%, 0.6%, 1.1%, and 1.8%, respectively.

Table 3. Bolt spacings and ultimate compressive bearing capacity of combined columns.

	Bolt Spacing/mm	Increasing Ratio	Increasing Bearing Capacity
C140-40-1.5-300	40, 60, 80, 100, 120, 150	275%	0.6%
C140-40-1.5-900	50-450, step: 50	800%	0.6%
C140-40-1.5-1500	50-200, step: 5; 200-600, step: 100	1100%	1.1%
C140-40-1.5-3000	50-200, step: 50; 200-600, step: 100	1100%	1.8%

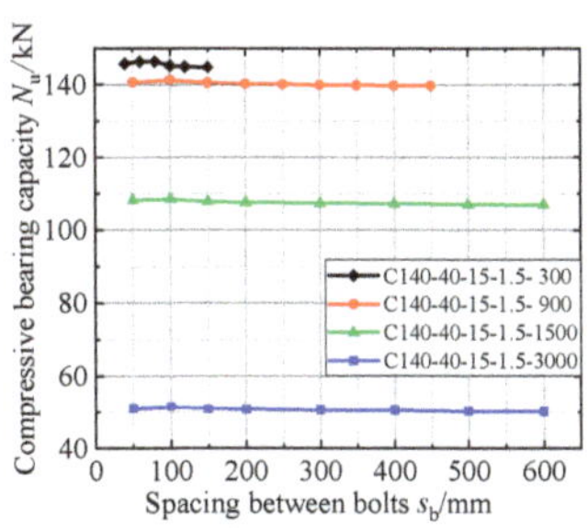

Figure 17. N_u-S_b relationship curves of the combined columns.

3.5. Effects of Opening in the Web

3.5.1. Effects of Opening Ratio of the Web

The openings in the web reduce the sectional area of the web and affect the failure mode of the combined columns. To investigate the effects of the opening ratios of the web on the ultimate compressive bearing capacity, parametric analyses were carried out on

C140-40-15-1.5 combined columns. The opening ratios were defined as the ratio of area of the openings to that of the web. In the parametric analyses, the oval openings were set in the web, and the spacing between the openings was fixed. The opening ratio increased from 0 to 60%, and the increasing ratio was about 8%.

With different lengths and opening ratios, the failure modes of the C-steel combined I-section columns are shown in Figure 18. With the increase in the opening ratio, the deformation of the columns was greater, and the local failure was more obvious in the middle of the combined columns. The relationship between the ultimate compressive bearing capacity and the opening ratios are shown in Figure 19. With the increase in the opening ratio, the compressive bearing capacity decreased. The decreasing ratios of the columns with 0~15% and 35~60% opening ratios were much greater than the columns with opening ratios of 15~35%. For the four series of combined columns with different length, the maximum increasing ratios of the opening ratio were 54.4%, 58.3%, 60.5%, and 62.9%, respectively, and the corresponding ultimate compressive bearing capacity decreased by 56.1%, 68.0%, 74.9%, and 81.7%, respectively. Except in the case that the openings reduced the sectional area of the web, the opening affected the half-wave length of the short column when the distortion happened, which reduced the compressive bearing capacity. For the medium and long combined columns, the openings aggravated the buckling of members and reduced the bearing capacity of the combined columns.

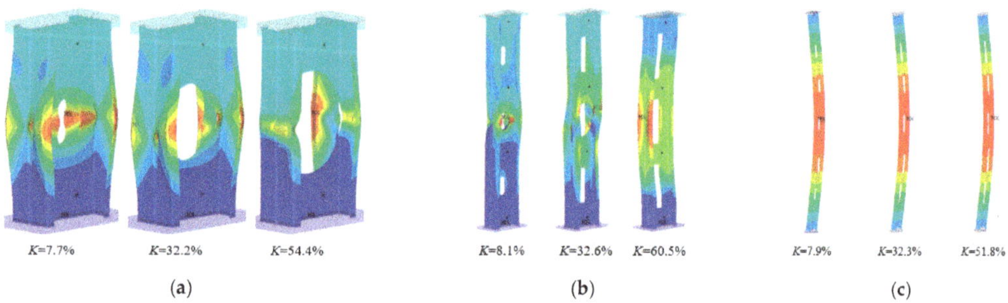

Figure 18. Failure modes of the combined columns with different opening ratios (**a**) L300; (**b**) L1500; (**c**) L3000.

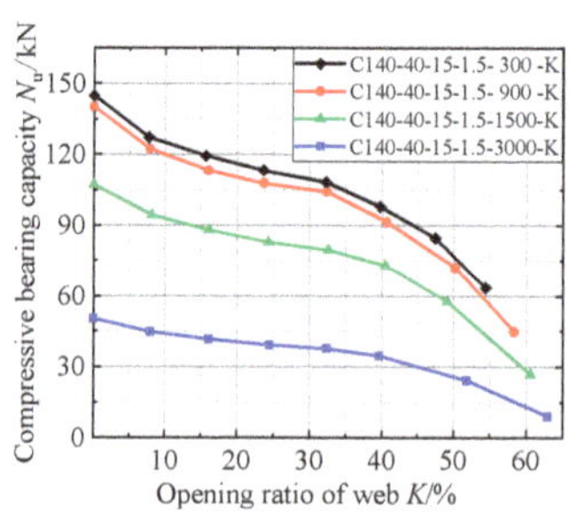

Figure 19. N_u-K relationship of combined columns.

3.5.2. Effects of Opening Spacing

In the last section, it was found that the location of the opening in the web affected the half-wave length of the columns. Therefore, in this section, parametric analyses were conducted to investigate whether the opening spacing has an influence on the ultimate compressive bearing capacity. Four series of C140-40-15-1.5 combined columns, with a length of 1500 mm, were modeled, and the opening ratios were set as 8.0%, 24.3%, 40.4%, and 62.9%, respectively. The opening spacings were in the range of 60~600 mm, with an increasing step of 60 mm. The relationship between the ultimate compressive bearing capacity and opening ratios are shown in Figure 20. According to Figure 20, the compressive bearing capacity increased with the increasing opening spacing, but the effects of the opening were not so obvious. When the opening spacing increased by 900%, the ultimate compressive bearing capacities of the four groups of combined columns increased by 6.5%, 6.9%, 6.8%, and 11.7%, respectively.

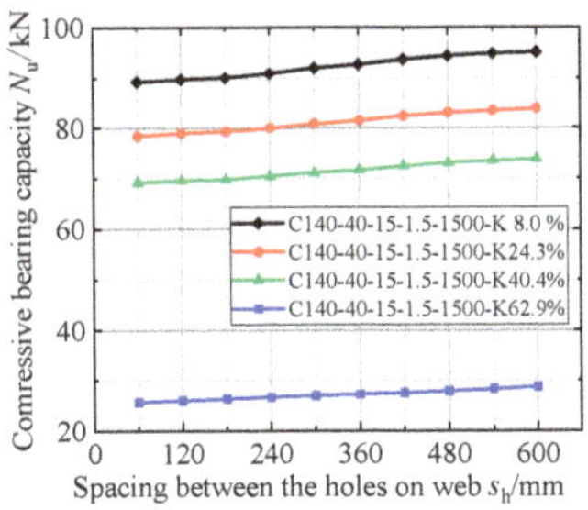

Figure 20. N_u-S_h relationship of combined columns.

4. Formulas for Ultimate Compressive Bearing Capacity

4.1. Formulas in Standards

4.1.1. Formulas in Chinese GB 50018 Standard

In the Chinese GB 50018 standard [46], there is no detailed recommended formula for the compressive bearing capacity of the combined column. In this standard, the C-steel combined I-section columns was seen as two individual C-section columns, and the method of effective width was adopted. According to the Chinese GB 50018 standard, the ultimate compressive bearing capacity of the combined column is:

$$N_u = \varphi A_e f_y \tag{1}$$

where φ is the compression stability factor. A_e (mm^2) denotes the effective area of the cross section and f_y (N/mm^2) is the yield strength of the steel. The compression stability factor φ can be found in Appendix B for Section B in this standard [46].

The effective area of the cross section is:

$$A_e = (b_{ew} + b_{ef} + b_{ec})t \tag{2}$$

In Equation (2), b_{ew}, b_{ef}, and b_{ec} are the effective width of the web, flange, and curling, respectively. When $b/t \leq 18\alpha\rho$,

$$b_e = b_c \tag{3}$$

In Equation (3), b and t are the width and thickness of the plate, respectively. Furthermore, when $18\alpha\rho < b/t < 38\alpha\rho$,

$$b_e = \left(\sqrt{\frac{21.8\alpha\rho}{b/t}} - 0.1\right) b_c \tag{4}$$

and when $b/t \geq 38\alpha\rho$,

$$b_e = \frac{25\alpha\rho}{b/t} b_c \tag{5}$$

In Equations (4) and (5), α is a coefficient, and

$$\alpha = 1.15 - 0.15\psi \tag{6}$$

In Equation (6), ψ is the coefficient of inhomogeneity of compressive stress, and

$$\psi = \sigma_{min} / \sigma_{max} \tag{7}$$

when $\psi < 0$, $\alpha = 1.15$. In Equations (4) and (5),

$$\rho = \sqrt{205 k_1 k_2 / \sigma_1} \tag{8}$$

In Equation (8), $\sigma_1 = \varphi f$ and k and k_1 denote the stability coefficient and constraint coefficient of the compressed plate, respectively. b_e is the width of the compressive area of the plate. When $\psi < 0$, $b_c = b$. Otherwise, $b_c = b/(1 - \psi)$.

4.1.2. Formulas in AISI S100-2016 Standard

In the AISI S100-2016 standard [47], the direct strength method was adopted to calculate the compressive bearing capacity of the combined column. Since there is relative deformation between the two connected components at the conditions, in this method, the slenderness ratio of the combined columns was modulated as follows:

$$\lambda_m = \sqrt{\lambda_0^2 + \left(s_a / i_y\right)^2} \tag{9}$$

where λ_m and λ_0 denote the slenderness ratios of the combined columns before and after the modulation, respectively. i_y is the radius of gyration of the individual component alone its weak axis, and s_a is the distance between two connected components.

According to AISI S100-2006 standard, first, the compressive bearing capacity P_{ne} with overall buckling should be obtained. When $\lambda_c \leq 1.5$,

$$P_{ne} = 0.658 \lambda_c^2 A_g f_y \tag{10}$$

Otherwise,

$$P_{ne} = \left(0.877 / \lambda_c^2\right) A_g f_y \tag{11}$$

In Equations (10) and (11), λ_c is the influence coefficient of flexibility with overall buckling, and $\lambda_c = \sqrt{f_y / f_{cre}}$. A_g denotes the area of the cross section, and f_{cre} is the elastic buckling stress.

Second, the compressive bearing capacity P_{nl} with local–overall coupled buckling should be determined. When $\lambda_1 \leq 0.776$,

$$P_{nl} = A_g f_{ne} \tag{12}$$

When $\lambda_1 > 0.776$,

$$P_{nl} = \left[1 - 0.15 \left(\frac{1}{\lambda_1^2}\right)^{0.4}\right] \left(\frac{1}{\lambda_1^2}\right)^{0.4} A_g f_{ne} \tag{13}$$

In Equations (12) and (13), λ_1 is the influence coefficient of flexibility with local buckling, and $\lambda_1 = \sqrt{f_{ne} / f_{o1}}$. f_{ne} and f_{ol} are the critical elastic buckling stress with overall and local buckling, respectively, and $f_{ne} = P_{ne} / A_g$.

Moreover, the compressive bearing capacity P_{nd} with distortion–overall coupled buckling can be calculated. When $\lambda_d \leq 0.561$,

$$P_{nd} = A_g f_y \tag{14}$$

Otherwise,

$$P_{nd} = \left[1 - 0.25 \left(\frac{1}{\lambda_d^2}\right)^{0.6}\right] \left(\frac{1}{\lambda_d^2}\right)^{0.6} A_g f_y \tag{15}$$

In Equations (14) and (15), λ_d is the influence coefficient of flexibility with distortion and f_{od} the critical stress of the distortion.

The ultimate bearing capacity of the combined columns is the minimum value of three values above, as shown in Equation (16).

$$P_n = \min(P_{ne}, P_{nl}, P_{nd}) \tag{16}$$

4.2. Comparisons between the Results of Numerical Analyses and Standards

In this section, the ultimate bearing capacities of the C-steel combined I-section columns obtained by the parametric analyses and formulas in standards were compared. The comparing results of all the combined columns in the parametric analyses are listed in Table A1 in Appendix A and some of the results are shown in Figure 21. In Table A1 and Figure 21, N_{FEM}, N_{CGB}, and N_{NAS} denote the ultimate compressive bearing capacity obtained by numerical analyses, the Chinese GB 50018 standard, and the AISI S100-2016 standard, respectively. According to Table A1 and Figure 21, the effective width method in the GB 50018 standard was conservative, and the maximum error was 60.26%, and the errors increased with greater slenderness ratio. Furthermore, the maximum error for the AISI S100-2016 standard was 21.49%.

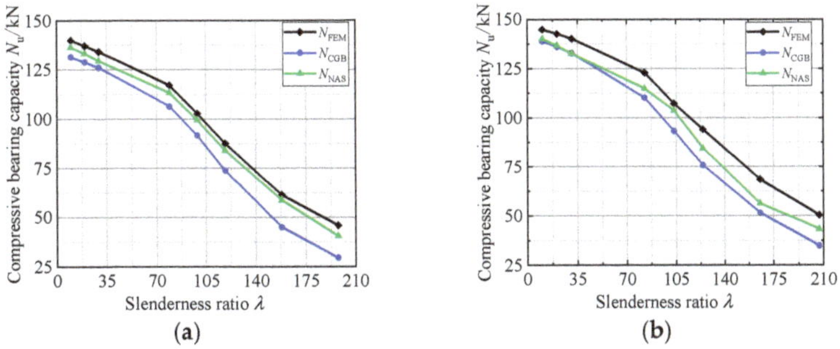

Figure 21. Comparisons of the bearing capacities between numerical analyses and standards (**a**) C90-40-15-1.5-L; (**b**) C140-40-15-1.5-L.

For the effective width method in the Chinese GB 50018 standard, the constraint effects between the plates, affecting the effective width, were considered. However, the interactions between the two C steels were ignored. With the increase in the slenderness ratio, the errors between the standard factors and slenderness ratios were greater. Therefore, with greater slenderness ratio, the formulas in Chinese GB 50018 standard were more conservative. Considering the influences of the bolt spacing on the slenderness ratio, the slenderness ratios in the AISI S100 standard were modulated. Thus, the accuracy of the method in the AISI S100 standard was higher than that of GB 50018 standard. However, with greater slenderness ratio, the influence of the bolt spacing on the slenderness ratio can be ignored. Additionally, the errors of the direct strength method in the AISI S100 standard increased.

4.3. Proposed Formulas for Compressive Bearing Capacity of C-Steel Combined I-Section Column

4.3.1. Combined Column without Opening

Since the slenderness ratio has the most significant effects on the ultimate bearing capacity of the combined column, and the combination's effects are not considered in the Chinese GB 50018 standard, the stability factor can be modulated to increase the accuracy of the formulas.

According to the Chinese GB 50018 standard, the stability factor φ and the effective area A_e of the cross section can be obtained. Thus, the stability factor in the numerical analyses φ_{FEM} should be:

$$\varphi_{FEM} = N_{FEM}/A_e f_y \tag{17}$$

and the correction factor for the stability factor is:

$$\xi = \varphi_{\mathrm{FEM}} / \varphi \tag{18}$$

The compression stability factors obtained by Equation (17) and the GB 50018 standard are shown in Figure 22 and Table A2 in Appendix B. Moreover, the relationship between the correction factors and slenderness ratios is shown in Figure 23. With the increase in the slenderness ratio, the correction factors increased rapidly. This is in accordance with the errors obtained by the numerical analyses and formulas in the GB 50018 standard. The formulas in the GB 50018 standard were more conservative with a greater slenderness ratio. The correction factor of the compression stability factor in Figure 23 was fitted as:

$$\xi = 1.0528 \times 10^{-5}\lambda^2 - 1.5318 \times 10^{-4}\lambda + 1.0661 \tag{19}$$

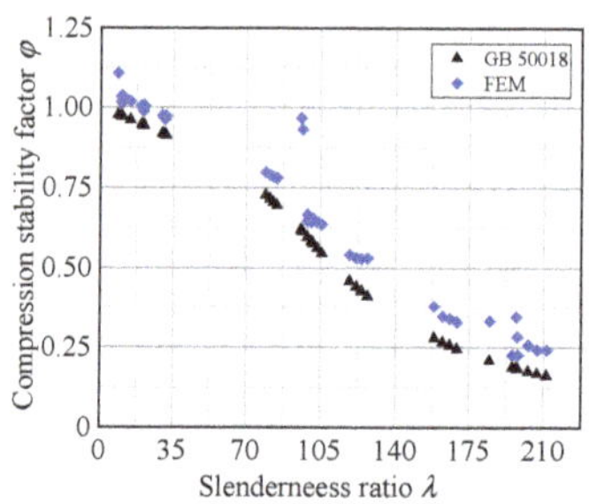

Figure 22. Stability factor of standard and numerical analyses.

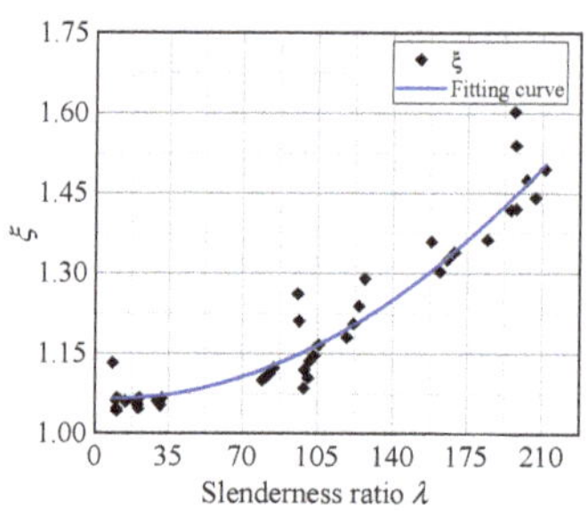

Figure 23. ξ–λ relationship.

Therefore, the ultimate compressive bearing capacity of the C-steel combined I-section column is:

$$N_{\mathrm{u}} = \xi \varphi A_{\mathrm{e}} f_{\mathrm{y}} \tag{20}$$

To validate the formulas proposed in this paper, the ultimate compressive bearing capacity obtained by Equation (20), numerical analyses, and formulas in the GB 50018 standard are compared in Figure 24. In Figure 24, the ultimate bearing capacity obtained by Equation (20) and numerical analyses were similar. The proposed formula can be used to predict the ultimate compressive bearing capacity of the C-steel combined I-section column.

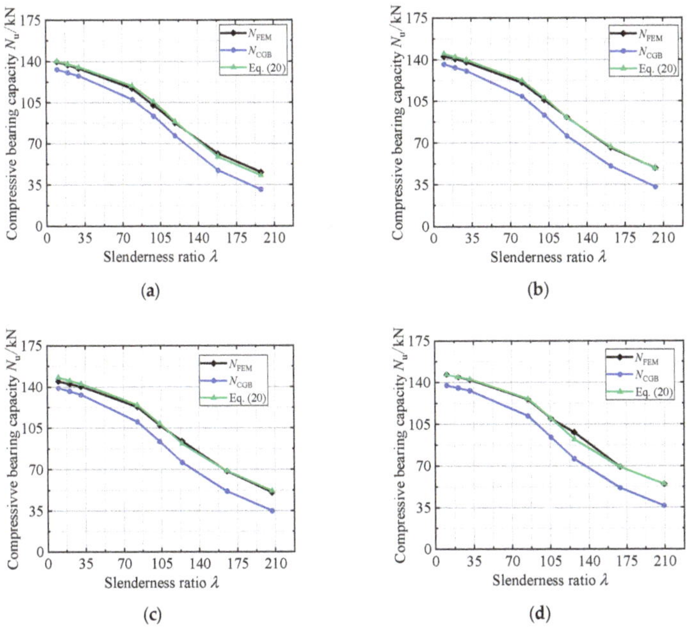

Figure 24. Comparisons of the ultimate compressive bearing capabilities (**a**) C90-40-15-1.5; (**b**) C120-40-15-1.5; (**c**) C140-40-15-1.5; (**d**) C160-40-15-1.5.

4.3.2. Combined Column with Openings

For the combined columns with openings, according to the parametric analyses, the slenderness ratio and opening ratio both have significant effects on the ultimate compressive bearing capacity. Comparing the ultimate compressive bearing capacities obtained by Equation (20) and numerical analysis, the reduction factor is:

$$\eta = \mathrm{N_u}/N_{\mathrm{FEM}} \tag{21}$$

and the ultimate bearing capacity of the C-steel combined I-section column is:

$$\mathrm{N_u} = \eta\xi\varphi A_{\mathrm{e}}f_{\mathrm{y}} \tag{22}$$

According to Equations (21) and (22) and the parametric analyses, the reduction factor η can be obtained and is shown in Figure 25 and Table A3 in Appendix B. In Figure 25, the reduction factor was fitted as:

$$\eta = -7.29K^3 + 5.2892K^2 - 1.7316K + 0.9748 \tag{23}$$

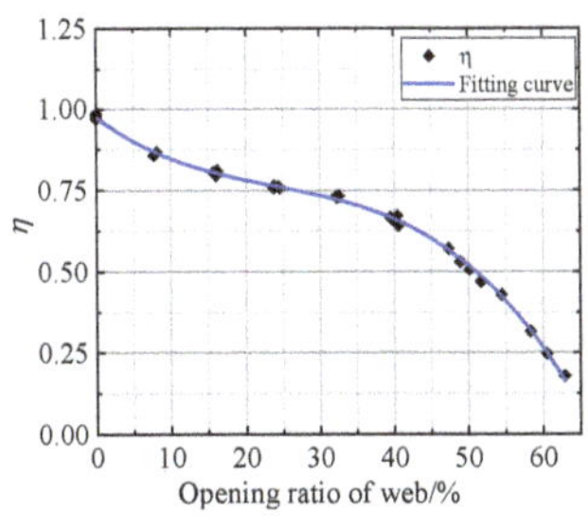

Figure 25. Relationship between reduction factor and opening ratio.

To validate Equation (22), the ultimate compressive bearing capacity obtained by the numerical analyses and Equation (22) are shown in Figure 26. In Figure 26a, the two curves are similar. In Figure 26b, N_u/N_{FEM} is in the range from 0.924 to 1.035, and the average value and variance are 0.999 and 0.018, respectively. Therefore, Equation (22) can be used to predict the ultimate compressive bearing capacity of the C-steel combined I-section column with opening in the web.

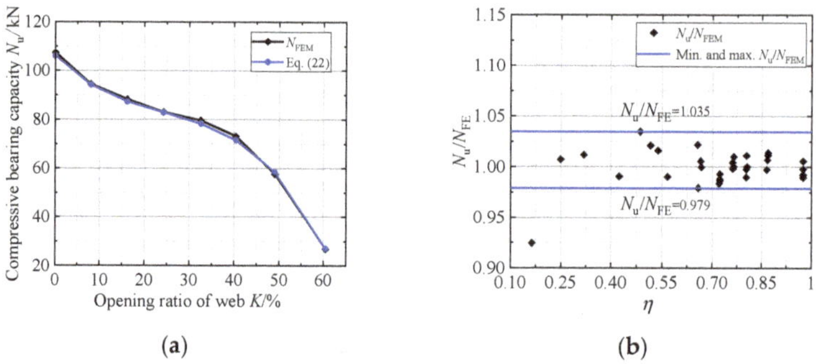

Figure 26. Validation of the proposed formulas for the combined column with opening (**a**) Comparisons of the bearing capacity of C-250-40-15-1.5-1900; (**b**) N_u/N_{FEM}.

5. Conclusions and Future Work

In this paper, the C-steel combined I-section columns were modeled. The numerical models were validated by experiments. Parametric analyses were carried out on the combined columns and the effects of slenderness ratio, height to thickness ratio, width to thickness ratio, bolt spacing, and opening were investigated. Afterward, the formulas for the ultimate compressive bearing capacity of the combined columns were proposed. The following conclusions can be drawn:

(1) The slenderness ratio has the most significant effect on the ultimate compressive bearing capacity of the combined column. With the increase in the height to thickness ratio and width to thickness ratio, the bearing capacity increases. The bolt spacing has less effect on the bearing capacity.
(2) The openings in the web decrease the ultimate compressive bearing capacity of the combined column. The opening in the web should be considered.
(3) The proposed formulas in this paper can predict the ultimate compressive bearing capacity of the combined columns with or without opening. The accuracy of the proposed formulas is higher than those of the Chinese and AISI S100 standard.

In our future work, experiments on the combined columns with openings in the web will be conducted. In addition, we will try to investigate the eccentric compressive behavior of the C-steel combined I-section columns. The combination effects of the column under eccentric compression will be investigated. The formulas for eccentric compressed C-steel combined I-section columns should be proposed. Furthermore, in our next step of research, we will try to take retrofitting countermeasures to increase the bearing capacity of the combined columns. Experiments and numerical analyses will be conducted, and the design method of the retrofitting countermeasures should be proposed.

Author Contributions: Conceptualization, C.H. and Y.C.; methodology, Y.C. and C.H.; validation, H.C.; formal analysis, C.H. and Y.C.; investigation, C.H.; resources, C.H.; data curation, H.C.; writing—original draft preparation, C.H.; writing—review and editing, Y.C.; visualization, C.H.; supervision, C.H.; project administration, Y.C. All authors have read and agreed to the published version of the manuscript.

Funding: This research was funded by the Natural Science Foundation of Hunan Province, grant number 2021JJ40737, and National Natural Science Foundation of China, grant number 52008406.

Institutional Review Board Statement: Not applicable.

Informed Consent Statement: Not applicable.

Data Availability Statement: The data provided in this study could be released upon reasonable request.

Conflicts of Interest: The authors declare no conflict of interest.

Appendix A Comparisons of the Bearing Capacities between Numerical Analyses and Standards

Table A1. Comparisons of the bearing capacities between numerical analyses and standards.

	λ_y	N_{FEM}/kN	N_{CGB}/kN	N_{NAS}/kN	N_{CGB}/N_{FEM}	N_{NAS}/N_{FEM}
C90-40-15-1.5-300	9.9	139.8	131.5	136.3	0.941	0.975
C90-40-15-1.5-600	19.8	137	128.9	133.1	0.941	0.971
C90-40-15-1.5-900	29.7	134	126.1	129.4	0.941	0.966
C90-40-15-1.5-1200	79.1	117.1	106.5	113.4	0.909	0.968
C90-40-15-1.5-1500	98.9	102.7	91.8	99.6	0.894	0.970
C90-40-15-1.5-1800	118.6	87.5	74.1	84.1	0.847	0.961
C90-40-15-1.5-2400	158.2	61.4	45.2	58.8	0.736	0.957
C90-40-15-1.5-3000	197.7	45.8	29.7	40.7	0.648	0.890
C120-40-15-1.5-300	10.1	142.6	136.2	139.6	0.955	0.979
C120-40-15-1.5-600	20.3	140.7	133.4	136.2	0.948	0.968
C120-40-15-1.5-900	30.4	137.9	130.5	132.3	0.946	0.959
C120-40-15-1.5-1200	81.1	120.6	109.1	114.3	0.905	0.948
C120-40-15-1.5-1500	101.4	105.8	93.2	101.7	0.881	0.961
C120-40-15-1.5-1800	121.7	91.6	76.0	87.3	0.830	0.953
C120-40-15-1.5-2400	162.3	65.7	50.4	60.2	0.767	0.917
C120-40-15-1.5-3000	202.8	48.9	33.2	41.4	0.679	0.847
C140-40-15-1.5-300	10.4	144.8	139.0	140.2	0.960	0.968
C140-40-15-1.5-600	20.7	142.6	136.2	136.8	0.955	0.959
C140-40-15-1.5-900	31.0	140.1	133.1	132.7	0.950	0.948
C140-40-15-1.5-1200	82.8	122.9	110.4	115.1	0.898	0.937
C140-40-15-1.5-1500	103.5	107.2	93.5	103.8	0.872	0.968
C140-40-15-1.5-1800	124.2	94.1	76.0	84.7	0.808	0.900
C140-40-15-1.5-2400	165.5	68.6	51.8	56.5	0.755	0.823
C140-40-15-1.5-3000	206.9	50.4	35.0	43.5	0.694	0.864
C160-40-15-1.5-300	10.6	146.7	137.5	142.4	0.937	0.971
C160-40-15-1.5-600	21.1	144.5	135.4	141.3	0.937	0.978
C160-40-15-1.5-900	31.7	142	133.0	137.0	0.937	0.965
C160-40-15-1.5-1200	84.5	125.7	111.9	119.9	0.890	0.954
C160-40-15-1.5-1500	105.7	109.5	93.9	96.0	0.858	0.877
C160-40-15-1.5-1800	126.8	98	76.0	81.0	0.776	0.827
C160-40-15-1.5-2400	169.1	69.2	51.7	67.0	0.747	0.969
C160-40-15-1.5-3000	209.3	54.7	36.6	51.6	0.669	0.944
C92.8-41.8-14.8-1.19-270	8.5	133.6	118.0	125.6	0.883	0.940
C93.8-41.8-14.7-1.19-270	8.5	133.5	117.9	124.9	0.883	0.936
C93.5-41.5-15-1.18-1533	96.5	102.3	84.5	97.6	0.826	0.954
C92.5-42-14.5-1.2-1531	95.7	106.7	84.6	98.1	0.793	0.919
C91.8-43.2-14.4-1.19-3033	184.6	44.5	32.7	40.8	0.734	0.916
C92.8-40.2-15-1.18-3038	197.2	44.9	28.0	41.3	0.624	0.920
C142.7-42.9-15.1-1.47-451	14.5	142.7	134.4	140.3	0.942	0.983
C144.2-42.8-14.8-1.48-451	14.6	143.1	135.0	137.1	0.943	0.958
C142-42.5-15.5-1.48-1532	98.8	106.2	97.8	102.7	0.921	0.967
C142-42-15-1.48-1533	100.7	105.1	95.2	102.2	0.906	0.972
C140.8-42-15.5-1.49-3034	198.0	48	33.8	43.7	0.704	0.910
C141.5-42.3-16-1.47-3033	195.4	47.7	33.6	42.9	0.705	0.899
Mean value	-	-	-	-	0.847	0.939
Variance	-	-	-	-	0.101	0.041

Appendix B Correction Factors of Compression Stability Factor

Table A2. Comparison of stability factor and correction factor of combined factor.

	λ_y	A/mm^2	A_e/mm^2	φ_{FEM}	φ	ξ
C90-40-15-1.5-300	9.9	562.2	442.3	0.974	1.035	1.063
C90-40-15-1.5-600	19.8	562.2	445.7	0.947	1.007	1.063
C90-40-15-1.5-900	29.7	562.2	449.5	0.919	0.976	1.062
C90-40-15-1.5-1200	79.1	562.2	479.9	0.726	0.799	1.100
C90-40-15-1.5-1500	98.9	562.2	504.7	0.595	0.666	1.119
C90-40-15-1.5-1800	118.6	562.2	529.2	0.459	0.541	1.180
C90-40-15-1.5-2400	158.2	562.2	529.5	0.279	0.380	1.360
C90-40-15-1.5-3000	197.7	562.2	529.5	0.184	0.283	1.540
C120-40-15-1.5-300	10.1	652.2	458.2	0.973	1.019	1.047
C120-40-15-1.5-600	20.3	652.2	461.9	0.946	0.997	1.054
C120-40-15-1.5-900	30.4	652.2	466.0	0.917	0.969	1.057
C120-40-15-1.5-1200	81.1	652.2	499.9	0.715	0.790	1.105
C120-40-15-1.5-1500	101.4	652.2	529.6	0.576	0.654	1.135
C120-40-15-1.5-1800	121.7	652.2	564.7	0.440	0.531	1.206
C120-40-15-1.5-2400	162.3	652.2	619.5	0.267	0.347	1.303
C120-40-15-1.5-3000	202.8	652.2	619.5	0.175	0.258	1.475
C140-40-15-1.5-300	10.4	712.2	468.1	0.973	1.013	1.042
C140-40-15-1.5-600	20.7	712.2	472.0	0.945	0.989	1.047
C140-40-15-1.5-900	31.0	712.2	476.3	0.915	0.963	1.053
C140-40-15-1.5-1200	82.8	712.2	513.0	0.705	0.784	1.113
C140-40-15-1.5-1500	103.5	712.2	545.9	0.561	0.643	1.146
C140-40-15-1.5-1800	124.2	712.2	583.4	0.426	0.528	1.239
C140-40-15-1.5-2400	165.5	712.2	659.8	0.257	0.340	1.325
C140-40-15-1.5-3000	206.9	712.2	679.5	0.169	0.243	1.440
C160-40-15-1.5-300	10.6	772.2	463.3	0.972	1.037	1.067
C160-40-15-1.5-600	21.1	772.2	469.8	0.944	1.007	1.067
C160-40-15-1.5-900	31.7	772.2	477.2	0.913	0.974	1.068
C160-40-15-1.5-1200	84.5	772.2	527.7	0.694	0.780	1.123
C160-40-15-1.5-1500	105.7	772.2	564.1	0.545	0.636	1.166
C160-40-15-1.5-1800	126.8	772.2	604.2	0.412	0.531	1.290
C160-40-15-1.5-2400	169.1	772.2	684.9	0.247	0.331	1.339
C160-40-15-1.5-3000	209.3	772.2	739.5	0.162	0.242	1.496
C92.8-41.8-14.8-1.19-270	8.5	466.5	318.7	0.978	1.107	1.132
C93.8-41.8-14.7-1.19-270	8.5	468.4	318.5	0.978	1.107	1.132
C93.5-41.5-15-1.18-1533	96.5	464.0	359.5	0.770	0.932	1.211
C92.5-42-14.5-1.2-1531	95.7	469.0	361.5	0.767	0.967	1.261
C91.8-43.2-14.4-1.19-3033	184.6	468.9	436.7	0.245	0.334	1.363
C92.8-40.2-15-1.18-3038	197.2	456.2	424.1	0.216	0.347	1.603
C142.7-42.9-15.1-1.47-451	14.5	724.3	457.8	0.962	1.021	1.061
C144.2-42.8-14.8-1.48-451	14.6	731.1	459.8	0.961	1.019	1.060
C142-42.5-15.5-1.48-1532	98.8	726.9	537.7	0.596	0.647	1.086
C142-42-15-1.48-1533	100.7	721.0	535.6	0.582	0.642	1.104
C140.8-42-15.5-1.49-3034	198.0	725.0	692.3	0.160	0.227	1.420
C141.5-42.3-16-1.47-3033	195.4	722.5	689.7	0.160	0.226	1.418

Table A3. Comparison of opening ratio and reduction factor of combined factor.

	λ_y	Opening Ratio K/%	N_{FEM}	N_u	η
C140-40-15-1.5-300-K0%	10.4	0.0	144.8	148.2	0.977
C140-40-15-1.5-300-K7.7%	10.4	7.7	127.1	148.2	0.858
C140-40-15-1.5-300-K15.7%	10.4	15.7	119.3	148.2	0.805
C140-40-15-1.5-300-K23.7%	10.4	23.7	113.2	148.2	0.764
C140-40-15-1.5-300-K32.2%	10.4	32.2	108.3	148.2	0.731

Table A3. *Cont.*

	λ_y	Opening Ratio $K/\%$	N_{FEM}	N_u	η
C140-40-15-1.5-300-K39.7%	10.4	39.7	98.0	148.2	0.661
C140-40-15-1.5-300-K47.4%	10.4	47.4	84.7	148.2	0.572
C140-40-15-1.5-300-K54.4%	10.4	54.4	63.5	148.2	0.429
C140-40-15-1.5-900-K0%	31.0	0.0	140.1	142.6	0.982
C140-40-15-1.5-900-K7.8%	31.0	7.8	122.4	142.6	0.859
C140-40-15-1.5-900-K16%	31.0	16.0	113.3	142.6	0.794
C140-40-15-1.5-900-K23.7%	31.0	23.7	108.0	142.6	0.757
C140-40-15-1.5-900-K32.3%	31.0	32.3	104.3	142.6	0.732
C140-40-15-1.5-900-K40.6%	31.0	40.6	91.5	142.6	0.642
C140-40-15-1.5-900-K50.2%	31.0	50.2	72.1	142.6	0.506
C140-40-15-1.5-900-K58.3%	31.0	58.3	44.9	142.6	0.315
C140-40-15-1.5-1500-K0%	103.5	0.0	107.2	108.8	0.985
C140-40-15-1.5-1500-K8.1%	103.5	8.1	94.4	108.8	0.868
C140-40-15-1.5-1500-K16.2%	103.5	16.2	88.2	108.8	0.810
C140-40-15-1.5-1500-K24.3%	103.5	24.3	83.0	108.8	0.763
C140-40-15-1.5-1500-K32.6%	103.5	32.6	79.6	108.8	0.732
C140-40-15-1.5-1500-K40.4%	103.5	40.4	73.1	108.8	0.672
C140-40-15-1.5-1500-K49%	103.5	49.0	57.7	108.8	0.530
C140-40-15-1.5-1500-K60.5%	103.5	60.5	26.9	108.8	0.247
C140-40-15-1.5-3000-K0%	206.9	0.0	50.4	52.0	0.969
C140-40-15-1.5-3000-K7.9%	206.9	7.9	44.8	52.0	0.862
C140-40-15-1.5-3000-K16.1%	206.9	16.1	41.8	52.0	0.805
C140-40-15-1.5-3000-K24.5%	206.9	24.5	39.4	52.0	0.758
C140-40-15-1.5-3000-K32.3%	206.9	32.3	37.8	52.0	0.727
C140-40-15-1.5-3000-K39.5%	206.9	39.5	34.7	52.0	0.667
C140-40-15-1.5-3000-K51.7%	206.9	51.7	24.4	52.0	0.470
C140-40-15-1.5-3000-K62.9%	206.9	62.9	9.2	52.0	0.178

References

1. Ding, F.X.; Ding, H.; He, C.; Wang, L.P.; Lyu, F. Method for flexural stiffness of steel-concrete composite beams based on stiffness combination coefficients. *Comput. Concr.* **2022**, *29*, 127–144.
2. He, C.; Xie, Q.; Jiang, L.Z.; Jiang, L.Q. Seismic terminal displacement of UHV post electrical equipment considering flange rotational stiffness. *J. Constr. Steel Res.* **2021**, *183*, 106701. [CrossRef]
3. He, C.; Wei, M.M.; Xie, Q.; Jiang, L.Z. Seismic responses of bundled conductor interconnected electrical equipment. *Structures* **2021**, *33*, 3107–3121. [CrossRef]
4. Hu, Y.; Jiang, L.Q.; Ye, J.H.; Zhang, X.; Jiang, L.Z. Seismic responses and damage assessment of a mid-rise cold-formed steel building under far-fault and near-fault ground motions. *Thin Walled Struct.* **2021**, *163*, 107690. [CrossRef]
5. Abou-Rayan, A.; Khalil, N.; Youssef, A.; Eldeib, M. Flexural Behavior of Encased Beam Flat or Perforated steel Cold Formed Sections. *Int. J. Steel Struct.* **2021**, *21*, 1465–1477. [CrossRef]
6. Zhang, J.F.; Li, B.; Li, A.Q.; Pang, S.Y. Critical stress determination of local and distortional buckling of lipped angle columns under axial compression. *Buildings* **2022**, *12*, 712. [CrossRef]
7. Thirunavukkarasu, K.; Kanthasamy, E.; Gatheeshgar, P.; Poologanathan, K.; Rajanayagam, H.; Suntharalingam, T.; Dissanayake, M. Sustainable performance of a modular building system made of built-up cold-formed steel beams. *Buildings* **2021**, *11*, 460. [CrossRef]
8. Dar, M.A.; Yusuf, M.; Dar, A.R.; Raju, J. Experimental study on innovative sections for cold formed steel beams. *Steel Compos. Struct.* **2015**, *19*, 1599–1610. [CrossRef]
9. Shaker, F.M.F.; Mamdooh, Z.; Deifalla, A.; Yehia, M.M. Experimental investigations of the behavior of stiffened perforated cold-formed steel sections subjected to axial compression. *Buildings* **2022**, *12*, 812. [CrossRef]
10. Jiang, L.Q.; Yu, K.; Ye, J.H.; Hu, Y.; Jiang, L.Z. Seismic Damage Assessment and Shaking-Table Test Validation of Midrise Cold-Formed Steel Composite Shear Wall Buildings. *J. Struct. Eng.* **2022**, *148*, 04022093. [CrossRef]
11. Jiang, L.Q.; Ye, J.H. Quantifying the effects of various uncertainties on seismic risk assessment of CFS structures. *Bull. Earthq. Eng.* **2019**, *18*, 241–272. [CrossRef]
12. Jiang, L.Q.; Ye, J.H. Redundancy of a mid-rise CFS composite shear wall building based on seismic response sensitivity analysis. *Eng. Struct.* **2019**, *200*, 109647. [CrossRef]
13. Zhou, X.H. Research progress on cold-formed steel structural framing. *Steel Constr.* **2020**, *35*, 1–19.

14. Li, Y.Q.; Li, Y.L.; Wang, S.K.; Shen, Z.Y. Investigation on ultimate capacity of built-up columns with double channel sections under axial compression. *J. Build. Struct.* **2014**, *35*, 104–113. [CrossRef]
15. Chen, M.; Huang, J.H.; Zhao, G.T. Research progress of compound section cold-formed thin-wall steel structures. *Eng. Mech.* **2016**, *33*, 1–11. [CrossRef]
16. Bae, S.-W.; Laboube, R.A.; Belarbi, A.; Ayoub, A. Progressive collapse of cold-formed steel framed structures. *Thin Walled Struct.* **2008**, *46*, 706–719. [CrossRef]
17. Whittle, J.; Ramseyer, C. Buckling capacities of axially loaded, cold-formed, built-up C-channels. *Thin Walled Struct.* **2009**, *47*, 190–201. [CrossRef]
18. Zhou, X.H.; Li, Z.; Liu, Y.J.; Shi, Y. Calculation method for bearing capacity of cold-formed steel built-up columns under axial compression. *J. Archit. Civil. Eng.* **2012**, *29*, 1–6.
19. Chen, M.; Sun, F.F.; Sun, Y. Analysis on load-carring behavior of axial compression short column with double cold-formed C steel. *Constr. Technol.* **2012**, *41*, 684–686.
20. Chen, M.; Bian, W.; Sun, F.F.; Chen, Y. Analysis on load-carrying behavior of axial compression short column with gusset plate between double back-to-back C steel. *Build. Struct.* **2013**, *43*, 1447–1450.
21. Chen, M.; Lu, W.W.; Liu, K. Aanlysis of bearing capacity of double cold-formed thin-walled C steel back-to-back composite column with plate coupled. *Ind. Constr.* **2014**, *44*, 124–127.
22. Li, Y.Q.; Li, Y.L.; Wang, S.K.; Shen, Z.Y. Ultimate load-carrying capacity of cold-formed thin-walled columns with built-up box and I section under axial compression. *Thin Walled Struct.* **2014**, *79*, 202–217. [CrossRef]
23. Dabaon, M.; Ellobody, E.; Ramzy, K. Experimental investigation of built-up cold-formed steel section battened columns. *Thin Walled Struct.* **2015**, *92*, 137–145. [CrossRef]
24. Dabaon, M.; Ellobody, E.; Ramzy, K. Nonlinear behaviour of built-up cold-formed steel section battened columns. *J. Constr. Steel Res.* **2015**, *110*, 16–28. [CrossRef]
25. Abbasi, M.; Khezri, M.; Rasmussen, K.J.R.; Schafer, B.W. Elastic buckling analysis of cold-formed steel built-up sections with discrete fasteners using the compound strip method. *Thin Walled Struct.* **2018**, *124*, 58–71. [CrossRef]
26. Rahnavard, R.; Craveiro, H.D.; Laím, L.; Simões, R.A.; Napolitano, R. Numerical investigation on the composite action of cold-formed steel built-up battened columns. *Thin Walled Struct.* **2021**, *162*, 107553. [CrossRef]
27. Zhou, T.H.; Wang, L.; Wang, Q. *Experimental research on bearing capacity of cold-formed steel double-shaft built-up long columns. The 6th National Civil Engineering Graduate Academic Forum*; Atlantis Press: Amsterdam, The Netherlands, 2008; p. 65.
28. Zhou, T.H.; Nie, S.F.; Liu, X.B. Experimental study on cold-formed steel three limbs built-up section members under axial compression. *J. Build. Struct.* **2012**, *33*, 22–29.
29. Zhou, T.H.; Yang, D.H.; Nie, S.F.; Wu, H.H. Experimental study and numerical analysis of the behavior of cold-formed steel quadruple-C built-up section members under axial compression. *China Civ. Eng. J.* **2012**, *45*, 77–85.
30. Zhou, T.H.; Li, Y.C.; Wu, H.H.; Lu, Y.; Ren, L. Analysis to determine flexural buckling of cold-formed steel built-up back-to-back section columns. *J. Constr. Steel Res.* **2020**, *166*, 105898. [CrossRef]
31. Stone, T.A.; Laboube, R.A. Behavior of cold-formed steel built-up I-sections. *Thin Walled Struct.* **2005**, *43*, 1805–1817. [CrossRef]
32. Yao, X.Y.; Li, X. Tests and direct strength method on the distortional buckling and interactive buckling of cold-formed thin-walled steel built-up I-section columns under axial compression. *Prog. Steel Build. Struct.* **2021**, *23*, 33–46.
33. Fratamico, D.C.; Torabian, S.; Zhao, X.; Rasmussen, K.J.; Schafer, B.W. Experimental study on the composite action in sheathed and bare built-up cold-formed steel columns. *Thin Walled Struct.* **2018**, *127*, 290–305. [CrossRef]
34. Fratamico, D.C.; Torabian, S.; Zhao, X.; Rasmussen, K.J.; Schafer, B.W. Experiments on the global buckling and collapse of built-up cold-formed steel columns. *J. Constr. Steel Res.* **2018**, *144*, 65–80. [CrossRef]
35. Southwell, R.V. On the analysis of experimental observations in problems of elastic stability. *Proc. R. Soc. Lond.* **1932**, *135*, 601–616. [CrossRef]
36. Roy, K.; Ting, T.C.H.; Lau, H.H.; Lim, J.B. Effect of thickness on the behaviour of axially loaded back-to-back cold-formed steel built-up channel sections-Experimental and numerical investigation. *Structures* **2018**, *16*, 327–346. [CrossRef]
37. Roy, K.; Ting, T.C.H.; Lau, H.H.; Lim, J.B. Nonlinear behaviour of back-to-back gapped built-up cold-formed steel channel sections under compression. *J. Constr. Steel Res.* **2018**, *147*, 257–276. [CrossRef]
38. Barszcz, A.M.; Giżejowski, M.A.; Stachura, Z. On elastic lateral-torsional buckling analysis of simply supported I-shape beams using Timoshenko's energy method. In *Modern Trends in Research on Steel, Aluminium and Composite Structures*; Routledge: Leiden, The Netherlands, 2021; pp. 92–98.
39. Kuś, J.; Maleska, T. Lateral torsional buckling of tapered steel I-beams with stiffner ribs. In *Modern Trends in Research on Steel, Aluminium and Composite Structures*; Giżejowski, M.A., Kozłowski, A., Chybiński, M., Rzeszut, K., Studziński, R., Szumigała, M., Eds.; Routledge: Leiden, The Netherlands, 2021; pp. 428–434.
40. Trahair, N.S. Bending and buckling of tapered steel beam structures. *Eng. Struct.* **2014**, *59*, 229–237. [CrossRef]
41. Tankova, T.; Martins, J.P.; da Silva, L.S.; Simões, R.; Craveiro, H.D. Experimental buckling behaviour of web tapered I-section steel columns. *J. Constr. Steel Res.* **2018**, *147*, 293–312. [CrossRef]
42. *Ansys®Academic Research Mechanical, Release 19.0, Help System, Coupled Field Analysis Guide*; ANSYS, Inc.: Canonsburg, PA, USA, 2018.

43. Lu, Y.; Zhou, T.H.; Li, W.C.; Wu, H. Experimental investigation and a novel direct strength method for cold-formed built-up I-section columns. *Thin Walled Struct.* **2017**, *112*, 125–139. [CrossRef]
44. *GB 50017–2017*; Ministry of Housing and Urban-Rural Development of the People's Republic of China. Standard for design of steel structures. Architecture and Building Press: Beijing, China, 2017.
45. Li, Y.C.; Zhou, T.H.; Ding, J.H.; Li, C.Y.; Zhang, X.C. Investigation on distortion buckling behavior of cold-formed thin-walled steel built-up box-section columns. *J. Hunan Univ. Nat. Sci.* **2021**, *48*, 10–21.
46. *GB 50018–2002;* Ministry of Housing and Urban-Rural Development of the People's Republic of China. Technical Code of Cold-Formed Thin-Wall Structures. Planing Press: Beijing, China, 2002.
47. *AISI-S100–16;* AISI. North American Specification for the Design of Cold-Formed Steel Structural Members. American Iron and Steel Institute: Washington, DC, USA, 2016.

buildings

MDPI

Article

Study on Parameters' Influence and Optimal Design of Tuned Inerter Dampers for Seismic Response Mitigation

Ruoyu Zhang [1,2], Jizhong Huang [2,*], Meigen Cao [3], Qingyang Luo [4] and Xiuwei Guo [1,2]

1 School of Mechanics and Engineering Science, Shanghai University, Shanghai 200444, China; 21820359@shu.edu.cn (R.Z.); guoxw@shu.edu.cn (X.G.)
2 Institute for Conservation of Cultural Heritage, Shanghai University, Shanghai 200444, China
3 School of Civil Engineering, North China University of Technology, Beijing 100144, China; caomeigen@ncut.edu.cn
4 School of Materials Science and Engineering, Shanghai University, Shanghai 200444, China; lqy4227@shu.edu.cn
* Correspondence: hjizhong@163.com

Abstract: In this paper, parameter analyses of a tuned inerter damper (TID) are carried out based on the displacement mitigation ratio. The optimal design of TID based on the closed-form solution method is carried out and compared with the fixed-point method. Meanwhile, applicable conditions of two methods are discussed in wider range of values of objective function under different inherent damping ratios. Finally, seismic responses of SDOF system with TID are carried out, which verifies the feasibility of the closed-form solution optimization method. Compared with the fixed-point method, the inherent damping ratio of the original structure is considered in the closed-form solution method, and the optimal damping ratio of a TID is smaller than that of the fixed-point method under same displacement mitigation ratio. The parameters' combination of a TID designed by the fixed-point method obtains a vibration mitigation effect with a larger damping ratio by cooperating with the deformation enhancement effect of the inerter, which may make the vibration mitigation effect of the TID lower than that of the VD in structures with large inherent damping ratios. However, the deformation enhancement effect on the damping element of the inerter can be fully used by using the closed-form solution method. Better applicability and robustness are shown in closed-form solution method. Under the same displacement mitigation ratio, the damping ratio of a TID obtained by using the closed-form solution method is about one tenth of that obtained by using the fixed-point method, which can realize the lightweight design of the TID.

Keywords: tuned inerter damper; stochastic response; closed-form solution; parameter optimization; displacement mitigation ratio; seismic response

Citation: Zhang, R.; Huang, J.; Cao, M.; Luo, Q.; Guo, X. Study on Parameters' Influence and Optimal Design of Tuned Inerter Dampers for Seismic Response Mitigation. *Buildings* **2022**, *12*, 558. https://doi.org/10.3390/buildings12050558

Academic Editor: Jorge Manuel Branco

Received: 11 April 2022
Accepted: 25 April 2022
Published: 27 April 2022

1. Introduction

Buildings, lifelines, bridges and other structures have high vulnerability to ground motions or fluctuating wind [1–3], especially for super high structures, steel-frame buildings [4], and transmission towers [5]. The main reason is that the vibration frequencies of the structures are close to the predominant frequency of external excitation [6], resulting in the vibration amplification of the structure [7]. Hence, much research has studied reinforcement measures or energy dissipation devices for vibration mitigation, as well as the design methods of these devices for the vibration control of different structures [8,9]. At present, tuned mass damper (TMD) is the most widely used [10] in the fields of electronics, machinery, building, and so on [11]. Other devices for vibration control used in practice, such as traditional isolation systems [12,13], include isolation bearings, e.g., elastomeric [14] or wire rope isolators [15], are installed at the bottom of the structures as in the case of traditional base-isolated buildings [16] or rigid blocks [17]. Some electronic and mechanical equipment has much higher requirements for the accuracy and robustness of vibration control than structures, and buildings and other structures are exposed to the external environment for a long time [18]. Many super-high-rise structures adopt the TMD design scheme. However, in order to achieve a certain vibration control effect, TMD requires larger size and volume [19]. However, from the perspective of economy and applicability [20], this is unrealistic for most structures, such as light space truss structures, because structures of this type are quite flexible and it is difficult to realize large-scale vibration mitigation measures based on TMD. Thus, it is necessary to design a lightweight and efficient damper which is easily designed and has little impact on the original structure, such as an inerter-based damper. A scheme that uses inerter-based dampers to suppress these vibrations [21–23]—including in some special structures, such as transmission lines [24], tall buildings [25], and transformer-bushing systems [26], will be satisfactory.

Vibration control technologies based on inerters have been developed based on electromechanical similarity theory [27]. Compared with TMDs, the inerter-based damper can control the inertial force at the two terminals directly. Moreover, the inerter element can effectively enlarge the small apparent mass through converting the translational motion into rotary motion such as with a ball screw. In 2001, Smith [28] put forward the concept of inerter elements and gave the basic forms of ball screw inerter elements and rack and pinion inerter elements and designed hydraulic inerter elements in 2013 [29]. Subsequently, shock absorbers such as tuned viscous mass dampers (TVMD) and tuned inerter dampers (TID) were proposed. The design method of inerter systems is also studied. Ikago et al. [30] derived a closed-form formula for TVMD optimization design based on fixed-point theory. Pan et al. [31] considered the natural damping of the original structure and the cost of the inerter-based damper and make up for the deficiency of fixed-point theory. Then, they proposed the design method of an SPIS-II inerter-based damper based on stochastic response mitigation ratio [32]. Hwang et al. [33] proposed a rotation inerter system connected with a toggle brace based on a ball screw. It is shown that the system can be effectively used in the structure with small drift. Zhang et al. [34,35] applied the inerter damper system to high-rise structures such as chimneys and wind power towers and proved the effectiveness of the inerter-based damper in high-rise structures. Gao et al. [36] put forward an optimum design method of viscous inerter damper (VID) based on the feedback control theory. De Domenico and collaborators [37–47] proposed the optimal design methods of inerter-based TMD systems for seismic response mitigation. Although some scholars have used the inerter-based damper in practical engineering [48], most of the research on the inerter-based damper is still in the stage of theoretical analysis, and only a few scholars have proposed the connection mode and design method of the inerter system be applied in building structures [49]. Xie et al. [50,51] put forward a cable-bracing inerter system (CBIS) and showed that it is easy to install and can effectively control displacement and acceleration of a structure. Wang et al. [52] put forward a new tuned inerter negative stiffness damper (TINSD) based on the fixed-point method which is more effective than the TID, TVMD, and INSD in reducing the dynamic response of structures.

At present, many design methods [31,53] are based on Den Hartog's tuning theory [54] of dynamic absorbers, which does not consider the inherent damping of the original structures and the excitation properties. Moreover, the design methods do not consider the target or demanded performance of the primary structure. Pan et al. [32] proposed a parameter optimization method that considers the inherent damping ratio and control cost for an T structure with an inerter system, such as TVMD. The vibration response mitigation ratio of a structure based on H_∞ norm or H_2 norm [16,55,56] is also used in parametric optimization of inerter systems. Meanwhile, the vibration mitigation effect of inerter systems, such as TIDs, is inferior to that of some traditional or ordinary dampers, such as viscous damper (VD), under some optimum parameters obtained based on the fixed-point method. Many parameter design methods of inerter systems have been proposed in previous research, such as using multi-objective optimization. However, these design methods are used in specific conditions; applicable conditions of these methods should be discussed in a wider range of values of objective functions or boundary conditions.

Therefore, it is necessary to discuss the applicability of current design methods of inerter systems.

In this paper, the motion control equations of the single-degree-of-freedom (SDOF) system with a typical inerter damper, the tuned inerter damper (TID), are established in Section 2. Parameter analyses of the TID are carried out based on the displacement mitigation ratio in Section 3. At the same time, the parameter optimization design of TIDs based on the closed-form solution method is carried out and compared with the fixed-point method. Applicable conditions of two methods are discussed in a wider range of values of objective function under different inherent damping ratios. Finally, seismic responses of the SDOF system with the TID are carried out, which verifies the feasibility of the closed-form solution optimization method in Section 4. Applicability of the two design methods to structures with different inherent damping ratios is discussed, as are the value ranges of invalid parameters of the TID by comparison with the mitigation performance of VD under identical damping ratios. The research in this paper can provide reference for the selection of design methods of inerter systems.

2. Motion Control Equation of TID System

Equation (1) is the output force of an inerter element when its two ends have different accelerations. Hence, the output of the inerter element is also directly proportional to the relative acceleration at two terminals, which can be shown as:

$$f_I = b(a_2 - a_1) \tag{1}$$

where f_I is the output force of the inerter element, a_1 and a_2 are the accelerations at terminals, and Figure 1a is the mechanical model of the inerter element. The inerter element is the same as the mass element and cannot dissipate energy by itself. It is generally used in combination with a damping element, such as a viscous damper. Figure 1b shows the mechanical model of the TID, and the TID is a series-parallel layout inerter system composed of inerter, damping element, and stiffness elements. Where k_d, c_d, and b are the stiffness, damping, and inertance of the TID. The mechanical model of the SDOF system with TID is shown in Figure 1c, and m, k, c are the mass, stiffness, and damping of the original structure.

Equation (2) is the motion control equation of SDOF with TID:

$$\begin{cases} m\ddot{u} + c\dot{u} + ku + k_d(u - u_d) + c_d(\dot{u} - \dot{u}_d) = -ma_g \\ b(\ddot{u}_d + a_g - a_g) = k_d(u - u_d) + c_d(\dot{u} - \dot{u}_d) \end{cases} \tag{2}$$

where u and u_d are the displacement responses of SDOF and the TID, and a_g is the ground acceleration. Dimensionless parameters are defined at the same time as follows:

$$\zeta_0 = \frac{c}{2m\omega_0}, \omega_0 = \sqrt{\frac{k}{m}}, \mu = \frac{b}{m}, \zeta = \frac{c_d}{2m\omega_0}, \kappa = \frac{k_d}{k} \tag{3}$$

where ζ_0 and ω_0 are the damping ratio and circular frequency of the original structure; κ, μ, and ζ are the stiffness ratio, inerter–mass ratio and damping ratio of the TID. Rewrite Equation (2) into dimensionless form as:

$$\begin{cases} \ddot{u} + 2\zeta_0\omega_0\dot{u} + {\omega_0}^2 u + \kappa{\omega_0}^2(u - u_d) + 2\zeta\omega_0(\dot{u} - \dot{u}_d) = -a_g \\ \mu(\ddot{u}_d + a_g - a_g) = \kappa{\omega_0}^2(u - u_d) + 2\zeta\omega_0(\dot{u} - \dot{u}_d) \end{cases} \tag{4}$$

The Laplace transform of Equation (4) can be obtained:

$$\begin{cases} s^2U + 2\zeta_0\omega_0 sU + {\omega_0}^2 U + \kappa{\omega_0}^2(U - U_d) + 2\zeta\omega_0 s(U - U_d) = -A_g \\ \mu s^2(U_d + A_g - A_g) = \kappa{\omega_0}^2(U - U_d) + 2\zeta\omega_0 s(U - U_d) \end{cases} \tag{5}$$

where $s = i\Omega$, Ω is the excitation frequency of ground motion, U, U_d, and A_g are the laplace transforms of u, u_d, and a_g. Then, the displacement response transfer function (frequency response function, FRF) of SDOF with TID H_U of can be obtained by solving the linear Equation (5):

$$H_U = \frac{U}{A_g} = -\frac{s^2\mu + 2s\zeta\omega + \kappa\omega^2}{s^4\mu + 2s^3(\zeta + \zeta\mu + \zeta_0\mu)\omega + s^2(4\zeta\zeta_0 + \kappa\mu + \kappa + \mu)\omega^2 + 2s(\zeta + \kappa\zeta)\omega^3 + \kappa\omega^4} \tag{6}$$

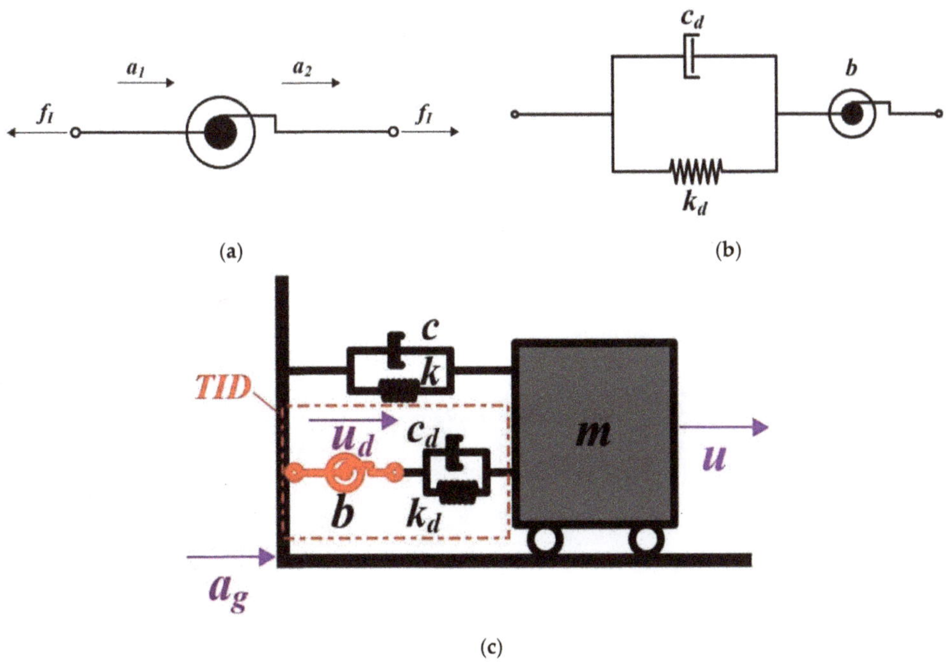

Figure 1. Mechanical model of a TID: (**a**) Inerter element; (**b**) TID; (**c**) SDOF with TID.

Assuming that the seismic excitation is a stationary white noise process, and according to the Parseval's theorem, the displacement response root mean square (RMS) σ_U of the original structure with a TID is:

$$\sigma_U = \sqrt{\int_{-\infty}^{+\infty} |H_U|^2 S_0 ds} \tag{7}$$

where S_0 is the power spectral density of white noise. The mitigation ratio of the RMS of displacement response of the original system with and without a damper can be compared to measure the effect of the shock absorber, namely, the displacement mitigation ratio. For an SDOF system with a viscous damper (VD) and a TID, the closed-form solutions of the structural displacement mitigation ratios J_{VD} and J can be obtained by using the James integral formula:

$$J_{VD} = \frac{\sigma_{U_{VD}}(\zeta_0)}{\sigma_{U_0}(\zeta_0)} = \sqrt{\frac{\zeta_0}{\zeta_0 + \zeta_0}}, J = \frac{\sigma_U}{\sigma_{U_0}} = \frac{I_4}{I_2} \tag{8}$$

$$J = \sqrt{\frac{\zeta_0^2(\kappa^2\mu^2 + 4\mu\zeta^2 + 4\kappa(\mu+1)\zeta^2) + \zeta_0\zeta\left(\mu^2 + \kappa^2(\mu+1)^2 - \kappa\mu(\mu+2) + 4\zeta^2(\mu+1)\right) + 4\kappa\mu\zeta\zeta_0^3}{\zeta\zeta_0\left(-2\kappa\mu + \mu^2 + \kappa^2(\mu+1)^2 + 4\zeta^2(\mu+1)\right) + \zeta_0^2(\kappa^2\mu^2 + 4\mu\zeta^2 + 4\kappa(\mu+1)\zeta^2) + \mu^2\zeta^2 + 4\kappa\mu\zeta\zeta_0^3}} \quad (9)$$

where σ_{U0} and σ_{UVD} are the RMS of displacement responses of the original structure with and without VD, I_4 and I_2 are the fourth-order and second-order James formula analytical solutions. It can be seen from Equation (9) that J_{VD} and J have nothing to do with the white noise excitation amplitude. The smaller the displacement mitigation ratio, the better the vibration mitigation effect of the TID, so $\boldsymbol{J}$ can be used as the main performance index for parametric analysis and optimization.

3. Parametric Analysis

3.1. Parametric Analysis of TID Based on Closed-Form Solution J

The closed-form solution of the displacement mitigation ratio based on stochastic response is taken as the parametric analysis index, and the inerter–mass ratio (μ), the damping ratio (ζ), and the stiffness ratio (κ) are selected to carry out the parametric analysis of the TID. The inherent damping ratio (ζ_0) of the original structure is assumed to be 0.02. The displacement mitigation ratio of the SDOF system with TID is related to three parameters of the TID, and it is quite difficult to see the trends of J with variations of three parameters simultaneously. In order to determine the optimal design parameters quickly, one parameter should be fixed and the variable trends of the other two parameters can be observed. In the parametric analysis, the ranges of parameters of the inerter–mass ratio (μ), the damping ratio (ζ), and the stiffness ratio (κ) are all 0.001 to 1. The values of μ, κ, and ζ are fixed as 0.02, 0.05, 0.1, 0.2, 0.5, and 1, respectively, and the continuous variable trends of the other two parameters except the fixed values within 0.001 to 1 are studied. At the same time, the contour plots are drawn as shown in Figures 2–4, and the area of the dark blue area in the Figure is defined as the optimal solution range.

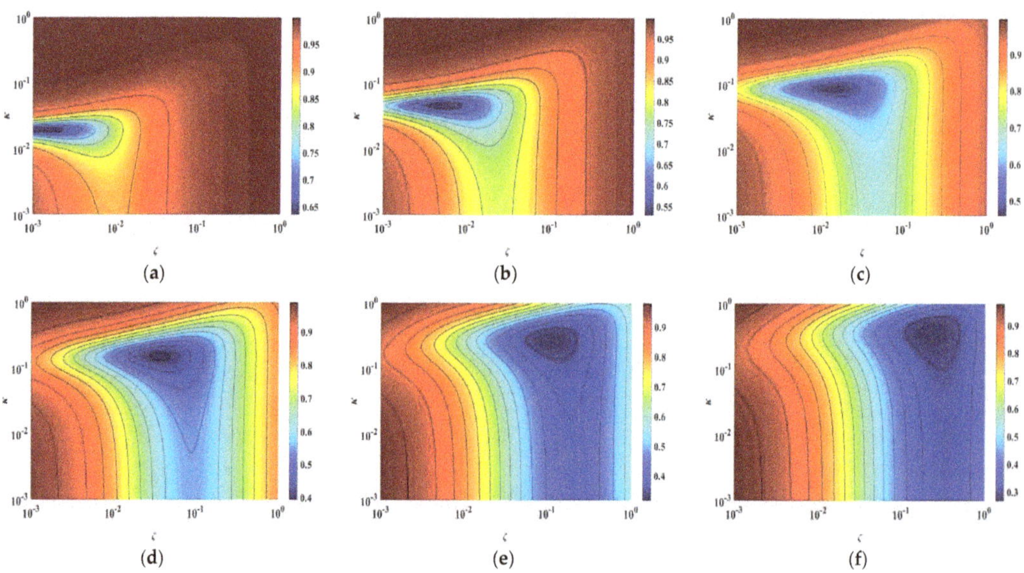

Figure 2. Contour plots of J with different inerter–mass ratios: (**a**) μ = 0.1; (**b**) μ = 0.2; (**c**) μ = 0.3; (**d**) μ_d = 0.1; (**e**) μ = 0.2; (**f**) μ = 0.3.

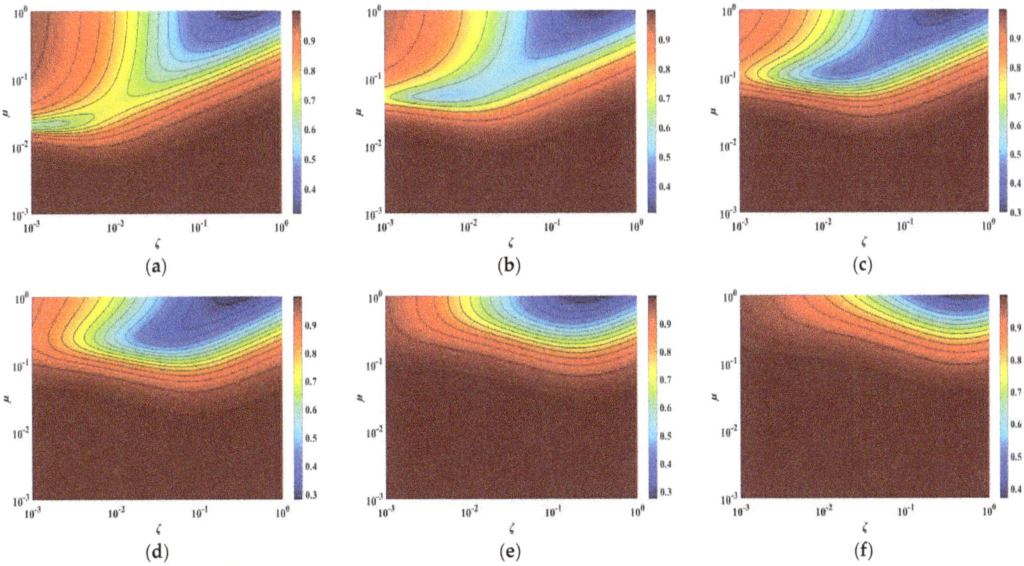

Figure 3. Contour plots of J with different stiffness ratios: (**a**) κ = 0.02; (**b**) κ = 0.05; (**c**) κ = 0.1; (**d**) κ = 0.2; (**e**) κ = 0.5; (**f**) κ = 1.

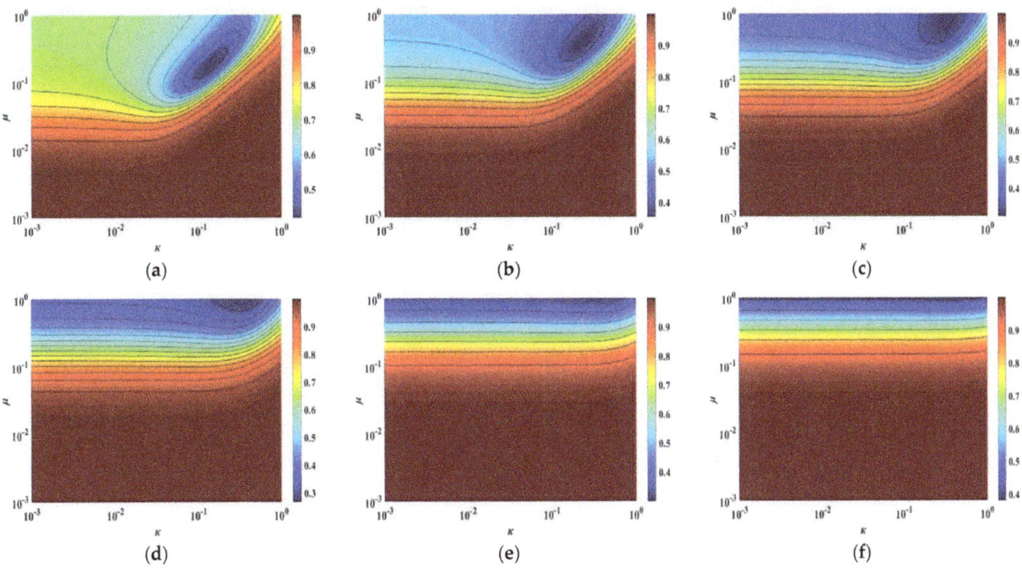

Figure 4. Contour plots of J with different damping ratios: (**a**) ζ = 0.02; (**b**) ζ = 0.05; (**c**) ζ = 0.1; (**d**) ζ = 0.2; (**e**) ζ = 0.5; (**f**) ζ = 1.

It can be seen from Figure 2 that with increasing μ, the lowest point moves to the upper-right part of the overall contour, indicating that the optimal stiffness ratio (κ) and damping ratio (ζ) are positively correlated with the inerter–mass ratio (μ). In addition, no matter how much the inerter–mass ratio (μ) is fixed, the lowest point will appear in the contour plot, indicating that there always exists an optimal combination of stiffness ratio (κ) and damping ratio (ζ) within the parameter range. Figure 3 shows that the position of the lowest point is basically stable at the upper-right of the contour with increasing κ, but the optimal solution range gradually decreases (the dark blue area). When κ is fixed as 0.02, 0.2, and 0.1, the minimum of J (J_{min}) is 0.35, 0.30, and 0.39, respectively, indicating that J_{min} first decreases and then increases with the increasing κ. However, the fluctuation of J_{min} is not obvious in general. As shown in Figure 4, when the value of ζ is within [0, 0.2], the optimal solution range increases gradually. When ζ = 0.2, J_{min} reaches the minimum value of 0.29—but when the value of ζ is in [0.2, 1]—the optimal solution range becomes smaller, and J_{min} also increases, indicating that the vibration mitigation effect of the TID decreases at the same time. With increasing ζ, the variable trends of the contour become gentle and are only correlated with μ but have no connection to the value of κ. At the same time, with increasing μ, κ, and ζ, the cost of the TID will increase simultaneously. Figures 2–4 are only the analytical solution of Equation (9). It is difficult to achieve a damping ratio (ζ) greater than 0.2 in practical engineering. The smaller values of J, μ, κ, and ζ should be determined simultaneously.

Figure 5 shows the maximum frequency response function (FRF) of original structural displacement with a TID under different μ. It can be seen from Figures 2 and 5 that the optimal solution ranges of the contour will be small under small inerter–mass ratios (μ) whether it is the response under harmonic excitation or the stochastic response, and the difficulty of optimization increases accordingly.

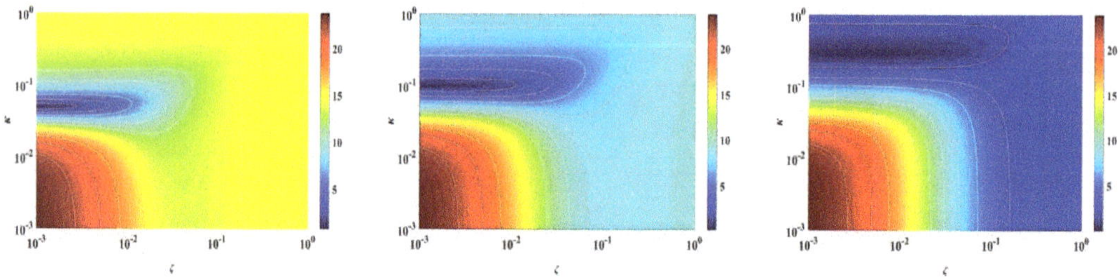

Figure 5. Contour plots of the maximum frequency response function (FRF) of original structural displacement with TID: (**a**) μ = 0.05; (**b**) μ = 0.1; (**c**) μ = 0.3.

Moreover, compared with the viscous damper (VD), the different topological connection forms of the inerter elements in the SDOF system can enhance the deformation of the viscous damping element and improve the energy dissipation ratio. Therefore, a new index of the damping effect of the TID is proposed by comparing J_{VD} and J, namely, inerter-enhanced energy dissipation coefficient η:

$$\eta = \frac{\sigma_{U0}(\zeta_0) - \sigma_U(\zeta_0, \zeta, \kappa, \mu)}{\sigma_{U0}(\zeta_0) - \sigma_{UVD}(\zeta_0, \zeta)} \tag{10}$$

where η is under the same additional damping ratio, the damping element with inerter element has higher deformation effect. When η is greater than 1, it shows that the inerter can strengthen the energy dissipation of the structure, and it can be used as an index to evaluate the economy and robustness of the TID. Figure 6 shows the 3D contour plots of η under different inerter–mass ratio μ.

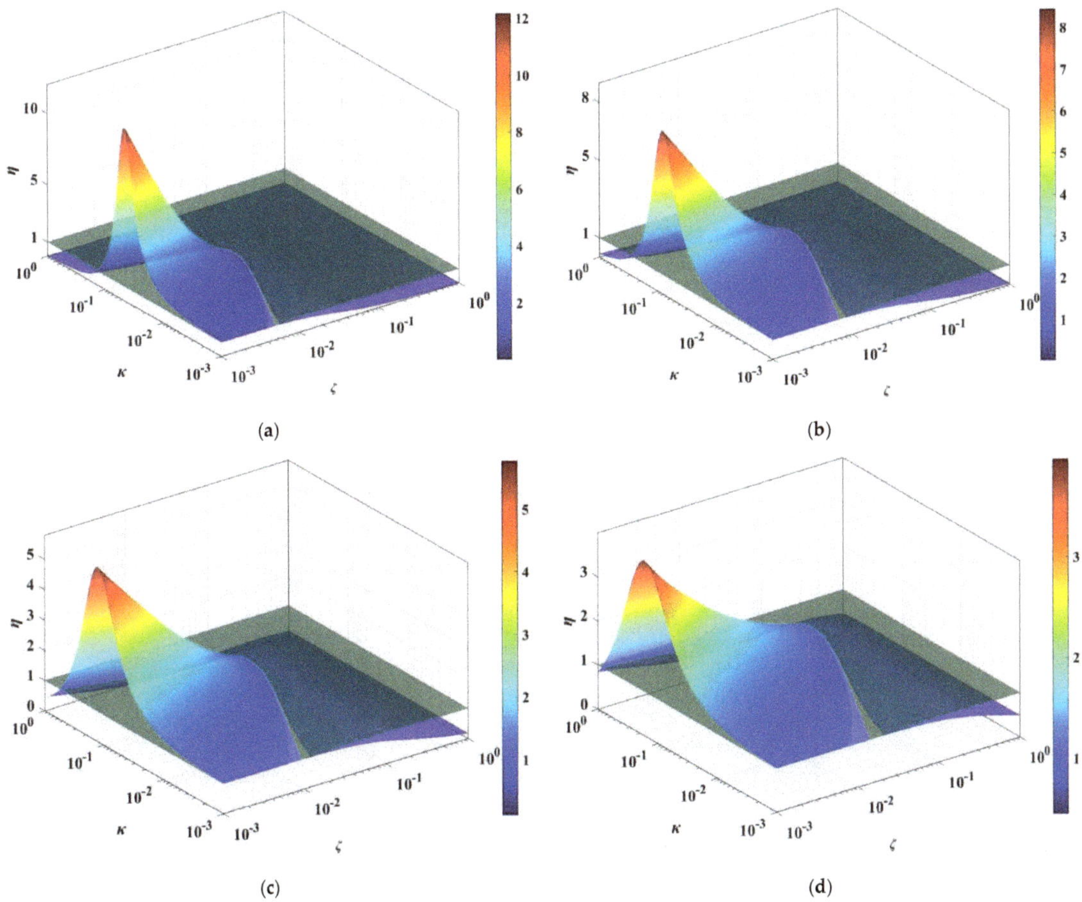

Figure 6. Contour plots 3D contour plots of η under different inerter–mass ratio μ: (**a**) μ = 0.05; (**b**) μ = 0.1; (**c**) μ = 0.2; (**d**) μ = 0.5.

It can be seen from Figure 6 that there always exist ranges of less than 1 (below the light green surface) in the contour plot. This means that the damping effect of the TID may be worse than VD under some specific value of optimal design parameter (κ, μ, and ζ) combinations. Figure 7 shows that the vibration mitigation effect of the TID is lower than VD under some specific ranges of damping ratio ($J > J_{VD}$) (invalid parameters in the shaded part). This will cause loss of the value of adding an inerter element. Hence, the vibration mitigation effect of the designed parameters of the TID should be compared with the VD after parameter optimization.

3.2. Parameter Optimization of TID by Using Closed-Form Solution Method

The design parameters of the TID can be determined by the fixed-point method. Based on the Den Hartog fixed-point theory [54], Pan et al. [16] deduced the design equation of the SDOF structure with TID. After the inerter–mass ratio is determined, the stiffness ratio and damping ratio can be obtained by the following formulas:

$$\kappa = \frac{\mu}{(1+\mu)2}, \zeta = \frac{\mu}{2(1+\mu)}\sqrt{2+\frac{1}{2+\mu-\sqrt{\mu 2+\mu}}} \tag{11}$$

However, the inherent damping ratio and external excitation characteristics of the original structure are not considered in the fixed-point theory, and other performance demands of the original structure, such as displacement of the original structure, cannot be considered. At the same time, all three parameters cannot be optimized simultaneously by using the fixed-point method.

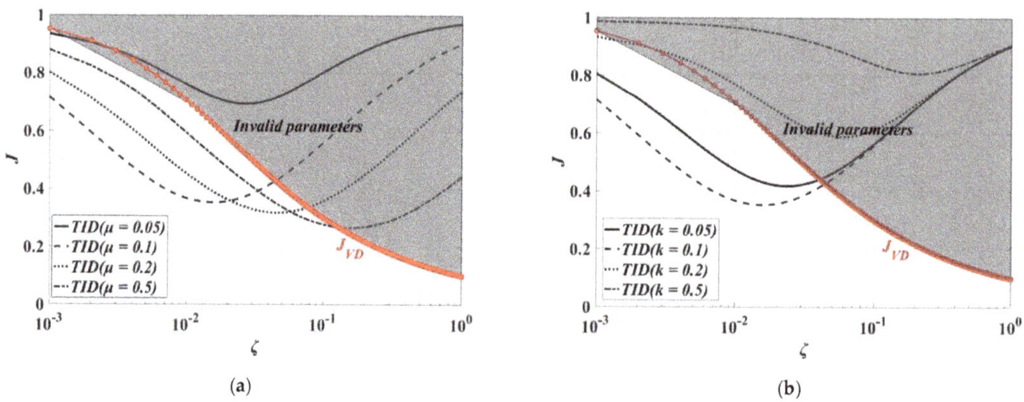

Figure 7. Displacement mitigation ratio of TID and VD with same damping ratio ($\zeta_0 = 0.01$): (**a**) $k = 0.1$; (**b**) $\mu = 0.1$.

For this reason, when TID parameters are designed, the main performance of the original structure should be considered first, and the target mitigation ratio can be determined according to the performance demands of different structures. The extreme conditions are used for parameter optimization at the same time, namely:

$$J = J_t, \frac{\partial J}{\partial \kappa} = 0, \frac{\partial J}{\partial \zeta} = 0 \tag{12}$$

where J_t is the target displacement mitigation ratio of the original structure. It is difficult to realize the installation of a TID in practical engineering under a high damping ratio. Therefore, the constraint condition of Equation (12) is to meet the target mitigation ratio (J_t) and the damping ratio is simultaneously determined to be as small as possible to meet the practical engineering requirements. This is called the closed-form solution method in this paper. After the optimal solution of the damping ratio is obtained, J_t should also be smaller than the displacement mitigation ratio of the VD system (J_{VD}) under the same damping ratio, namely, Formula (13). If Formula (13) is not satisfied, that is $J_t > J_{VD}$, the vibration mitigation effect of the TID is lower than the VD, and the design parameters are invalid parameters correspondingly.

$$J_t \leq J_{VD} \tag{13}$$

The inerter–mass ratio obtained by the closed-form solution method is also used in the fixed-point method as additional tuning condition, and then the optimal damping ratio and optimal stiffness ratio of the fixed-point method are obtained by using Formula (11).

3.3. Applicability Analysis of Closed-Form Solution Method and Fixed-Point Method

In this paper, the optimization designs of different original structures with inherent damping ratios (ζ_0 is considered to be 0.01, 0.02, and 0.05) are carried out. Table 1 shows the optimization parameter values of the TID and the J_{VD} of the VD under the same damping ratio. It can be seen from Table 1 that the damping ratio (ζ) of the TID obtained by the closed-form solution method is always smaller than the fixed-point method, and the J_{VD} obtained is always greater than J_t, indicating that the TID design parameters obtained by the closed-form solution method are effective parameters which can achieve the effect of enhancing damping. However, when the inherent damping ratio (ζ_0) of the original structure is equal to 0.05, the design parameters of the TID obtained by using the fixed-point method are invalid, indicating that the fixed-point method is not suitable for structures with large inherent damping ratios (blue font in Table 1).

Figure 8 shows the displacement transfer function of the SDOF structure for the two optimization methods. It can be seen from Figure 8 that the TID vibration mitigation effect obtained by using the fixed-point method is significantly lower than that of the closed-form solution method, and only the resonance response of the original structure can be controlled under the fixed-point method, which is similar to the VD. The optimal parameters of the TID by using the closed-form solution method can expand the frequency domain of displacement control effect and that it is no longer limited to narrow-band control. The displacement response in the near-circular frequency of the original structure or even a wider range can be controlled, which is more obvious when considering a larger inherent damping ratio.

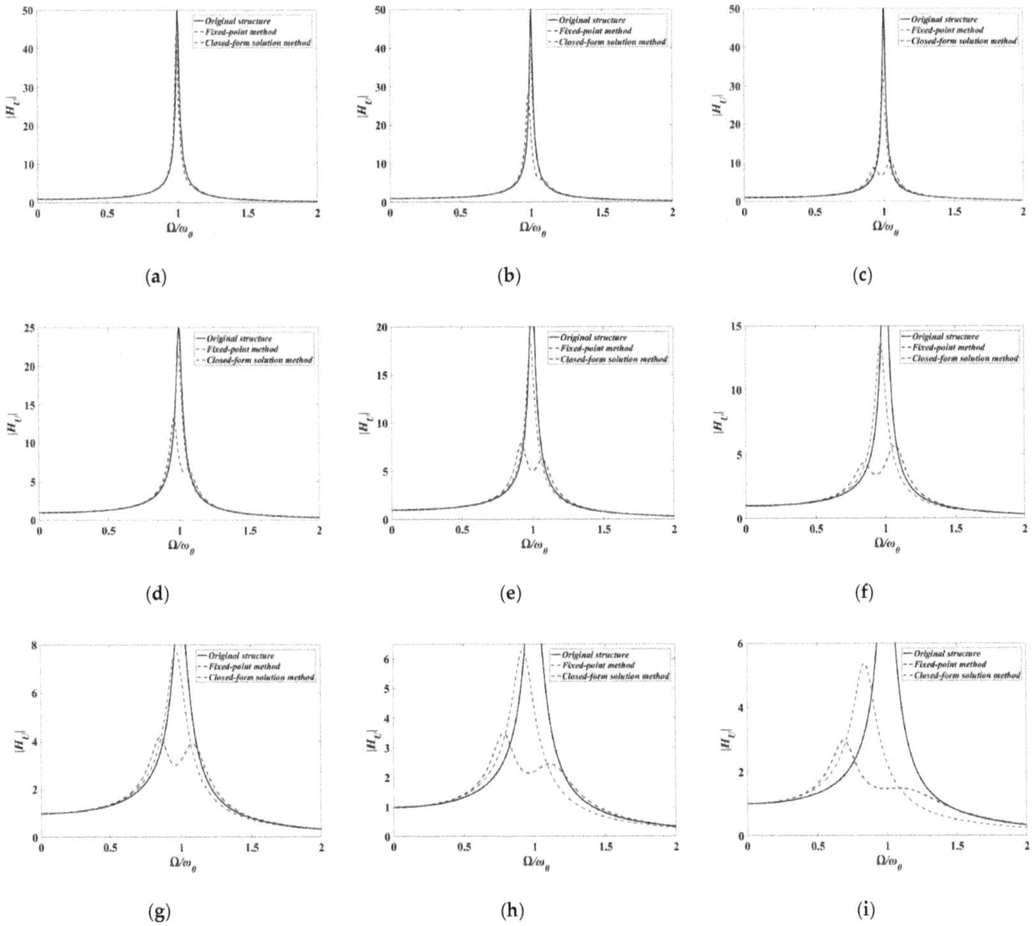

Figure 8. Displacement transfer function of original structural with TID with different optimal methods: (**a**) $J = 0.7$, $\zeta_0 = 0.01$; (**b**) $J = 0.6$, $\zeta_0 = 0.01$; (**c**) $J = 0.5$, $\zeta_0 = 0.01$; (**d**) $J = 0.7$, $\zeta_0 = 0.02$; (**e**) $J = 0.6$, $\zeta_0 = 0.02$; (**f**) $J = 0.5$, $\zeta_0 = 0.02$; (**g**) $J = 0.7$, $\zeta_0 = 0.05$; (**h**) $J = 0.6$, $\zeta_0 = 0.05$; (**i**) $J = 0.5$, $\zeta_0 = 0.05$.

Figure 9 shows the range of the TID invalid parameters under the two optimization methods, and the shaded part in Figure 8 represents the ranges of invalid parameters ($J_t > J_{VD}$). It can be seen from Figure 8 that when the inherent damping ratio (ζ_0) of the original structure is 0.01 and 0.02, the TID design parameters obtained by the closed-form solution are all effective. When ζ_0 is 0.05, the closed-form solution fails when J is within [0.25, 0.4]. When J is within [0.25, 0.4], as the inherent damping ratio of the original structure increases, the effect of the TID becomes closer to that of the VD, indicating that the vibration mitigation effect of the TID is worse. The range of invalid parameters increases to [0.17, 0.6] as ζ_0 increases to 0.1. The TID design parameter combination obtained by the fixed-point method has no invalid parameters only when ζ_0 is 0.01. The invalid ranges of parameters under other inherent damping ratios all exist, and the invalid ranges are larger than those found by using the closed-form solution method. Within the range of effective design parameters of the TID, the positions of curves of equivalent J_{VD} obtained by the closed-form solution method are always above those of the fixed-point method, indicating that the cost of the closed-form solution method is lower than that of the fixed-point method. That is, the optimal damping ratio (ζ_{opt}) of TID obtained by the closed-form solution method is smaller than that by the fixed-point method.

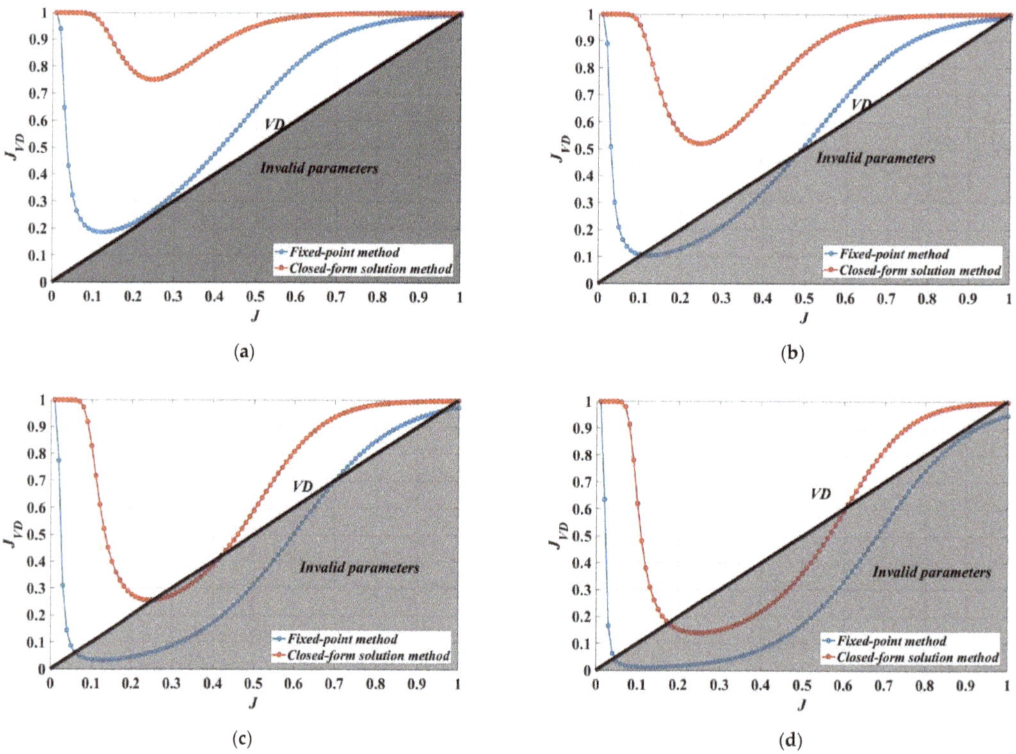

Figure 9. Range of invalid designed parameters of TID with different optimal method: (**a**) $\zeta_0 = 0.01$; (**b**) $\zeta_0 = 0.02$; (**c**) $\zeta_0 = 0.05$; (**d**) $\zeta_0 = 0.1$.

Table 1. Designed parameters of TID and J_{VD} with different optimal methods.

ζ_0		*Closed-Form Solution Method*				*Fixed-Point Method*			
	J_t	μ	κ	ζ	J_{VD}	μ	κ	ζ	J_{VD}
0.01	**0.7**	0.0026	0.0031	0.0001	**0.9901**	0.0026	0.0025	0.0020	**0.9117**
	0.6	0.0065	0.0073	0.0003	**0.9860**	0.0065	0.0073	0.0052	**0.8104**
	0.5	0.0165	0.0156	0.0010	**0.9528**	0.0165	0.0160	0.0134	**0.6541**
0.02	**0.7**	0.0100	0.0105	0.0005	**0.9873**	0.0100	0.0098	0.0080	**0.8445**
	0.6	0.0256	0.0250	0.0020	**0.9532**	0.0256	0.0243	0.0210	**0.6984**
	0.5	0.0647	0.0533	0.0071	**0.8589**	0.0647	0.0570	0.0554	**0.5151**
0.05	**0.7**	0.0608	0.0533	0.0068	**0.9381**	0.0608	0.0541	0.0521	**0.6991**
	0.6	0.1559	0.1267	0.0264	**0.8089**	0.1559	0.1167	0.1462	**0.5048**
	0.5	0.3935	0.2705	0.0934	**0.5905**	0.3935	0.2027	0.4508	**0.3160**

4. Seismic Responses

In order to verify the vibration mitigation effect of the TID and parameter optimization results, 0.5 is taken as the target displacement mitigation ratio J_t, the seismic responses of the SDOF system with TID under ground motions are carried out, and the TID parameters obtained by the two optimization methods were compared. The original structural mass is $m = 1 \times 10^4$ kg, stiffness $k = 1.57 \times 10^3$ kN/m, and the inherent damping ratio ζ_0 is taken as 0.01, 0.02, and 0.05.

EL Centro record, Taft record, Chi-Chi record, and Kobe record are selected, and Figure 10 is the acceleration response spectrum of the four records. At the same time, different dominant frequencies of records are selected four including low frequency, high frequency and many components such as El Centro record. Figure 11 is the displacement time history response of the SDOF system under four records. It can be seen from Figure 11 that the vibration control effects of the TID parameters obtained by using the closed-form solution on the structure are better than those obtained by using the fixed-point method under the ground motions with different predominant frequencies, indicating that the design method of the TID based on the closed-form solution method has better robustness. As the seismic response of the structure progresses, the vibration control effects of the TID in the later stage become better, indicating that a stable vibration state of the original structure can be quickly brought under control by the TID. Figure 12 shows the Fourier amplitude spectrum of the displacement response of the SDOF system under the excitation of the EL Centro record. It can be seen from Figure 12 that the TID parameters obtained by the closed-form solution method can be tuned to match all high-amplitude regions of ground motion. The control frequency domain is wider than that of the fixed-point method, and the controlled displacement amplitude decreases significantly at the fundamental frequency of the original structure. The spectral curve has no obvious peaks or corresponding frequencies, which also corresponds to the steady and low seismic response in the time history curves.

Figure 13 shows the hysteresis loops of damping elements of the TID and VD with the same damping ratio under the excitation of EL Centro record. The corresponding TID design parameters are shown in Table 1. It can be seen from Figure 13 that, under the same damping ratio, the deformation and damping force of the damping element of the TID are both greater than those of the VD, and the overall energy dissipation effect is better than that of the VD. Compared with the hysteresis loops of the TID and VD with the same damping ratio under the two optimization methods, it can be seen that the hysteresis loop areas of the VD corresponding to the fixed-point method increase with the increase of the inherent damping ratio, while the corresponding to closed-form solution method decreases. These show that the design parameters of the fixed-point method are relatively conservative, a larger damping ratio is selected for energy dissipation, and the damping force is 4–5 times that of the closed-form solution method. However, the damping deformation enhancement characteristics of the inerter system are not obvious by using the combination of designed parameters obtained by the fixed-point method. At the same time, the inherent damping ratio of the original structure is not considered in the fixed-point method, which will cause a detuning effect. The inherent damping ratio of the original structure is not considered in the closed-form solution method, so the damping deformation enhancement characteristics of the inerter system can be brought into full play and the same vibration control effect as the fixed-point method is achieved with a smaller damping coefficient.

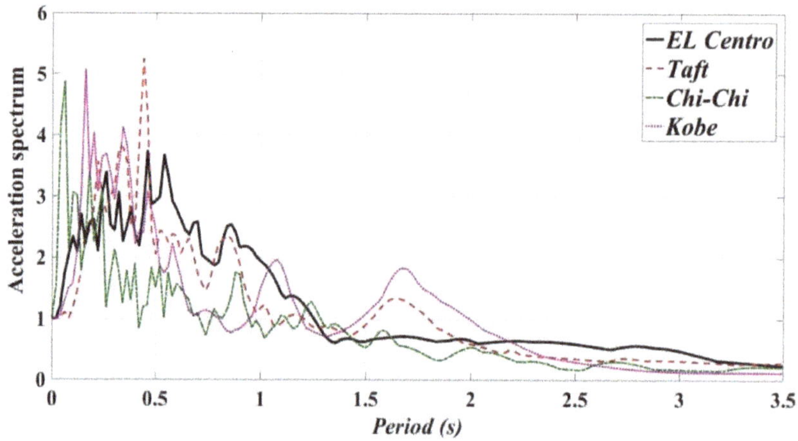

Figure 10. Normalized acceleration spectra of ground motion records.

Figure 14 shows the hysteresis loops of the damping element of the TID by using different optimization methods under the excitation of EL Centro record. It can be seen from Figure 14a that the damping coefficient of the TID obtained by using the fixed-point method is much higher than that obtained by the closed-form solution method. Although a high damping coefficient will increase the damping force and increase the energy dissipation effect, the cost of the TID will increase dramatically. However, the damping deformation enhancement characteristics of the inerter system found using the closed-form solution method are about three times higher than those found using the fixed-point method. Hence, the damping element of the TID obtained by the closed-form solution method still has effective energy dissipation. Figure 14b shows the energy dissipation effect of inherent damping ratio and the equivalent VD obtained by the two methods. The optimal damping ratio of the TID obtained by the closed solution method is much smaller, which is 11% of the damping ratio obtained by using the fixed-point method and 8.3% of the inherent damping ratio. From Figure 14a, we can see that the maximum displacements of the main structure with a TID are nearly identical using two optimal design methods; however, the additional damping ratio of the TID designed by the closed-form solution method is considerably smaller compared to that of the fixed-point method. The damping enhancement effect of the inerter element can be fully utilized when employing the proposed performance-demand-based design approach. The designed parameters of the TID obtained by the closed-form solution method can not only meet the performance demands of the structures with different inherent damping ratio, but the engineering cost is also greatly reduced, achieving a lightweight design of the TID at the same time.

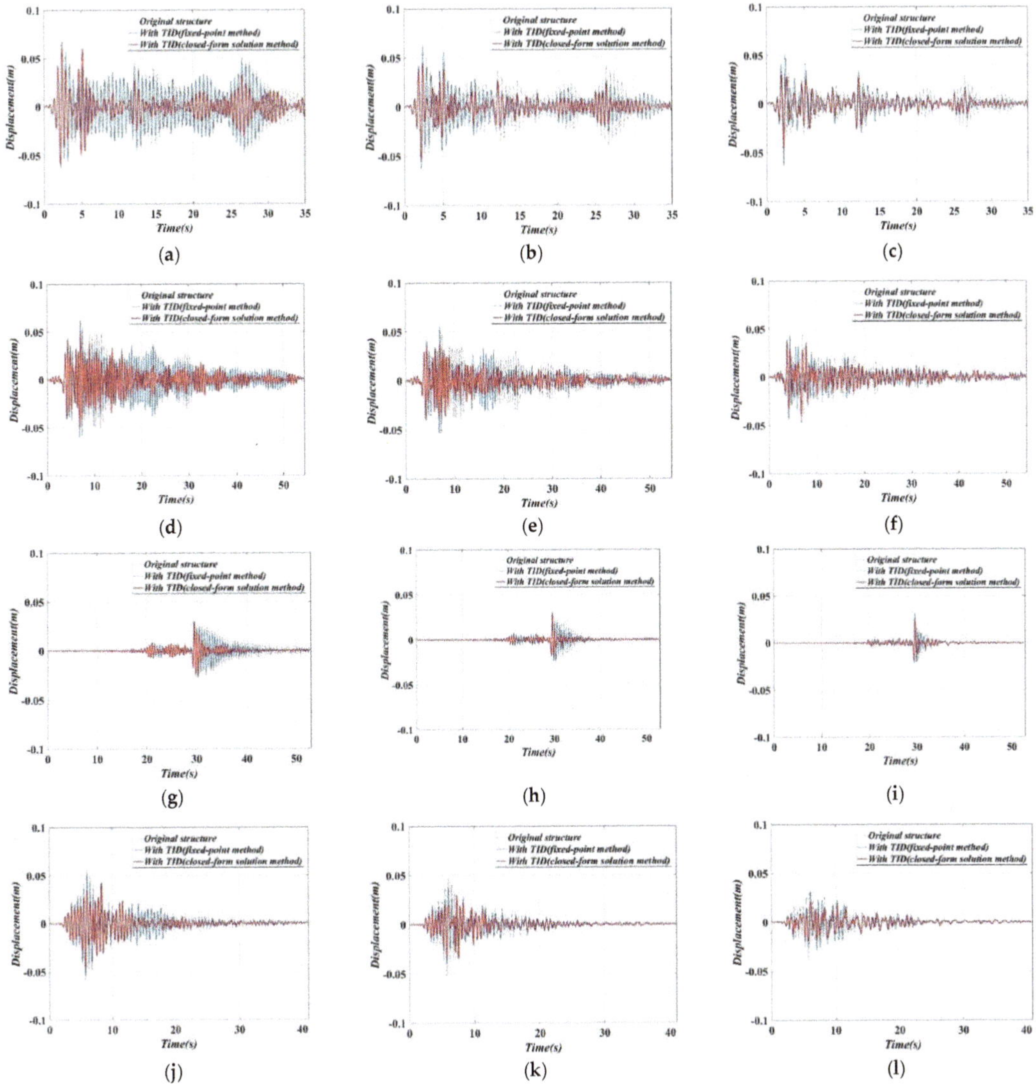

Figure 11. Displacement responses of SDOF ($J = 0.5$): (**a**) EL Centro, $\zeta_0 = 0.01$; (**b**) EL Centro, $\zeta_0 = 0.02$; (**c**) EL Centro, $\zeta_0 = 0.05$; (**d**) Taft, $\zeta_0 = 0.01$; (**e**) Taft, $\zeta_0 = 0.02$; (**f**) Taft, $\zeta_0 = 0.05$; (**g**) Chi-Chi, $\zeta_0 = 0.01$; (**h**) Chi-Chi, $\zeta_0 = 0.02$; (**i**) Chi-Chi, $\zeta_0 = 0.05$; (**j**) Kobe, $\zeta_0 = 0.01$; (**k**) Kobe, $\zeta_0 = 0.02$; (**l**) Kobe, $\zeta_0 = 0.05$.

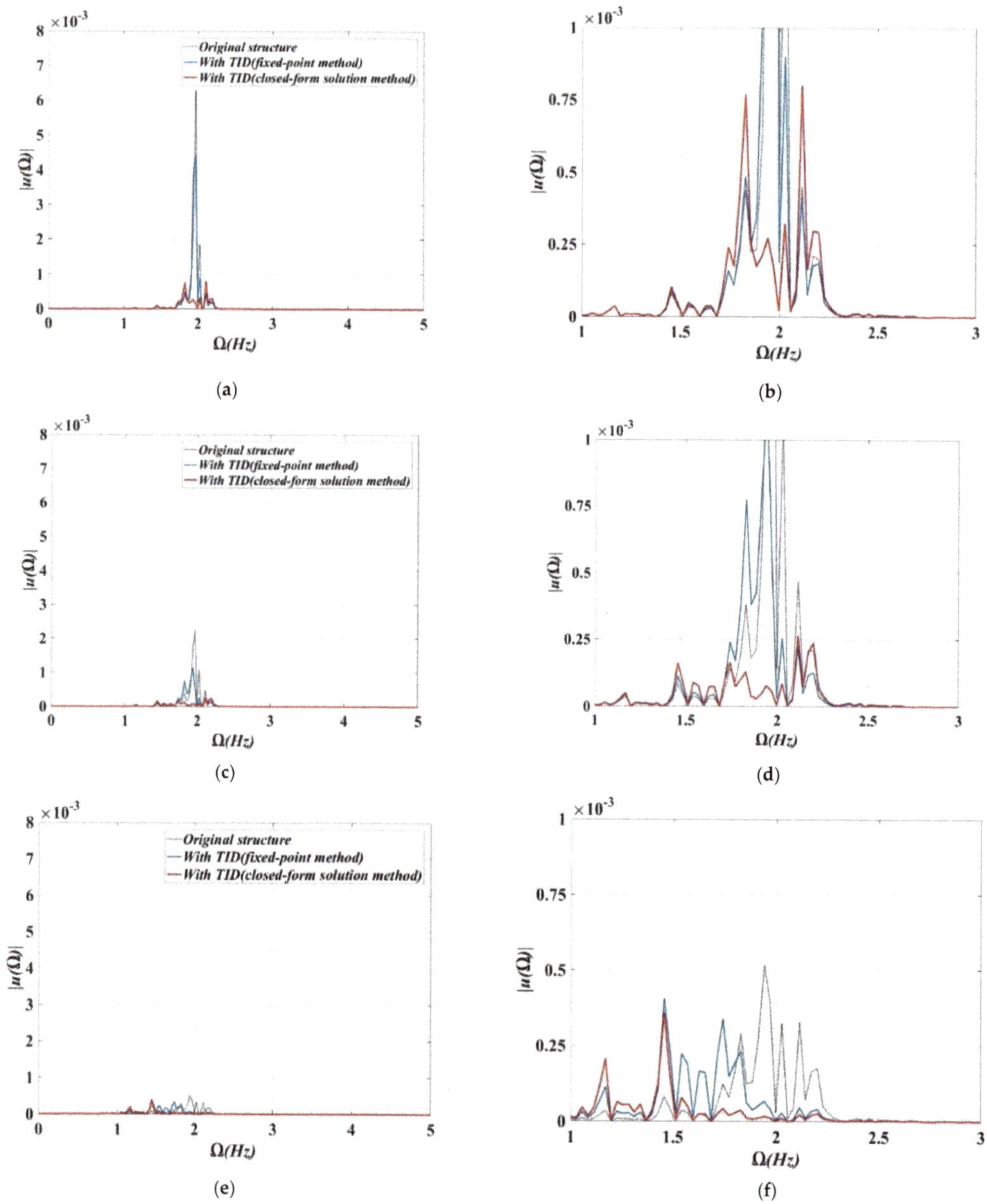

Figure 12. Single-sided Fourier spectra of displacement responses of SDOF under EL Centro record (J = 0.5): (**a**) ζ_0 = 0.01; (**b**) Zoom-in, ζ_0 = 0.01; (**c**) ζ_0 = 0.02; (**d**) Zoom-in, ζ_0 = 0.02; (**e**) ζ_0 = 0.05; (**f**) Zoom-in, ζ_0 = 0.05.

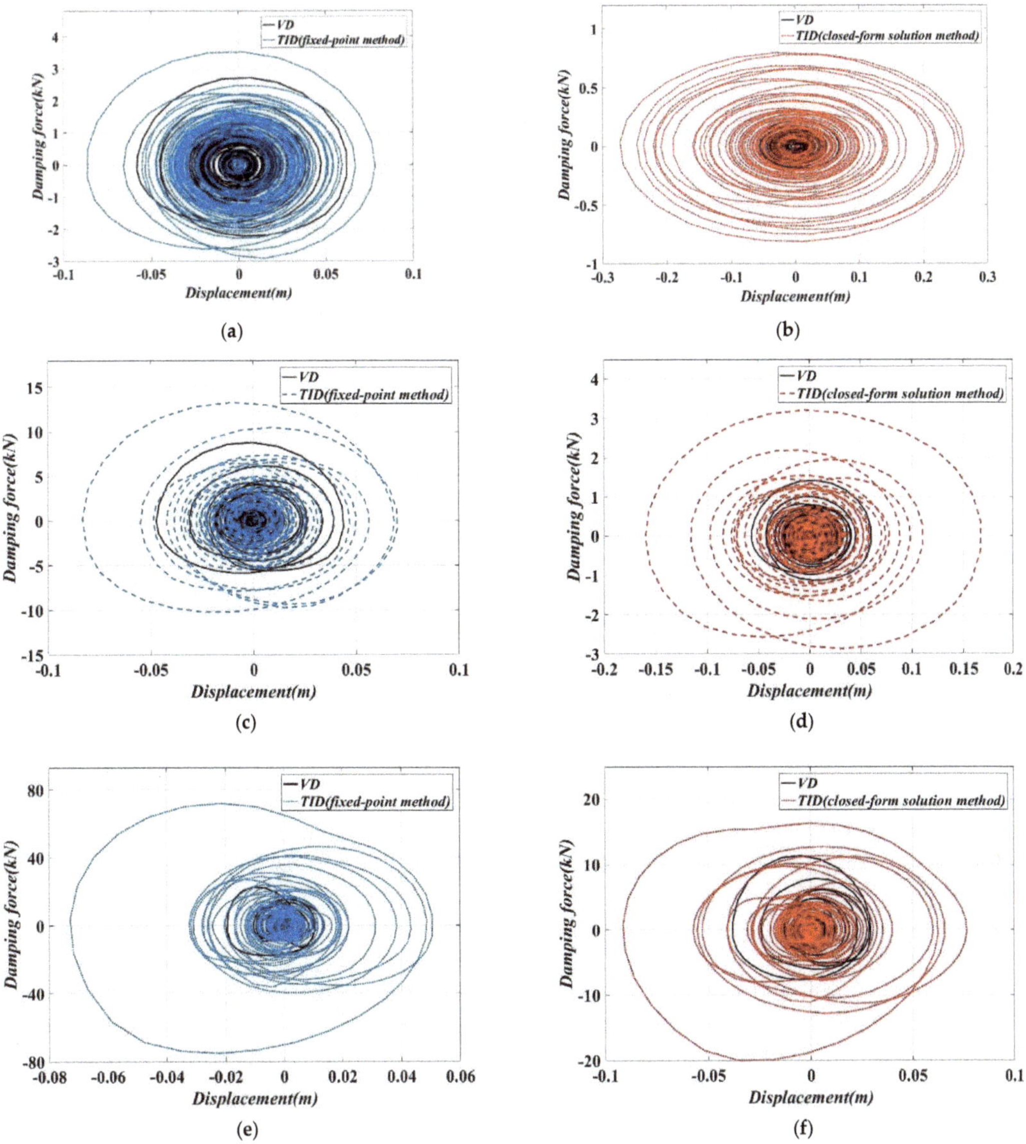

Figure 13. Hysteretic loops of the damper under EL Centro record (J = 0.5): (**a**) Fixed-point method, ζ_0 = 0.01; (**b**) Closed-form solution method, ζ_0 = 0.01; (**c**) Fixed-point method, ζ_0 = 0.02; (**d**) Closed-form solution method, ζ_0 = 0.02; (**e**) Fixed-point method, ζ_0 = 0.05; (**f**) Closed-form solution method, ζ_0 = 0.05.

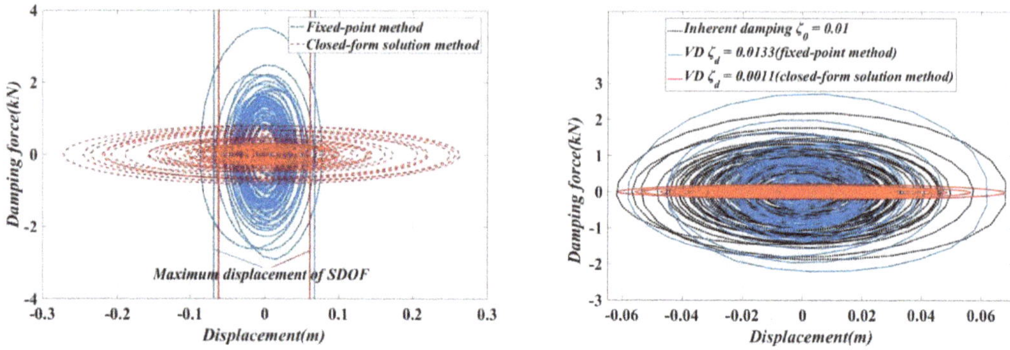

(a) (b)

Figure 14. Hysteretic loops of the damping element compared with two optimal methods with same inerter–mass ratio: (**a**) TID, $\zeta_0 = 0.01$; (**b**) VD.

5. Conclusions

In this paper, the motion control equations of the SDOF system with TID are established. The inerter–mass ratio, damping ratio, and stiffness ratio are analyzed based on the displacement mitigation ratio of stochastic response, and the parameter optimization design of the TID based on the closed-form solution method are also carried out compared with the fixed-point method. Additionally, applicable conditions of the two methods are discussed in a wider range of values of objective function under different inherent damping ratios. Finally, seismic responses are carried out on the SDOF system with TID, which verifies the feasibility of the closed-form solution optimization method. The main conclusions are as follows:

- The investigation on the variation patterns of invalid designed parameters of TIDs under different inherent damping ratios indicates that compared with the fixed-point method, better applicability and robustness are shown in closed-form solution method.
- The displacement mitigation ratio could be used as a target during the determination of design parameters of the TID, which meets the structural design concept based on performance demand.
- Nearly identical performances of main structures with TID can be realized using two optimal design methods; however, the additional damping ratio of the TID designed by the closed-form solution method is considerably smaller compared to that of the fixed-point method. The damping enhancement effect of the inerter element can be fully utilized when employing the proposed performance demand-based design approach.
- Considering the lower additional damping ratio and inerter–mass ratio, applications of the TID designed by the closed-form solution method can be extended and its installation made more flexible. The corresponding verification experiment study must be conducted in the near future.

Author Contributions: R.Z.: methodology, conceptualization, software, investigation, writing—original draft, writing—review and editing; J.H.: supervision, funding acquisition, methodology, investigation, validation, writing—review and editing. M.C.: conceptualization, writing—review and editing; Q.L., validation, writing—review and editing. X.G., validation, writing—review and editing. All authors have read and agreed to the published version of the manuscript.

Funding: This work is supported in part by the National Key R&D Program of China (Grant no. 2019YFC1520500, 2020YFC1523004).

Institutional Review Board Statement: Not applicable.

Informed Consent Statement: Not applicable.

Data Availability Statement: The data used to support the findings of this study are included within the article.

Conflicts of Interest: The authors declare no conflict of interest.

References

1. Cao, M.; Zhou, F.; Tan, P.; Zhang, G.; Zhengguo, G. Seismic response of transformer and bushing isolation system and parameter analysis of isolation layer. *J. Proc. CSEE* **2012**, *32*, 166–174.
2. Maleska, T.; Nowacka, J.; Beben, D. Application of EPS Geofoam to a Soil–Steel Bridge to Reduce Seismic Excitations. *Geosciences* **2019**, *9*, 448. [CrossRef]
3. Rahmani, M.; Todorovska, M.I. Structural health monitoring of a 32-storey steel-frame building using 50 years of seismic monitoring data. *Earthq. Eng. Struct. Dyn.* **2021**, *50*, 1777–1800. [CrossRef]
4. Liang, J.; Jin, L.; Todorovska, M.I.; Trifunac, M.D. Soil–structure interaction for a SDOF oscillator supported by a flexible foundation embedded in a half-space: Closed-form solution for incident plane SH-waves. *Soil Dyn. Earthq. Eng.* **2016**, *90*, 287–298. [CrossRef]
5. Zhang, R.; Cao, M. Study on vibration control and parameters influence of cable inerter viscous damping system. *Shock. Vib.* **2022**, *2022*, 2983700. [CrossRef]
6. Meirovitch, L. *Dynamics and Control of Structures*; John Wiley & Sons: Hoboken, NJ, USA, 1990.
7. Soong, T.T.; Dargush, G.F. *Passive Energy Dissipation Systems in Structural Engineering*; Wiley: Chichester, UK, 1997.
8. Housner, G.; Bergman, L.A.; Caughey, T.K.; Chassiakos, A.G.; Claus, R.O.; Masri, S.F.; Skelton, R.E.; Soong, T.; Spencer, B.; Yao, J.T. Structural control: Past, present, and future. *J. Eng. Mech.* **1997**, *123*, 897–971. [CrossRef]
9. Kareem, A.; Kijewski, T.; Tamura, Y. Mitigation of motions of tall buildings with specific examples of recent applications. *Wind Struct. Int. J.* **1999**, *2*, 201–251. [CrossRef]
10. McNamara, R.J. Tuned Mass Dampers for Buildings. *J. Struct. Div.* **1977**, *103*, 1785–1798. [CrossRef]
11. Hwang, J.S.; Lee, S.H.; Min, K.W.; Kim, J. Equivalent damping of a structure with vibration control devices subjected to wind loads. *Wind. Struct.* **2003**, *6*, 249–262. [CrossRef]
12. Losanno, D.; Calabrese, A.; Madera-Sierra, I.E.; Spizzuoco, M.; Marulanda, J.; Thomson, P.; Serino, G. Recycled versus natural-rubber fiber-reinforced bearings for base isolation: Review of the experimental findings. *J. Earthq. Eng.* **2020**, *1*, 1–20. [CrossRef]
13. Losanno, D.; Palumbo, F.; Calabrese, A.; Barrasso, T.; Vaiana, N. Preliminary Investigation of Aging Effects on Recycled Rubber Fiber Reinforced Bearings (RR-FRBs). *J. Earthq. Eng.* **2021**, *1*, 1–18. [CrossRef]
14. Kelly, J.M. Seismic isolation for earthquake-resistant design. In *Earthquake-Resistant Design with Rubber*; Springer: London, UK, 1997; pp. 1–18.
15. Vaiana, N.; Spizzuoco, M.; Serino, G. Wire rope isolators for seismically base-isolated lightweight structures: Experimental characterization and mathematical modeling. *Eng. Struct.* **2017**, *140*, 498–514. [CrossRef]
16. Nagarajaiah, S.; Reinhorn, A.M.; Constantinou, M.C. Nonlinear dynamic analysis of 3-D-base-isolated structures. *J. Struct. Eng.* **1991**, *117*, 2035–2054. [CrossRef]
17. Pellecchia, D.; Lo Feudo, S.; Vaiana, N.; Dion, J.L.; Rosati, L. A procedure to model and design elastomeric-based isolation systems for the seismic protection of rocking art objects. *Comput.-Aided Civ. Infrastruct. Eng.* **2021**, *1–18*, 1093–9687. [CrossRef]
18. Simiu, E.; Scanlan, R. *Wind Effects on Structures*; Wiley: Hobeken, NJ, USA, 1996.
19. Tian, L.; Ma, R.; Pan, H.; Qiu, C.; Li, W. Wind-induced progressive collapse analysis of long span transmission tower-line system considering the member buckling effect. *Adv. Struct. Eng.* **2019**, *22*, 30–41. [CrossRef]
20. Jiang, Y.Y.; Zhao, Z.P.; Zhang, R.F.; De Domenico, D.; Pan, C. Optimal design based on analytical solution for storage tank with inerter isolation system. *Soil Dyn. Earthq. Eng.* **2020**, *129*, 105924. [CrossRef]
21. Sadeghian, M.A.; Yang, J.; Wang, F.; Wang, X. Structural vibration control using novel adaptive tuned mass inertance damper (ATMID) with adjustable inertance. *Appl. Sci.* **2022**, *12*, 4028. [CrossRef]
22. Brzeski, P.; Lazarek, M.; Perlikowski, P. Influence of variable damping coefficient on efficiency of TMD with inerter. *Energies* **2020**, *13*, 6175. [CrossRef]
23. Yu, N.; Ouyang, H.; Zhang, J.; Ye, J.; Yan, Q.; Zheng, X. Performance analyses of passive vibration isolator with parallel connection of quasi-zero stiffness and inerter dampers. *Appl. Sci.* **2020**, *10*, 6894. [CrossRef]
24. Liu, X.; Yang, Y.; Sun, Y.; Zhong, Y.; Zhou, L.; Li, S.; Wu, C. Tuned-mass-damper-inerter performance evaluation and optimal design for transmission line under harmonic excitation. *Buildings* **2022**, *12*, 435. [CrossRef]
25. Weber, F.; Huber, P.; Borchsenius, F.; Braun, C. Performance of TMDI for tall building damping. *Actuators* **2020**, *9*, 139. [CrossRef]
26. Zhang, R.; Cao, M.; Huang, J. Study on seismic response and parameter influence in a transformer-bushing with inerter isolation system. *Buildings* **2022**, *12*, 530. [CrossRef]
27. Lee, S.H.; Min, K.W.; Hwang, J.S.; Kim, J. Evaluation of equivalent damping ratio of a structure with added dampers. *Eng. Struct.* **2004**, *26*, 335–346. [CrossRef]
28. Smith, M.C.; Wang, F.C. Performance benefits in passive vehicle suspensions employing inerters. *Veh. Syst. Dyn.* **2004**, *42*, 235–257. [CrossRef]
29. Smith, M.C. Synthesis of mechanical networks: The inerter. *IEEE Trans. Autom. Control* **2002**, *47*, 1648–1662. [CrossRef]

30. Ikago, K.; Saito, K.; Inoue, N. Seismic control of single-degree-of-freedom structure using tuned viscous mass damper. *Earthq. Eng. Struct. Dyn.* **2012**, *41*, 453–474. [CrossRef]
31. Pan, C.; Zhang, R.; Luo, H.; Li, C.; Shen, H. Demand-based optimal design of oscillator with parallel-layout viscous inerter damper. *Struct. Control Health Monit.* **2018**, *25*, e2051. [CrossRef]
32. Pan, C.; Zhang, R. Design of structure with inerter system based on stochastic response mitigation ratio. *Struct. Control Health Monit.* **2018**, *25*, e2169. [CrossRef]
33. Hwang, J.S.; Kim, J.; Kim, Y.M. Rotational inertia dampers with toggle bracing for vibration control of a building structure. *Eng. Struct.* **2007**, *29*, 1201–1208. [CrossRef]
34. Zhang, R.F.; Zhao, Z.P.; Dai, K.S. Seismic response mitigation of a wind turbine tower using a tuned parallel inerter mass system. *Eng. Struct.* **2019**, *180*, 29–39. [CrossRef]
35. Zhao, Z.P.; Zhang, R.F.; Wierschem, N.E.; Jiang, Y.Y.; Pan, C. Displacement mitigationoriented design and mechanism for inerter-based isolation system. *J. Vib. Control* **2021**, *27*, 1991–2003. [CrossRef]
36. Gao, H.; Wang, H.; Li, J.; Wang, Z.; Liang, R.; Xu, Z.; Ni, Y. Optimum design of viscous inerter damper targeting multi-mode vibration mitigation of stay cables. *Eng. Struct.* **2021**, *226*, 111375. [CrossRef]
37. De Domenico, D.; Impollonia, N.; Ricciardi, G. Soil-dependent optimum design of a new passive vibration control system combining seismic base isolation with tuned inerter damper. *Soil Dynam. Earthq. Eng.* **2018**, *105*, 37–53. [CrossRef]
38. De Domenico, D.; Ricciardi, G. An enhanced base isolation system equipped with optimal tuned mass damper inerter (TMDI). *Earthq. Eng. Struct. Dyn.* **2018**, *47*, 1169–1192. [CrossRef]
39. De Domenico, D.; Ricciardi, G. Optimal design and seismic performance of tuned mass damper inerter (TMDI) for structures with nonlinear base isolation systems. *Earthq. Eng. Struct. Dyn.* **2018**, *47*, 2539–2560. [CrossRef]
40. Zhao, Z.; Zhang, R.; Jiang, Y.; Pan, C. Seismic response mitigation of structures with a friction pendulum inerter system. *Eng. Struct.* **2019**, *193*, 110–120. [CrossRef]
41. Nakamura, Y.; Fukukita, A.; Tamura, K.; Yamazaki, I.; Matsuoka, T.; Hiramoto, K.; Sunakoda, K. Seismic response control using electromagnetic inertial mass dampers. *Earthq. Eng. Struct. Dyn.* **2014**, *43*, 507–527. [CrossRef]
42. Wang, F.-C.; Chen, C.-W.; Liao, M.-K.; Hong, M.-F. Performance analyses of building suspension control with inerters. In Proceedings of the 2007 46th IEEE Conference on Decision and Control, New Orleans, LA, USA, 12–14 December 2007; pp. 3786–3791.
43. Lazar, I.F.; Wagg, D.J.; Neild, S.A. An inerter vibration isolation system for the control of seismically excited structures. In Proceedings of the International Conference on Urban Earthquake Engineering, Tokyo, Japan, 1–2 March 2013.
44. Luo, H.; Zhang, R.; Weng, D. Weng, Mitigation of liquid sloshing in storage tanks by using a hybrid control method. *Soil Dynam. Earthq. Eng.* **2016**, *90*, 183–195. [CrossRef]
45. Marian, L.; Giaralis, A. Optimal design of a novel tuned mass-dampereinerter (TMDI) passive vibration control configuration for stochastically support-excited structural systems. *Probabilist. Eng. Mech.* **2014**, *38*, 156–164. [CrossRef]
46. Hashimoto, T.; Fujita, K.; Tsuji, M.; Takewaki, I. Innovative base-isolated building with large mass-ratio TMD at basement for greater earthquake resilience. *Future Cities Environ.* **2015**, *1*, 9. [CrossRef]
47. Pietrosanti, D.; De Angelis, M.; Basili, M. Optimal design and performance evaluation of systems with tuned mass damper inerter (TMDI). *Earthq. Eng.Struct. Dyn.* **2017**, *46*, 1367–1388. [CrossRef]
48. Sugimura, Y.; Goto, W.; Tanizawa, H.; Saito, K.; Nimomiya, T. Response control effect of steel building structure using tuned viscous mass damper. In Proceedings of the 15th World Conference on Earthquake Engineering, Lisbon, Portugal, 24–28 September 2012.
49. Kurata, M.; Leon, R.T.; Desroches, R. Rapid seismic rehabilitation strategy: Concept and testing of cable bracing with couples resisting damper. *J. Struct. Eng.* **2012**, *138*, 354–362. [CrossRef]
50. Xie, L.; Ban, X.; Xue, S.; Ikago, K.; Kang, J.; Tang, H. Theoretical study on a cable-bracing inerter system for seismic mitigation. *Appl. Sci.* **2019**, *9*, 4096. [CrossRef]
51. Xue, S.; Kang, J.; Xie, L.; Zhang, R.; Ban, X. Cross-layer installed cable-bracing inerter system for MDOF structure seismic response control. *Appl. Sci.* **2020**, *10*, 5914. [CrossRef]
52. Wang, H.; Gao, H.; Li, J.; Wang, Z.; Ni, Y.; Liang, R. Optimum design and performance evaluation of the tuned inerter-negative-stiffness damper for seismic protection of single-degree-of-freedom structures. *Int. J. Mech. Sci.* **2021**, *212*, 106805. [CrossRef]
53. Hu, Y.; Chen, M.Z.; Shu, Z.; Huang, L. Analysis and optimisation for inerter-based isolators via fixed-point theory and algebraic solution. *J. Sound Vib.* **2015**, *346*, 17–36. [CrossRef]
54. Den Hartog, J.P. *Mechanical Vibrations*, 4th ed.; McGraw-Hill: New York, NY, USA, 1956.
55. Baduidana, M.; Kenfack-Jiotsa, A. Optimal design of inerter-based isolators minimizing the compliance and mobility transfer function versus harmonic and random ground acceleration excitation. *J. Vib. Control* **2020**, *27*, 107754632094017. [CrossRef]
56. Garrido, H.; Curadelli, O.; Ambrosini, D. Improvement of tuned mass damper by using rotational inertia through tuned viscous mass damper. *Eng. Struct.* **2013**, *56*, 2149–2153. [CrossRef]

buildings

MDPI

Article

Investigation on Load Path of a Latticed Shell Structure under Localized Fire Based on Member Sensitivity

Jing Cui [1,2,*], Gang Hu [1], Yang Zhan [3] and Rui Pang [1]

[1] School of Civil Engineering, Henan University of Technology, Zhengzhou 450001, China
[2] Henan Key Laboratory of Grain and Oil Storage Facility & Safety, Henan University of Technology, Zhengzhou 450001, China
[3] School of Civil Engineering and Architecture, Nanjing Institute of Technology, Nanjing 211167, China
[*] Correspondence: jcui@haut.edu.cn

Abstract: Thermal expansion of a member can affect structural bearing capacity when one or more members are subjected to fire. To explore the developmental rule of member internal force in latticed shell structure, a computation method of member sensitivity is presented based on the alternate path method. Taking a K8 single-layer latticed shell structure as the analysis object, finite element models are established by ANSYS, applying temperature loads on radial and circumferent ribs, calculating the sensitivity of each member during heating, and exploring the rule of member sensitivity at different temperatures. It is revealed from the numerical results that when one member is in fire, member sensitivity is proportional to temperature and inversely proportional to the distance from the member subjected to fire. Taking the member sensitivity coefficient as an index, the internal force will be transmitted along the member with high sensitivity, the rule of load path and internal force redistribution is given when the latticed shell structure is under elevated temperature.

Keywords: member sensitivity; temperature; latticed shell structure; load path

Citation: Cui, J.; Hu, G.; Zhan, Y.; Pang, R. Investigation on Load Path of a Latticed Shell Structure under Localized Fire Based on Member Sensitivity. *Buildings* **2022**, *12*, 1938. https://doi.org/10.3390/buildings12111938

Academic Editors: Liqiang Jiang, Jihong Ye, Wei Guo and Nerio Tullini

Received: 4 October 2022
Accepted: 2 November 2022
Published: 10 November 2022

1. Introduction

Due to its lightweight, good rigidity, and large span, a spatial structure is mostly used in public buildings, such as stadiums and exhibition halls. Steel is the main material in a spatial structure, which has poor fire resistance; the material's mechanical properties decrease as the temperature rises. Most members of the large-span space structure are axially compressed, and structural members are more likely to be damaged when the structure is in a fire. As one of the main secondary disasters of a building fire, progressive collapse is generally difficult to predict.

Progressive collapse of a building structure is defined as a local failure and unbalanced force caused by unexpected loads; it can cause member failure because internal force changes dramatically. Building fire is one of the disasters with high frequency and serious harm; it is also an occasional effect that seriously endangers the safety of spatial structures. Many scholars have carried out a series of studies on the progressive collapse resistance of building structures [1–6], especially for the mechanical response of large-span space structures subjected to fire; a number of conclusions have been obtained through fire experiments and FE analysis. Guo [7,8] carried out eight destructive fire tests on a scale-reticulated shell model. The experimental results show the shell specimen performed arching under vertical load and the fire effect due to thermal expansion, the global deformation patterns and the axial force distributions of the specimen were approximately symmetric when the fire was located at the center of the test model, and the ultimate bearing capacity of the specimen gradually decreased with the combustion of the fire. Yu [9] uses fire dynamics software FDS to set different fire source positions to generate the fire temperature field of the single-layer cylindrical reticulated shell and used the finite element software ABAQUS to conduct a numerical simulation of the behavior of the cylindrical shell structure under different fire conditions. The influence of several parameters, such as the steel pipe section size, rise span ratio of the structure, roof load, and thickness of the fireproof coating, on the single-layer cylindrical reticulated shell in fire is analyzed.

Sensitivity analysis is an analytical method to evaluate the parameter change of structural response while the structure is in particular load conditions; sensitivity index is commonly used in the analysis of the progressive collapse-resisting performance of building structures. Based on member sensitivity, many scholars have established a method to evaluate the progressive collapse-resisting performance of structures, such as the member importance index [10,11], structural redundancy index [12,13], and robustness analysis [14–16]. The load path analysis is an important problem of structural resistance to progressive collapse. Wang [17] investigates a test of a scaled Kiewitt-6 dome; the effects of force transmission paths and initial imperfections on the progressive collapse resistance of large-span single-layer latticed domes were analyzed. The results indicate that the failure mode of the tested dome was local collapse when the two force transmission paths failed. The dome relies mainly on the compression mechanism to resist progressive collapse. Zhao [18] proposes a new expression of internal stiffness to visualize the load paths, instead of measuring strain energy caused by displacement at the loading point. An L-bracket test is carried out to validate the index. However, these indexes are not applicable to the structure in fire because material properties attenuation at elevated temperature is not considered.

In the condition of a building fire, steel members are prone to fail, but it is a dynamic process of member failure as the temperature increases. The failure of a single member can cause damage to multiple adjacent members in the local zone; it is a great danger to structural safety. In order to clarify the internal forces redistribution rule and prevent progressive collapse when a single-layer latticed shell structure is subjected to fire, this study proposes a member sensitivity index considering material properties reduction, quantitative analyses of the member sensitivity at different temperatures through finite element analysis; the transmission path is determined through the member sensitivity change with temperature. The current work aims to evaluate the structure's progressive collapse-resisting performance when a latticed shell structure is under fire conditions.

2. Member Sensitivity Index

Fire can cause serious adverse effects on the structure. The temperature of a building fire generally is greater than 800 °C. However, the mechanical properties of steel significantly decline at 500 °C. With the temperature rising, the thermal effect on the structure becomes more significant. This may cause two security concerns. On the one hand, the elevated temperature will reduce the member bearing capacity, which is easy to cause local failure; on the other hand, mutual support exists between members in a spatial structure, so the failure of local members may cause the progressive collapse of the structure. Therefore, in the process of heating up, exploring member sensitivity to temperature, and clarifying the force transmission path, ensure not only the fire resistance design safety of the structure but also provide a basis for the fire occurrence performance evaluation.

The sensitivity coefficient is an index used to quantitatively evaluate the vulnerability of a certain member; it reflects which member is more likely to fail than others under a specific working condition. The sensitivity coefficient of member SI is generally defined as Equation (1) [19].

$$\mathrm{SI} = \frac{\lambda - \lambda_0}{\lambda_0} \tag{1}$$

where λ_0 and λ are the mechanical parameters of the initial state and the final state of the structure, respectively.

Equation (1) is suitable for structure sensitivity analysis at room temperature. However, steel is a thermosensitive material, and its mechanical properties change significantly with temperature. The elastic modulus and yield strength should be considered for materials at elevated temperatures, as temperatures can significantly affect the member sensitivity.

When the structure is in fire, the fire-retardant coating of some members is burnt out; these members are directly exposed to fire and rapidly heat up; in contrast, other members heat slowly because of the protection of the fire-retardant coating. There is a temperature difference and stress difference between the two types, resulting in an uneven temperature distribution of the members in the structure. To quantitatively analyze the member sensitivity of the temperature effect, the attenuation of material properties at elevated temperatures is considered. The yield stress of steel at temperature T is introduced into Equation (1), and the sensitivity formula of the member at a high temperature can be expressed as Equation (2):

$$\mathrm{SI}_{ij} = \frac{\sigma_j^{(k+1)} - \sigma_j^{(k)}}{f_s^T - \left|\sigma_j^{(k)}\right|} \tag{2}$$

where σ is the maximum principal stress of the member section, which is a function varying with the temperature T; subscript i is the number of the member at elevated temperature, j is the number of other members at room temperature, and $i \neq j$; k is the number of load steps; $\sigma_j^{(k)}$ is the stress calculated up to the kth step while member j at temperature T, f_s^T is the yield stress at temperature T. $f_s^T - \left|\sigma_j^{(k)}\right|$ represents the stress margin of member j at temperature T, which is a constant positive value. $\sigma_j^{(k+1)} - \sigma_j^{(k)}$ represents the stress increment of member j from step k to step $k+1$.

SI is a ratio of the stress increment to the stress margin, and the value range is (−1.0, 1.0); the larger the value is, the greater the member tends to be in a fully stressed state, and the smaller the capacity to continue bearing. When SI is greater than 1.0, it indicates the strength failure of member j. It is a mathematical certainty that the value of SI is a slope, which is the change rate of sensitivity with temperature. For the calculation results, SI can be either positive or negative; when SI > 0, it indicates that the internal force of the member will bear more load in the process of internal force redistribution; when SI < 0, it is a process of internal force release.

3. Finite Element Implementation

3.1. Model Design of Latticed Shell Structure

To find the changeable rule of member sensitivity with temperature when a structure is subjected to fire and then determine the influence of temperature on the progressive collapse-resisting performance of the spatial structure, a K8 single-layer latticed shell structure is taken as the research object, as shown in Figure 1. The structural span D = 32 m, the height h = 4 m, the members use a circular steel tube ϕ 114 × 3, the live load is 0.5 kN/m^2, and the dead load is 0.5 kN/m^2.

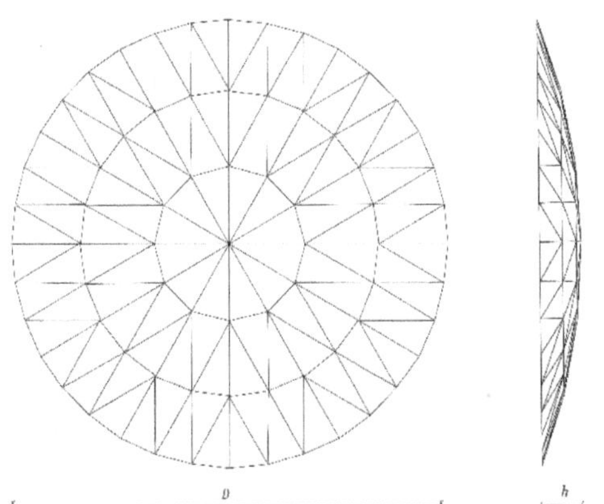

Figure 1. K8 single-layer latticed shell model.

3.2. Material Properties

The material is steel Q235B, and the values of the material parameters are taken according to the Code for fire safety of steel structures in buildings (GB512479-2017) [20]. The density is 7850 kg/m^3, Poisson's ratio is 0.3, coefficient of thermal expansion is 1.4×10^{-5}, and the elastic modulus at room temperature is 206 GPa. The reduction coefficient ξ_T of the elastic modulus when the temperature ranges from 20 to 600 °C is shown in Equation (3).

$$\xi_T = \frac{7T_s - 4780}{6T_s - 4760} \tag{3}$$

where T_s is the temperature of the steel. The stress-strain relationship of steel at the elevated temperature is an elastic-plastic isotropic strengthening model; strengthening and neck shrinkage are ignored [21]. The multilinear kinematic hardening model is selected in ANSYS. The stress-strain relationships of steel at different temperatures are shown in Figure 2.

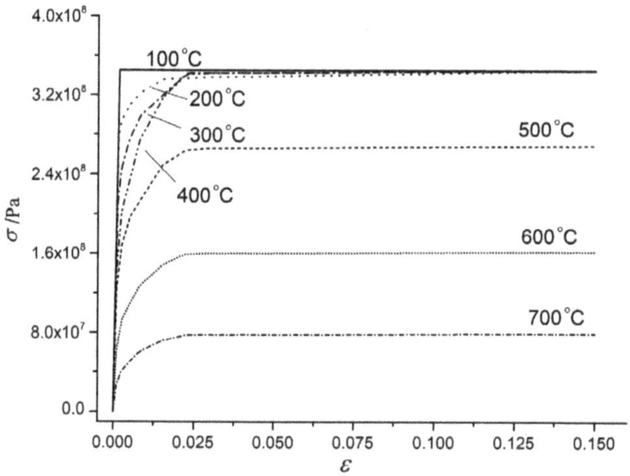

Figure 2. Stress-strain relationship of Q235B steel at different temperatures.

3.3. Loading System

The temperature action is according to the ISO 834 standard heating curve, and the expression is

$$T_g - T_0 = 345\lg(8t + 1) \tag{4}$$

where T_g is the member temperature at time t; T_0 is the initial temperature of the member; t is the fire duration time.

According to Reference [20], the section shape factor of circle tube ϕ 114 × 3 is greater than 300; therefore, the steel temperature can be equal to the air temperature.

3.4. Finite Element Model

The finite element model is established by ANSYS; Beam188 is used to simulate the members, which is based on Timoshenko beam theory; it is well-suited for simulating the members of the latticed shell structure. The nodes are rigid connections; the bottom nodes are simply supported. The initial geometric defect is applied by the first-order mode of the structure. The large geometric deformation is turned on in the calculation. Temperature loads are applied to the structure in the form of body loads. Set multiple load steps with temperature changes, and the temperature value is set every minute according to ISO834. The FE model is shown in Figure 3.

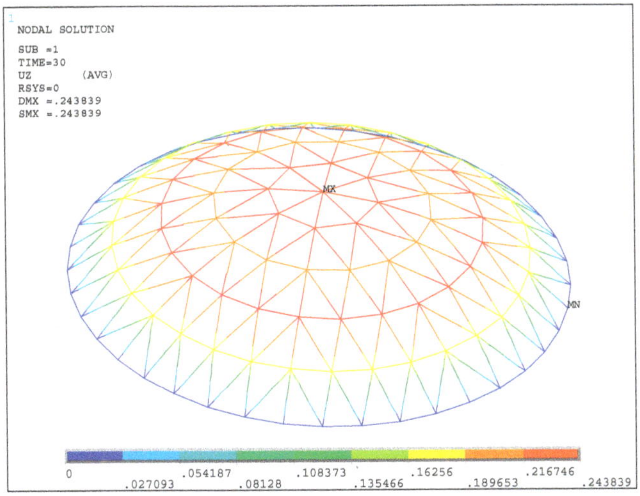

Figure 3. FE model of latticed shell structure.

4. Test Results

The positions of accidental loads are random, and the affected members are also uncertain. Due to the large number of members of the lattice shell structure, it is very important to traverse and calculate all members. In this study, the characteristic members are selected, and a temperature load is applied to one member to analyze the response law of other members of the structure. According to the symmetry of the latticed shell structure, the half-side structure is taken as a schematic, and the number of each member is shown in Figure 4.

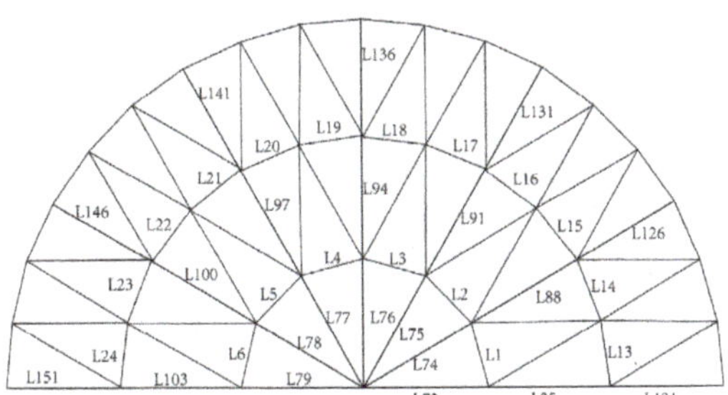

Figure 4. Schematic diagram of member numbering.

4.1. Effect of Temperature on Radial Ribs

Taking the radial rib as the analysis object, which is the main force transmission member in a single-layer latticed shell structure, the sensitivity of each member of the structure when one radial rib is heated is calculated. Three analysis conditions were designed in which radial ribs L73, L85, and L121 were heated. The first condition is that L73 is subjected to temperature, and the temperature variation range is from 20 °C to 500 °C. The other conditions are similar to explore the influence of heating of different radial ribs on the whole structure.

Take condition 1 for analysis. Applying a temperature load on member L73, the temperature function is according to Equation (4); the sensitivity coefficient of each member at the temperature load of each step is calculated using Equation (2). The calculation results of member sensitivity at different temperatures are marked in Figure 5.

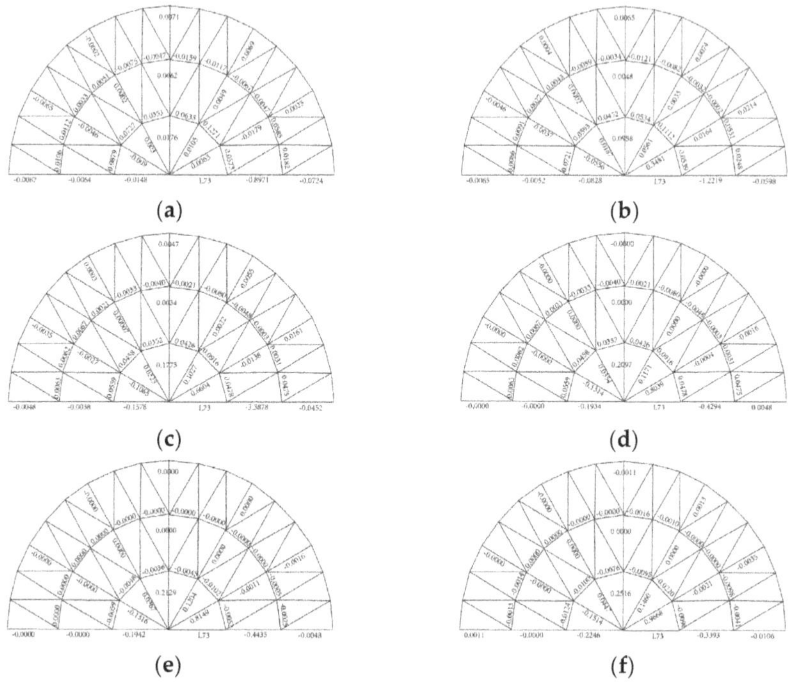

Figure 5. Sensitivity coefficient of L73 main members under fire ($\times 10^{-1}$). **(a)** t = 20 °C, **(b)** t = 100 °C, **(c)** t = 200 °C, **(d)** t = 300 °C, **(e)** t = 400 °C, **(f)** t = 500 °C.

As shown in Figure 5, it can ignore the effect of member sensitivity marked 0.0000, which is the calculated value is less than 10^{-5}. When member L73 is subjected to a temperature load, from the radial point, the farther away from the elevated temperature member, the larger the value of the sensitivity coefficient. This indicates that members L85 and L121 bear the additional force generated during L73 heating up, and their own compressive stress decreases. From the circumferential point, the sensitivity values of radial ribs L74 and L76 change greatly, and L74 is the main forced member when the L73 temperature rises over 200 °C. The value of the circular ribs is tiny, which indicates that the circular ribs have less effect in the process of internal force redistribution. The above analysis shows that the radial members are the main forced members when radial ribs are subjected to fire.

The variation law of sensitivity with the temperature of circumferential ribs is shown in Figure 6.

As shown in Figure 5, the change trend of each circumferential rib is the same, and the sensitivity decreases with increasing temperature. The curve can be divided into three stages. In the first stage, the temperature is between 20 °C and 270 °C, the member sensitivity coefficient decreases linearly, and the stress increment of the adjacent member decreases. The main reason is that the axial force caused by the external load decreases due to thermal stress. In the second stage, between 270 °C and 400 °C, the curve drops sharply, and there are multiple inflection points in the fluctuation curve. The main reason is that the elastic modulus of steel decreases at the elevated temperature, and the stress increment decreases from step k to step $k + 1$, indicating that the internal force flow of the structure is the most significant when the member temperature rises in this range. In the third stage, the value of member sensitivity changes from positive to negative, and the value is small, which indicates that other members are not sensitive to L73 when the temperature increases over 400 °C. It can be considered that internal force redistribution has been completed.

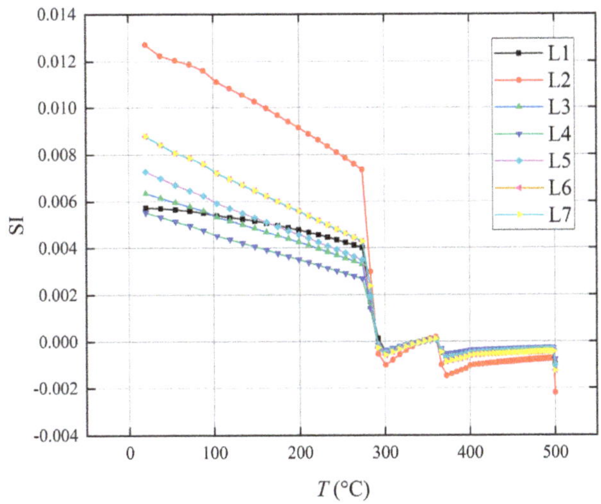

Figure 6. Variation rule of sensitivity coefficient of circumferential ribs with temperature (L73).

To study the different fire-receiving elements on the longitudinal ribs, radial ribs L73, L85, and L121 were selected to heat up for analysis. Taking member L136 as the analysis object, the change curve of member sensitivity is shown in Figure 7.

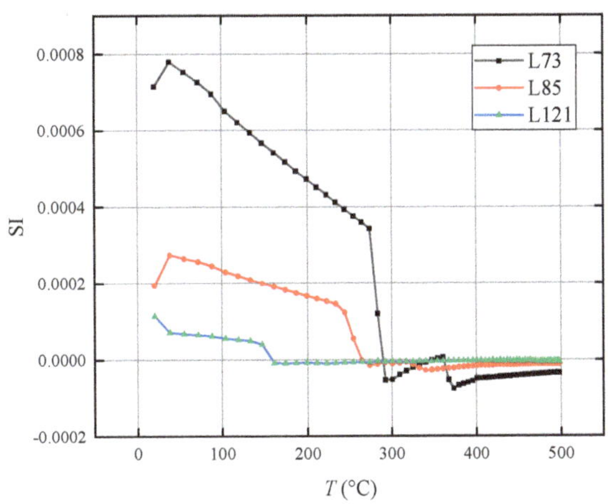

Figure 7. Sensitivity coefficient of member L136 under different fire conditions.

As shown in Figure 7, the member near the vertex is more sensitive than the member near the supports. The reason is that the force of member L121 can be directly transmitted to supports, so member sensitivity is a kind of member with less influence in the transmission process of internal force.

4.2. *Effect of Temperature on Ring Ribs*

From the geometric point of view, the latticed shell structure is a perfectly symmetric structure, and it is representative of analyzing any circumferential member. Heating member L1, which is located at the first circular ribs, is heated to 500 °C. The sensitivity coefficients of the circumferential ribs, radial ribs, and diagonal braces are calculated and marked in Figure 8.

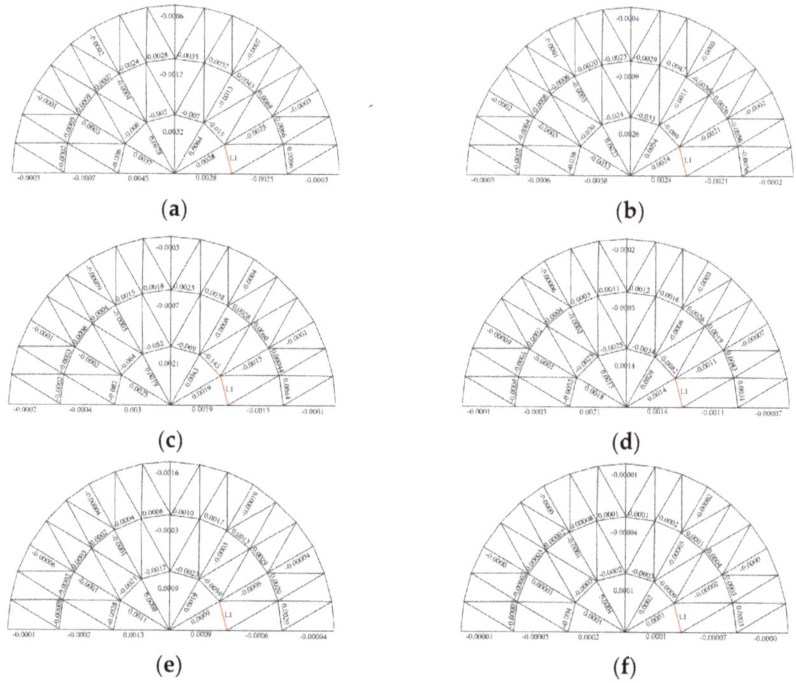

Figure 8. Sensitivity coefficient of the main members when the ring ribs are under different temperatures. (**a**) t = 20 °C, (**b**) t = 100 °C, (**c**) t = 200 °C, (**d**) t = 300 °C, (**e**) t = 400 °C, (**f**) t = 500 °C.

As shown in Figure 8, there is an inverse correlation with the distance between members and member L1, which heats up. The member is much closer to L1, the greater the sensitivity value, which indicates that the sensitivity value in the first quadrant is greater than members in the second quadrant; the sensitivity of radial ribs in the first ring is greater than members in the second ring.

The member sensitivity coefficients with temperatures of L2~L7 in the first ring are shown in Figure 9.

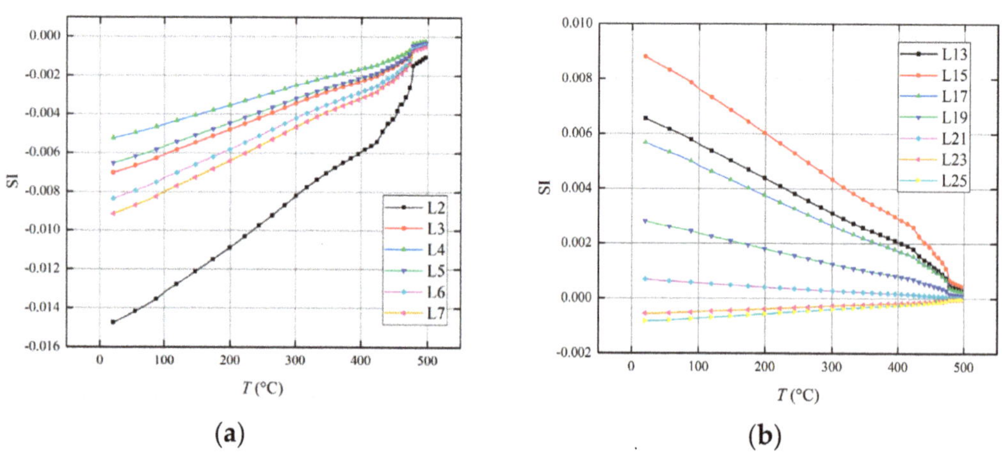

Figure 9. Law of the sensitivity coefficient with temperature. (**a**) Circumferent ribs; (**b**) radial ribs.

As shown in Figures 8 and 9, the following conclusions can be drawn:

(1) The sensitivity of members L2~L7 in the first ring increases with temperature. The sensitivity of member L2 adjacent to L1 has a maximum value, and member L4, which is perpendicular to L1, has the smallest sensitivity. This indicates that in the process of internal force redistribution, the elastic modulus of member L1 keeps decreasing with increasing temperature, the bearing capacity of member L1 decreases, member L2 adjacent to member L1 needs to bear more internal force, and is greatly affected by temperature and more dangerous under the elevated temperature. There is a rule for the sensitivity of radial ribs that the sensitivity is larger with the angle between the member and member in fire from 0° to 90°; the sensitivity decreases from 90° to 180°. This shows that during the process of inertial force redistribution, not only the members in the same direction as the load-bearing member but also members in other directions are affected.

(2) Compared with circumferential members, the sensitivity coefficient of the radial rib is smaller, which indicates that the members on the same ring are the main force transmission member when one circumferential member is heated up, and the distal members are less affected.

(3) Between 400 °C and 500 °C, there are inflection points in the curve of the sensitivity coefficient with increasing temperature, and the inflection points are consistent with the inflection point of the mechanical properties of steel.

5. Load Path of Member under Elevated Temperature

The form of the single-layer latticed shell structure is an arch, and the members mainly bear axial compression under the action of external loads. The local structure is taken as the analysis component, which is composed of one node and six connected members, as shown in Figure 10.

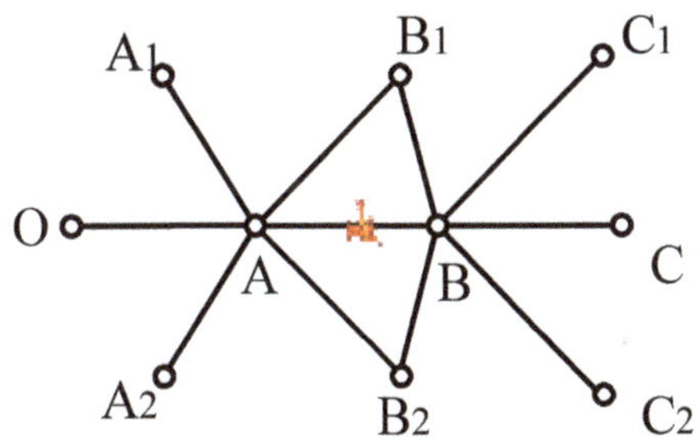

Figure 10. Schematic diagram of the analysis component.

As shown in Figure 10, assuming that member AB is in fire, the unbalanced forces F and F' are generated by the thermal effect at both ends of the member nodes, as shown in Figure 11a. The deformation of points A and B is constrained by other members, and the unbalanced forces F and F' are transmitted to adjacent members through the nodes. A schematic diagram of force decomposition is shown in Figure 11b; the force component $F_y = F\ \cos\theta$ causes an increase in inertial force in OA, and $F_x = F\ \sin\theta$ causes a shear force and a bending moment. The shear force F_y causes the upward displacement of point A. According to the same analysis, member BC can generate additional internal forces, and point B produces upward displacement. This process is the redistribution of local internal forces under temperature effects.

The unbalanced force generated by member AB is transmitted to adjacent members, and the radial component of force transfers to the next ring along BC, BC1, and BC2. The internal force change of a single member causes internal force flow in a local zone, and the change effect decreases with the increase in participating members. According to the analysis in Section 3, there is a certain range of the internal force redistribution of the latticed shell structure rather than the participation of all members of the structure, which can result from the latticed shell structure having a number of indeterminateness and high structural redundancy.

Figure 11. Schematic diagram of radial member force. (**a**) Unbalanced force diagram; (**b**) decomposition of force.

Generally, only one member failure is considered in current domestic and foreign design codes and guidelines for progressive collapse-resistant building structures. Different from static loads, members are more likely to buckle or be damaged in the condition of extreme fire disasters. When a structure is exposed to fire for a long time, one or more members of a latticed shell structure are prone to failure, resulting in progressive collapse because local members fail. This study analyses the flow direction of the internal force when the structure is in fire, determines the main forced transition member and load path, and analyses the most sensitive member. The results can provide a method to analyze the progressive collapse resistance of a single-layer latticed shell structure under fire disasters.

6. Discussion and Conclusions

A novel method to explore the law of redistribution of internal forces of Kiewitt reticulated shell structure under fire, the member sensitivity index, and the load path of members in the condition of a single member subjected to fire are studied. The following conclusions are obtained:

- A member sensitivity formula is proposed, which considers the temperature effect by introducing the yield stress of steel at elevated temperature; it can estimate the sensitivity of members to temperature by calculating the stress increment to evaluate the allowance of the members bearing capacity.
- The rule of member sensitivity with temperature is explored in the condition of radial ribs and circumferent ribs subjected to fire; the analysis reveals that member sensitivity increases with temperature; the farther away from the member subjected to fire, the smaller the member sensitivity. The additional internal force generated by one member subjected to fire is mainly borne by adjacent members, and the force will transmit along the member with higher sensitivity by changing the internal force of members in a local zone to make the structure reach a new equilibrium condition, the law of internal force redistribution is clarified when the structure is in fire.
- We proposed a prediction method proposed in this study; the sensitivity index is the difference calculation of stress from step k to step $k + 1$. The smaller the sensitivity coefficient is, the smaller the allowable bearing allowance of the member. In the condition of part of the structure subjected to fire, the maximum load path of the latticed shell structure can be determined, and the members that are prone to damage can be identified. This work provides a new idea for the analysis of resisting progressive collapse performance of a structure subjected to fire.

Author Contributions: Conceptualization, J.C. and G.H.; methodology, J.C. and Y.Z.; software, G.H.; validation, J.C. and R.P.; formal analysis, G.H. and Y.Z.; investigation, R.P.; data curation, G.H.; writing—original draft preparation, G.H.; writing—review and editing, J.C. and Y.Z.; funding acquisition, J.C. and R.P. All authors have read and agreed to the published version of the manuscript.

Funding: The work described in this paper was fully supported by Henan key scientific and technological project (182102210090); the Open Research Subject of Henan Key Laboratory of Grain and Oil Storage Facility & Safety (2020KF-1303, 2021KF-02).

Institutional Review Board Statement: Not applicable.

Informed Consent Statement: Not applicable.

Data Availability Statement: Data is contained within the article.

Conflicts of Interest: The authors declare no conflict of interest.

References

1. Ye, J.; Jiang, L. Collapse mechanism analysis of a steel moment frame based on structural vulnerability theory. *Arch. Civ. Mech. Eng.* **2018**, *3*, 833–843.
2. Elsanadedy, H.M.; Al-Salloum, Y.A.; Almusallam, T.H.; Ngo, T.; Abbas, H. Assessment of progressive collapse potential of special moment resisting RC frames—Experimental and FE study. *Eng. Fail. Anal.* **2019**, *105*, 896–918. [CrossRef]
3. Gernay, T.; Gamba, A. Progressive collapse triggered by fire induced column loss: Detrimental effect of thermal forces. *Eng. Struct.* **2018**, *172*, 483–496. [CrossRef]
4. Iribarren, B.S.; Berke, P.; Bouillard, P.; Vantomme, J.; Massart, T.J. Investigation of the influence of design and material parameters in the progressive collapse analysis of RC structures. *Eng. Struct.* **2011**, *33*, 2805–2820. [CrossRef]
5. Kwasniewski, L. Nonlinear dynamic simulations of progressive collapse for a multistory building. *Eng. Struct.* **2010**, *32*, 1223–1235. [CrossRef]
6. Azghandi, R.R.; Shakib, H.; Zakersalehi, M. Numerical simulation of seismic collapse mechanisms of vertically irregular steel high-rise buildings. *J. Constr. Steel Res.* **2020**, *166*, 105914. [CrossRef]
7. Guo, X.; Zhu, S.; Jiang, S.; Zhang, C.; Chen, C. Fire tests on single-layer aluminum alloy reticulated shells with gusset joints. *Structures* **2020**, *28*, 1137–1152. [CrossRef]
8. Jiang, S.; Zhu, S.; Guo, X.; Li, Z. Full-Scale Fire Tests on Steel Roof Truss Structures. *J. Constr. Steel Res.* **2020**, *169*, 106025. [CrossRef]
9. Yu, Z.; Lu, C.; Zhong, Y. Performance-based analysis of single-layer cylindrical steel reticulated shells in fire. *Appl. Sci. -Basel* **2020**, *10*, 3099. [CrossRef]
10. Li, Y.; Nie, Q.; Luo, Y.; Liu, X.; Guo, X. Classification Method of Member Importance Based on Strain Energy Theory for Lattice Shell Structure. *J. South China Univ. Technol. (Nat. Sci. Ed.)* **2020**, *48*, 22–31.
11. Zhu, N.; Ye, J. Development and verification of redundancy analysis method for single-layer dome based on response sensitivity. *J. Build. Struct.* **2014**, *35*, 85–93.
12. Han, Q.H.; Liu, M.J.; Lu, Y.; Wang, C.X. Progressive collapse analysis of large-span reticulated domes. *Int. J. Steel Struct.* **2015**, *15*, 261–269. [CrossRef]
13. Jiang, S.; Yuan, X.; Ma, S. An evaluation method for component importance of pin-jointed structures considering structural redundancy. *J. Harbin Inst. Technol.* **2018**, *50*, 187–192.
14. Tavakoli, H.R.; Afrapoli, M.M. Robustness analysis of steel structures with various lateral load resisting systems under the seismic progressive collapse. *Eng. Fail. Anal.* **2018**, *83*, 88–101. [CrossRef]
15. Li, L.L.; Li, G.Q.; Jiang, B.; Lu, Y. Analysis of robustness of steel frames against progressive collapse. *J. Constr. Steel Res.* **2018**, *143*, 264–278. [CrossRef]
16. Elsanadedy, H.M.; Abadel, A.A. High-Fidelity Fe Models for Assessing Progressive Collapse Robustness of Rc Ordinary Moment Frame (Omf) Buildings. *Eng. Fail. Anal.* **2022**, *136*, 1062258. [CrossRef]
17. Wang, Y.; Nie, X.; Tian, L. Influence Factors on Progressive-Collapse Resistance of Large-span Single-layer Latticed Domes. *J. Disaster Prev. Mitig. Eng.* **2022**, *42*, 464–474.
18. Zhao, S.; Wu, N.; Karnaoukh, S. A new expression of internal stiffness for load path analysis in structures. *Int. J. Appl. Mech.* **2022**, *14*, 2250030. [CrossRef]
19. Japanese Society of Steel Construction; Council on Tall Buildings and Urban Habitat. *Guidelines for Collapse Control Design: Construction of Steel Building with High Redundancy*; Iron and Steel Federation: Tokyo, Japan, 2005.
20. Tongji University, Sub-Association for Fire Protection & Corrosion Prevention for Steel Structures. *Code for Fire Safety of Steel Structures in Buildings, GB51249-2017*; China Planning Press: Beijing, China, 2017.
21. *EN 1993-1-2*; Eurocode 3: Design of Steel Structures Part 1–2: General Rules-Structural Fire Design. Committee for Standardization: Brussels, Belgium, 2001.

MDPI
St. Alban-Anlage 66
4052 Basel
Switzerland
www.mdpi.com

Buildings Editorial Office
E-mail: buildings@mdpi.com
www.mdpi.com/journal/buildings

www.ingramcontent.com/pod-product-compliance
Lightning Source LLC
Chambersburg PA
CBHW040802250926
51850CB00007BA/1057
* 9 7 8 3 7 2 5 8 0 2 4 4 9 *